Professional Mathematics
for
Polytechnics
(for Third Semester)

Second Edition

H.K. DASS M.Sc.

Diploma in Specialist Studies (Maths),
University of Hull, England

CBS

CBS Publishers & Distributors Pvt. Ltd.

New Delhi • Bengaluru • Chennai • Kochi • Kolkata • Mumbai
Hyderabad • Nagpur • Patna • Pune • Vijayawada

ISBN: 978-93-85915-17-8

First Edition: 2012
Reprint: 2012, 2014, 2015
Second Edition: 2016
Reprint: 2017

Published by **Satish Kumar Jain** and produced by **Varun Jain** for
CBS Publishers & Distributors Pvt. Ltd.,
4819/XI Prahlad Street, 24 Ansari Road, Daryaganj, New Delhi - 110002
delhi@cbspd.com, cbspubs@airtelmail.in • www.cbspd.com
Ph.: 23289259, 23266861, 23266867 • Fax: 011-23243014

Corporate Office: 204 FIE, Industrial Area, Patparganj, Delhi - 110 092
Ph: 49344934 • Fax: 011-49344935
E-mail: publishing@cbspd.com • publicity@cbspd.com

Branches:
• *Bengaluru:* 2975, 17th Cross, K.R. Road, Bansankari 2nd Stage,
  Bengaluru - 70 • Ph: +91-80-26771678/79 • Fax: +91-80-26771680
  E-mail: cbsbng@gmail.com, bangalore@cbspd.com
• *Chennai:* No. 7, Subbaraya Street, Shenoy Nagar, Chennai - 600030
  Ph: +91-44-26681266, 26680620 • Fax: +91-44-42032115
  E-mail: chennai@cbspd.com
• *Kochi:* Ashana House, 39/1904, A.M Thomas Road, Valanjambalam,
  Ernakulum, Kochi • Ph: +91-484-4059061-65
  Fax: +91-484-4059065 • E-mail: cochin@cbspd.com
• *Kolkata:* 6-B, Ground Floor, Rameshwar Shaw Road, Kolkata - 700014
  Ph: +91-33-22891126/7/8 • E-mail: kolkata@cbspd.com
• *Mumbai:* 83-C, Dr. E. Moses Road, Worli, Mumbai - 400018
  Ph: +91-9833017933, 022-24902340/41 • E-mail: mumbai@cbspd.com

Representatives:

• Hyderabad: 0-9885175004   • Nagpur: 0-9021734563
• Patna: 0-9334159340   • Pune: 0-9623451994
• Vijayawada: 0-9000660880

Printed at:
J.S. Offset Printers, Delhi (India)

PREFACE TO THE SECOND EDITION

This book is revised according to the latest syllabus of Professional Mathematics for Third Semester of Govt. Polytechnics affiliated to the Bihar Board of Technical Education.

Question-papers of the examinations of Third Semester of Government Polytechnics, conducted by Bihar State Board of Technical Education are solved for the years 2008, 2009, 2010, 2011, 2012 and 2014 and their solutions are included in the body of the textbook, chapterwise. It will help the students to know the trends of the examination.

Emphasis has been laid on making the concept clear. Presentation is very simple and in logical manner. Simple and lucid language is used while writing the text. Lots of pain and concentration on the part of the author has gone in solving the examples in best possible way. Large number of solved examples are added in different chapters. In providing the solution of the problems care has been taken not to miss even minor step so that the students can follow the subject easily. Tabular explanation of specific topics are included in the solution of the problems.

I take this opportunity to express my deep sense of gratitude to Shri Satish Jain, CMD, CBS Publishers & Distributors Pvt. Ltd., for taking personal interest in this book.

Suggestions and costructive feed back from students and teachers to improve the book shall be personally acknowledged and deeply appreciated which will help to make it users friendly.

D-1/87, Janakpuri   **H.K. DASS**
New Delhi - 110058
Tele: 011-28525078, 32985078
Mob. 09350055078
hk_dass@yahoo.com

SYLLABUS

	Theory		No. of Periods in one session. 60		
	No. of Periods Per Week		Full Marks	:	100
	L	T	P/S	Annual Exam.	:	80
	06	00	00	Internal Exam.	:	20

RATIONALE:

A technical diploma holder is engaged generally as first line supervisor. He forms a bridge between workers and management. He has to understand the language of the modern management and communicate with the workers in their language. This subject will help accomplishment of the task in stipulated time, develop attitude towards cost effectiveness, selection of most effective alternative methods. This course will also help the student to tackle different numerical methods and computational techniques for problems solving in research organization as a programmer.

OBJECTIVE:

The course enables students to.

- Managerial skill based on mathematical footing
- The ability to find approximate solutions and/or answers to the problems where analytical methods become more complex.
- To choose correct numerical techniques for a given problem

S.No.	TOPICS	PERIODS
01	GROUP – A (Numerical Methods)	20
02	GROUP– B (Statistical Techniques)	20
03	GROUP– C (Management Techniques)	20
	Total	60

CONTENTS:

GROUP-A (NUMERICAL METHODS)

01.01   Introduction to Numerical methods: Approximation and errors (Truncation & Round off).

01.02   Numerical solutions of non-linear and Transcendental equations: Iterative methods. Newton-Raphson's method. Bisection method and Regula-Falsi method.

01.03   Solution of Linear Simultaneous Equations: Gaussian Elimination method and Gauss-Jordan method.

01.04   Finite difference: Backward and forward Differences. Finite Difference Interpolation Formula. Newton's Forward Difference formula and Newton's Backward Difference formula.

01.05   Numerical Differentiation & Integration: Newton's forward and backward differentiation formula. Trapezoidal Rule and Simpson's 1/3 rule for numerical integration.

01.06   Difference equations. Simple problem only.

GROUP-B (STATISTICAL TECHNIQUES)

02.01   Introduction to statistics: Measure of central tendencies: measures of dispersions: standard deviation and variance for discrete and grouped data: assumed mean and step deviation methods.

02.02   Theory of Probability: Random events and their types. Probability of Events. Definitions. Laws of Probability (Addition and Multiplication Laws)

02.03   Probability Distribution: Introduction to Arithmetic Mean and Standard Deviation of a probability distribution. Important probability distribution-Binomial distribution. Poisson's distribution & Their means and variance.

GROUP-C (MANAGEMENT TECHNIQUES)

03.01   Linear Models

03.01.01   Introduction to Operations Research (O.R) Steps of O.R.

03.01.02   Linear Programming Problems: Formulation of a LPP. Mathematical Modelling and Solution by graphical method.

03.01.03   Solution by Simplex Method: Basic Feasible Solution (Degenerator and Non-degenerator)

03.01.04   Transportation problem: Introduction and Solution Procedure-

   (*i*)   Finding the initial basic feasible solution by N-W Corner Rule, Least cost method and Vogel's approximation Method.

   (*ii*)   Test of optimality by u-v method only.

03.01.05   Assignment Problem: Introduction and Solution Procedure-Fundamental theory underlying Hungarian Method.

03.02   Network Analysis. CPM & PERT: introduction.

03.02.01   Basic concepts-Activities. Nodes. Edges. Networking of a project. Various times calculations. CPM to determine the optimal project schedule.

03.02.02   PERT-Definition, difference between CPM & PERT. Pessimistic times, optimistic times. Most likely times of various activities.

Contents

IMPORTANT FORMULAE
(Summary)

1. Absolute Error $= | X - X' |$

    where $X =$ True value, $X' =$ Approximate value

2. Relative Error $= \left| \dfrac{X - X'}{X} \right|$

3. Percentage Error $= \left| \dfrac{X - X'}{X} \right| \times 100$

4. **Significant figure:**

    The digits used to express a number are called significant digits.

    0.0163 contains 3 significant figures. 0.001027 contains four significant figures.

    Zeroes before decimal and just after decimal are not significant figures.

5. **Rounding off errors**

    The dropping of digits is called rounding off. For example, 3.14285143 is rounded off for three significant figure is 3.14.

6. **Truncation Error**

    Truncation Error arrises by dropping some numbers of a decimal fraction or on replacing an infinite series by a finite one.

    On truncation the remaining digits do not change but on rounding off the last digit changes according to the rounding off rules.

7. Error in approximate number $= \dfrac{1}{2} \times 10^{-n}$

    if a approximate number is correct upto n decimal places.

CHAPTER 2: NUMERICAL SOLUTION OF NON-LINEAR AND TRANSCENDENTAL EQUATIONS

1. Bisection Method (Bolzano Method)

If $f(x)$ is continuous in the interval (a, b) such that $f(a)$ and $f(b)$ are of opposite signs, then

$f(a) . f(b) < 0$ $[f(a) = -\text{ve and } f(b) = +\text{ve}]$

The curve crosses the *x*-axis between *a* and *b* then the first approximation to the root is

$$x_1 = \frac{1}{2}(a+b)$$

Now there are three cases, if

1. $f(x_1) = 0$, then x_1 is the root of $f(x) = 0$
2. $f(x_1) < 0$, then root lies between x_1 and *b*. [If $f(a) = -$ve, $f(b) = +$ve]
3. $f(x_1) > 0$, then root lies between *a* and x_1.

Suppose $f(x_1) > 0$, then $a < \text{Root} < x_1$

Second approximation to the root is $x_2 = \frac{1}{2}(a+x_1)$

and now we proceed as above.

2. Newton-Raphson formula

$$x_{n+1} = x_n - \frac{f(x_n)}{f'(x_n)}$$

The order of convergence in Newton-Raphson method = 2

3. Working Rule to solve f(x) = 0 by Newton-Raphson Method

Step 1. Choose two close numbers *b* and *c* such that $f(b)$ and $f(c)$ are of opposite signs, then the root α lies between *b* and *c*.

Step 2. Out of $f(b)$ and $f(c)$ choose which is nearer to zero. If $f(b)$ is nearer to zero, then *b* is an initial approximate root (x_0) of the given equation.

Step 3. Apply Newton-Raphson formula to find out better approximate root x_1.

$$x_1 = x_0 - \frac{f(x_0)}{f'(x_0)}$$

Repeat the process to get successive approximations.

Step 4. Stop when two approximate roots are equal.

4. Regula Falsi Method

Choose two roots *a*, *b* such that $f(a) \cdot f(b) < 0$.

Then $x = \dfrac{af(b) - bf(a)}{f(b) - f(a)}$

5. Iteration Method

Consider the equation $f(x) = 0$ ... (1)

We rewrite the equation in the form

$$x = \phi(x)$$ .... (2)

Let us draw two curves

$$y = x \text{ and } y = \phi(x)$$

The point of intersection of two curves is the root of (1).

Let $x = x_0$ be an initial approximate root, then first approximation x_1 is found by

$$x_1 = \phi(x_0)$$

Now taking x_1 as initial value, x_2 second value approximation is given by

$$x_2 = \phi(x_1) \quad \text{and so on.}$$

$$x_{n+1} = \phi(x_n)$$

This is also known as successive approximation method.

CHAPTER 3: SOLUTION OF LINEAR SIMULTANEOUS EQUATIONS

Working Rule to solve the equations by Gaussian Elimination method:

Step 1. We eliminate x_1 from 2nd, 3rd ....nth equation with the help of the first equation.

Step 2. We again eliminate x_2 from 3rd, 4th..... nth equation with the help of second equation.

Step 3. Continue this process.

In this way the given system is reduced to the triangular form.

Backward Substitution

We first find out the value of x_n from the last equation, then substitute the value of x_n in the $(n-1)$th equation to get the value of x_{n-1}. Again substitute the value of x_{n-1} in $(n-2)$th equation to get the value of x_{n-2}. By this backward substitution we can find the values of all the unknowns.

Gauss-Jordan method

This is modification of the Gauss elimination method.

By this method we eliminate unknowns not only from the equations below but also from the equations above. In this way the system is reduced to a diagonal matrix.

Finally each equation consists of only one unknown and thus, we get the solution. Here, the labour of backward substitution for finding the unknowns is saved.

CHAPTER 4: FINITE DIFFERENCES

1. Forward difference: $\Delta f(x) = f(x+h) - f(x)$

2. Backward difference: $\nabla f(x) = f(x) - f(x-h)$

3. Shifting operator $E : E f(x) = f(x+h)$

$$\boxed{E = \Delta + 1} \qquad \boxed{E^n \, f(x) = f(x+nh)}$$

4. Factorial Polynomial: $x^{(n)} = x(x-h)(x-2h)(x-3h) ..... [x-(n-1)h]$

5. Differences of a factorial polynomial:

$$\Delta x^{(n)} = n \, x^{(n-1)}$$

$$\frac{1}{\Delta} x^{(n)} = \frac{x^{(n+1)}}{(n+1)}$$

6. Relations:

(i) $\qquad\qquad x^{(n)} = n! \, {}^x C_n$

(ii) $\qquad\qquad \Delta^r \, {}^x C_n = {}^x C_{n-r}$

(iii) $\qquad\qquad \Delta^n \, {}^x C_n = 1$

CHAPTER 5: INTERPOLATION

1. **Interporlation:** Interpolation is the process of finding the value of $f(x)$ for any value of $f(a + nh)$ in the interval $[a, b]$.

2. Newton's Gregory forward interpolation formula:

$$f(a + ph) = f(a) + p\Delta f(a) + \frac{p(p-1)}{2!}\Delta^2 f(a) + \ldots\ldots$$

3. $\Delta^3 f(x) = (E - 1)^3 f(x)$
   $$= (E^3 - 3E^2 + 3E - 1) f(x)$$
   $$= E^3 f(x) - 3E^2 f(x) + 3E f(x) - f(x)$$
   $$= f(x + 3h) - 3 f(x + 2h) + 3 f(x + h) - f(x)$$

4. **Extrapolation:** This is the process for finding the value of $f(x)$ for a value of x lying outside the interval $[a, b]$.

CHAPTER 6: NUMERICAL DIFFERENTIATION

1. **Newton's Forward difference formula for differentiation**

$$f'(x) = f'(a + ph) = \frac{1}{h}\left[\Delta f(a) + \frac{2p-1}{1}\Delta^2 f(a) + \frac{3p^2 - 6p + 2}{6}\Delta^3 f(a) + \ldots\ldots\right]$$

2. **Newton's backward formula for differentiation**

$$f'(x) = f'(x_n + ph) = \frac{1}{h}\left[\nabla f(x_n) + \frac{2p+1}{2}\Delta^2 f(x_n) + \frac{3p^2 + 6p + 2}{6}\nabla^3 f(x_n) + \ldots\ldots\right]$$

CHAPTER 7: NUMERICAL INTEGRATION

1. **Trapezoidal Rule:**

$$\int_{x_0}^{x_0 + nh} f(x)\, dx = \frac{h}{2}\left[(y_0 + y_n) + 2(y_1 + y_2 + y_3 + \ldots + y_{n-1})\right]$$

2. **Simpson's $\frac{1}{3}$ Rule:**

$$\int_{x_0}^{x_0 + nh} f(x)\, dx = \frac{h}{3}\left[(y_0 + y_n) + 2(y_2 + y_4 + y_6 + \ldots y_{n-2}) + 4(y_1 + y_3 + y_5 + \ldots y_{n-1})\right]$$

$$\int f(x)\, dx = \frac{h}{3}\left[(y_0 + y_n) + 2\Sigma y_e + 4\Sigma y_o\right]$$

In applying Simpson's $\frac{1}{3}$ Rule the integration range must be divided into even sub intervals.

CHAPTER 8: DIFFERENCE EQUATIONS

1. **Difference Equation:** Difference equation is the equation between the differences of an unknown function.

    For example; $\Delta^2 y_n + 5 \Delta y_n + 6 y_n = 0$

    Second way to express the difference equation.

    $$E^2 y_n + 3 E y_n + 2 y_n = 0$$

    Third way $y_{n+2} + 3 y_{n+1} + 2 y_n = 0$

2. Order of difference equation $= \dfrac{\text{Largest argument} - \text{smallest argument}}{\text{Unit of interval}}$

3. **Formation of a difference equation** by eliminating the ordinary constant.

4. **Complementary function**

    $$a_0 E^2 y_n + a_1 E y_n + a_2 y_n = 0$$

5. **Auxiliary Equation:** $a_0 m^2 + a_1 m + a_2 = 0$ $\Rightarrow$ $m = m_1, m_2$

    Case 1. *Roots are real and different* say m_1, m_2

    Then $C.F. = C_1 (m_1)^n + C_2 (m_2)^n$

    Case 2. *Roots are equal* m_1, m_1.

    Then $C.F. = (C_1 + C_2 n) (m_1)^n$

    Case 3. *Roots are imaginary* say $\alpha \pm i\beta$

    Then $C.F. = r^n [C_1 \cos n\theta + C_2 \sin n\theta]$

    where $r = \sqrt{\alpha^2 + \beta^2} , \theta = \tan^{-1} \dfrac{\beta}{\alpha}$

6. **Particular Integral**

    (*a*)   P.I. $= \dfrac{1}{f(E)} a^n = \dfrac{a^n}{f(a)}$

    (*b*)   P.I. $= \dfrac{1}{E-a} . a^n = n a^{n-1}$

    (*c*)   P.I. $= \dfrac{1}{(E-a)^2} . a^n = \dfrac{n(n-1)}{2!} a^{n-2}$

    (*d*)   P. I. $= \dfrac{1}{f(E)} \sin kn = \dfrac{1}{f(E)} \left(\dfrac{e^{ikn} - e^{-ikn}}{2i} \right) = \dfrac{1}{2i} \left[\dfrac{1}{f(E)} . e^{ink} - \dfrac{1}{f(E)} . e^{-ikn} \right] = \dfrac{1}{2i} \left[\dfrac{e^{ink}}{f(e^{ik})} - \dfrac{e^{-ink}}{f(e^{-ik})} \right]$

    Particular integral $= \dfrac{1}{f(E)} n^2$

    E is to replaced by $\Delta + 1$ and n^2 is converted into factorial notation $n (n-1) + n = n^{(2)} + n^{(1)}$

    P.I. $= \dfrac{1}{(E-1)^2} 2^n . n^2 = 2^n \dfrac{1}{(2E-1)^2} n^2 = 2^n \dfrac{1}{(2+2\Delta-1)^2} . n^2$

    $= 2^n \dfrac{1}{(1+2\Delta)^2} . n^2 = 2^n . (1+2\Delta)^{-2} [n(n-1) + n]$

CHAPTER 9: INTRODUCTION TO STATISTICS

A.M. $= \dfrac{\Sigma f(x)}{\Sigma f}$

A.M. $= a + \dfrac{\Sigma f d}{\Sigma f}$, a is assumed mean, $d = x - a$

A.M. $= a + \dfrac{\Sigma f D}{\Sigma f} i$, where $D = \dfrac{x-a}{i}$

Median is the value of the middle item.

$$\text{Median} = l + \dfrac{\dfrac{1}{2}N - C}{f} \cdot i$$

l = lower limit of the median class

$N = \Sigma f$, f = frequency of the median class

C = Cumulative freq. above the median class

not cut = width of the class interval.

Lower Quartile: $\quad Q_1 = l + \dfrac{i}{f}\left(\dfrac{N}{4} - C\right)$

Upper Quartile: $\quad Q_3 = l + \dfrac{i}{f}\left(\dfrac{3N}{4} - C\right)$

First Decile: $\quad D_1 = l + \dfrac{i}{f}\left(\dfrac{N}{10} - C\right)$

Second Decile: $\quad D_2 = l + \dfrac{i}{f}\left(\dfrac{2N}{10} - C\right)$

Third Decile: $\quad D_3 = l + \dfrac{i}{f}\left(\dfrac{3N}{10} - C\right)$

Mode = the most frequented item

$$\text{Mode} = l + \left(\dfrac{f - f_{-1}}{2f - f_{-1} - f_{+1}}\right) i$$

where l is lower limit, i is the class length, f is the frequency of the modal class, f_{-1} and f_1 are the frequencies of the classes preceding and succeeding the modal class respectively.

Geometric Mean

$$G = (x_1 \times x_2 \times x_3 \times x_4 \times \ldots \times x_n)^{1/n}$$

Harmonic Mean

$$\dfrac{1}{H} = \dfrac{1}{n}\left[\dfrac{1}{x_1} + \dfrac{1}{x_2} + \ldots + \dfrac{1}{x_n}\right]$$

Standard Deviation

Standard deviation is defined as the square root of the mean of the square of the deviations from the arithmetic mean.

$$S.D. = \sigma = \sqrt{\frac{\Sigma f (x - \bar{x})^2}{\Sigma f}}$$

Note. 1. The square of the standard deviation *i.e.*; σ^2 is called variance.

    2. σ^2 is called the second moment about the mean and is denoted by μ_2.

$$S.D. = \sigma = \sqrt{\frac{\Sigma f d^2}{N} - \left(\frac{\Sigma f d}{N}\right)^2}$$ where d is the deviation from the assumed mean.

Symmetry

A distribution is said to be symmetrical when its mean, median and mode are identical. *i.e.*;

    Mean = Median = Mode.

Skewness

Skewness denotes the opposite of symmetry. It is lack of symmetry.

Skew symmetrical Distribution

A distribution which is not symmetrical is said to be skew symmetrical distribution. In skew symmetrical distribution the left tail and the right tail are not of equal length. One tail will be longer than the other.

(*a*) **Negatively skew distribution.** In negatively skew distribution the left tail is longer than the right tail.

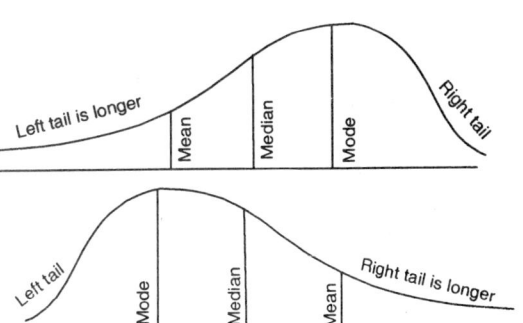

(*b*) **Positively skew distribution.** In positively skew distribution the right tail of the curve will be longer than the left.

Karl Pearson's Coefficient of Skewness

$$\text{Karl Pearson's Coefficient of Skewness} = \boxed{\frac{\text{Mean} - \text{Mode}}{\text{Standard deviation}}}$$

$$= \boxed{\frac{3\,(\text{Mean} - \text{Median})}{\text{Standard deviation}}}$$

1. Measure of skewness

$$\boxed{\beta_1 = \frac{\mu_3^2}{\mu_2^3}}$$

2. Measure of Kurtosis is given by β_2, where

$$\boxed{\beta_2 = \frac{\mu_4}{\mu_2^2}}$$

    *r*th moment $\mu_r = \dfrac{1}{N} \displaystyle\sum_{i=1}^{} f_i\,(x_i - \bar{x})^r$

3. Gamma Coefficients

$$\gamma_1 = \pm \sqrt{\beta_1}$$

$$\gamma_2 = \beta_2 - 3$$

A Mesokurtic   B Leptokurtic   C Platykurtic

If $\beta_2 = 3$, then the curve is normal or mesokurtic.

If $\beta_2 > 3$, then the curve is peaked or leptokurtic.

If $\beta_2 < 3$, then the curve is flat topped or platykurtic.

$$\gamma_2 = \beta_2 - 3$$

CHAPTER 10: PROBABILITY

1. **Exhaustive Events or Sample Space :** The set of all possible outcomes of a single performance of an experiment is exhaustive events or sample space. Each outcome is called a sample point. In case of tossing a coin once, $S = (H,T)$ is the *sample space*. Two outcomes Head and Tail constitute an exhaustive event because no other outcome is possible.

2. **Random Experiment :** There are experiments, in which results may be altogether different, even though they are performed under identical conditions. They are known as random experiments. Tossing a coin or throwing a die is random experiment.

3. **Continuous Random Variables:** A *continuous random variable* is one which can assume any value within a number. *i.e.,* all values of continuous scale. For example (*i*) the weights (in kg) of a group of individuals, (*ii*) the heights of a group of individuals.

4. **A discrete random variable** is one which can assume only isolated values. For example, (*i*) the number of heads in 4 tosses of a coin is a discrete random variable as it con not assume values other than 0, 1, 2, 3, 4.

5. **Trail and Event :** Performing a random experiment is called a trial and outcome is termed as event. Tossing of a coin is a trial and the turning up of head or tail is an event.

6. **Equally likely events:** Two events are said to be '*equally likely*', if one of them cannot be expected in preference to the other. For instance, if we draw a card from well-shuffled pack, we may get any card, then the 52 different cases are equally likely.

7. **Independent event :** Two events may be *independent*, when the actual happening of one does not influence in any way the probability of the happening of the other.

8. **Mutually Exclusive events:** Two events are known as *mutually exclusive*, when the occurrence of one of them excludes the occurrence of the other. For example, on tossing of a coin, either we get head or tail, but not both.

9. **Compound Event :** When two or more events occur in composition with each other, the simultaneous occurrence is called a compound event. When a die is thrown, getting a 5 or 6 is a compound event.

10. **Favorable Events :** The events, which ensure the required happening, are said to be favourable events. For example, in throwing a die, to have the even numbers, 2, 4 and 6 are favourable cases.

11. **Conditional Probability :** The probability of happening an event A, such that event B has already happened, is called the conditional probability of happening of A on the condition that B has already happened. It is usually denoted by $P(A/B)$.

12. **Odds in favour of an event and odds against an event**

If number of favourable ways = m, number of not favourable events = n

(*i*) Odds in favour of the event = $\dfrac{m}{n}$, Odds against the event = $\dfrac{n}{m}$.

13. **Classical Definition of Probability.** If there are N equally likely, mutually, exclusive and exhaustive events of an experiment and m of these are favourable, then the probability of the happening of the

event is defined as $\dfrac{m}{N}$.

14. **Expected value.** If $p_1, p_2, p_3, ..., p_n$ are the probabilities of the events $x_1, x_2, x_3 ... x_n$ respectively the expected value

$$E(x) = p_1 x_1 + p_2 x_2 + p_3 x_3 + ... + p_n x_n = \sum_{r=1}^{n} p_r x_r$$

15. Probability $= \dfrac{\text{Number of favourable ways}}{\text{Total number of equally likely ways}}$

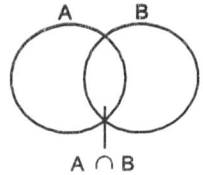

$$P(A \cup B) = P(A) + P(B) - P(A \cap B)$$

16. **Additional law of probability**

$$P = p_1 + p_2 + p_3 + ... + p_n$$

17. **Conditional Probability**

Let A and B be two events of a sample space S and let $P(B) \neq 0$. Then conditional probability of the event A, given B, denoted by $P(A/B)$, is defined by

$$P(A/B) = \dfrac{P(A \cap B)}{P(B)}$$

18. **Baye's Theorem**

If $B_1, B_2, B_3, ... B_n$ are mutually exclusive events with $P(B_i) \neq 0$, $(i = 1, 2, ... n)$ of a random experiment then for any arbitrary event A of the sample space of the above experiment with $P(A) > 0$, we have

$$P(B_2/A) = \dfrac{P(B_2)P(A/B_2)}{P(B_1)P(A/B_1) + P(B_2)P(A/B_2) + P(B_3)P(A/B_3)}$$

CHAPTER 11: PROBABILITY DISTRIBUTION

1. **Probability Distribution**

   The probability distribution of the number of heads in the simultaneous toss of four coins.

   ∴ The probability distribution of the random variable X is :

Number of heads (X)	0	1	2	3	4
Probability $P(X)$	$\frac{1}{16}$	$\frac{1}{4}$	$\frac{3}{8}$	$\frac{1}{4}$	$\frac{1}{16}$

CHAPTER 12: BINOMIAL DISTRIBUTION

1. $P(r) = {}^{n}C_{r}\ p^{r} \cdot q^{n-r}$
2. Mean $= np$
3. Variance $= npq$

CHAPTER 13: POISSON DISTRIBUTION

1. $P(r) = \dfrac{m^{r} e^{-m}}{r!}$
2. Mean $= m$
3. S.D. $= \sqrt{m}$
4. Mean deviation $= \dfrac{2}{e}$ S.D.
5. Recurrence relation: $P(r+1) = \dfrac{m}{r+1} P(r)$

CHAPTER 15: LINEAR PROGRAMMING (GRAPHICAL METHOD)

Some Definitions

1. **Linear Programming Problem**

   Here, we have to optimize the linear function Z subject to certain conditions. Such problems are called linear programming problems. As example 1 on page 200.

2. **Objective functions**

   Objective function is a linear function of several variables, subject to the conditions that $Z = 3x + 4y$ in the example 1 page 200.

3. **Optimal Value**

   Optimal value is a maximum or minimum value of a objective function to be calculated in a linear programming problem.

4. **Non-negative Constants**

   Production of any item is always non-negative, so we write $x \geq 0, y \geq 0$.

5. Linear Relations

All the mathematical relations used in L.P.P. are linear relations.

6. Programming

Programming is the method of determining a particular programme.

7. Decision Variables

Decision variables are x and y which denote the required number of items/products.

8. Constraints

Constraints are linear inequalities or equations involved in linear programming problem. In the example (1) page 200; (2), (3), (4) are constraints.

9. Optimization Problem

Optimization problem is a problem in which a objective function is to be maximize or minimize subject to the certain conditions. In the previous example we have to maximize the profit, so it is optimization problem.

Mathematical Formulation of Linear Programming Problems

In the previous section we have defined certain technical terms of L.P.P. Conversion of the verbal description of L.P.P. into algebraic equations/inequations is known as Mathematical Formultion.

Working Rule to formulate the L.P.P.

Step 1. Identify the decision variables to be determined and expressed them as x, y etc.

Step 2. Identify all the limitations or constraints in the given problem and then express them as linear inequalities or equations in terms of x, y etc.

Step 3. Identify the objective function (Z) which is to optimize (maximize or minimize) and express Z in terms of x, y.

Procedure:

The solution of the given L.P.P. should be divided under the following heads:

1. Preparation of *table of information.* 2. Write down the *decision variables.*

3. Form the *objective function.* 4. Write down the *constraints.*

5. Write down Mathematical Formulation.

Corner Point Method

This method is based on the fundamental extreme point theorem.

In previous class we have learnt how to formulate a system of linear inequalities involving two variables x and y mathematically.

Working Rule to Solve a L.P.P Problem

Step 1. Formulate the given L.P.P. in mathematical form.

Step 2. The inequations are converted into equations.

In the equation on putting $y = 0$ we get x-coordinate on x-axis. Similarly, putting $x = 0$ we get y-coordinate on y-axis. Join these two points to get the graph of the equation.

Step 3. The inequation of a line divides the plane into two half planes, to choose the plane of the inequation we put $x = 0$ and $y = 0$ in the inequation. If origin satisfies the inequation then the region containing the origin is the region represented by the given inequation. Otherwise the half plane not containing the origin is the region represented by the given inequation.

Step 4. The region satisfying all the inequations is the feasible region.

Step 5. The vertices (corner points) of the required region are known as extreme points of the set of all feasible solutions of the L.P.P.

Step 6. By putting the values of x and y of each corner point in the objective function we get the values of the objective function at each of the vertices of the feasible region. Out of all the values of the objective function, we get a point at which the objective function is optimum (maximum or minimum).

Solution of Linear Programming Problems

Here we will solve the linear programming problems.

Working Rule

Step 1. Define the problem mathematically.

Step 2. Graph the constraint inequalities by converting them into equations. Find out their respective intercept on both the axes and connect them by straight lines.

Step 3. Find out the vertices of the feasible region.

Step 4. Find out the value of the objective function on the vertices.

Step 5. Find out the optimum value of the objective function.

Procedure . The solution of the given LPP should be divided under the following heads:

1. Prepare a *table* of the data given in the problem.   2. Write down the *decision variables*.

3. Form the *objective function*.   4. Write down the constraints.

5. Mathematical formulation.   6. Region represented by inequations.

7. Apply Corner point method/Iso-cost or ISO-profit method.

Transportation Problems

The main problem of commercial organisation is to take decision that how much amount of an item should be transported from factories (sources) to various destinations (ware houses) in a minimum cost. This is known as transportation problem.

Diet Problem:

The cost of food items containing different quantity of proteins viamins, fats required for a normal man should be minimum.

CHAPTER 16: SIMPLEX METHOD

Theory of simplex method

The basis of the complex method consists of two fundamental conditions :-

(*i*) The feasibility condition

It ensures that the starting solution is basic feasible, only basic feasible solution will be obtained during computation.

(*ii*) Optimality condition

It guarantees only better solution (as compared to the current solution)

Some important definitions

Consider the following problem

Maximize $\quad Z = c_1x_1 + c_2x_2 + \ldots\ldots\ldots c_nx_n,$ Objective function

Subject to $\quad \left.\begin{array}{l} a_{11}x_1 + a_{12}x_2 + \ldots\ldots a_{1n}x_n \le b_1, \\ a_{21}x_1 + a_{22}x_2 + \ldots\ldots a_{2n}x_n \le b_2, \\ a_{m1}x_1 + a_{m2}x_2 + \ldots\ldots, + a_{mn}x_n \le b_m, \end{array}\right\}$ Constraints

where $x_1, x_2\ldots\ldots, x_n \ge 0.$

Introducing slack vairables $s_1, s_2, s_3\ldots\ldots\ldots s_m$ in m constraint equations. It can be put in the following standard form:

Maximize $Z = c_1x_1 + c_2x_2 + \ldots\ldots c_nx_n + s_1 + s_2 \ldots\ldots + s_m.$ .....(1)

Subject to
$$a_{11}x_1 + a_{12}x_2 + \ldots\ldots c_{1n}x_n + s_1 = b_1$$
$$a_{21}x_1 + a_{22}x_2 + \ldots\ldots a_{2n}x_n + s_2 = b_2$$
$$a_{m1}x_1 + a_{m2}x_2 + \ldots a_{mn}x_n + s_m = b_m$$
.....(2)

where $x_1, x_2, \ldots\ldots, x_n, s_1, s_2, \ldots\ldots s_m \geq 0$ .....(3)

1. **Solution :** To find out the values of $x_1, x_2 \ldots\ldots x_n, s_1, \ldots\ldots s_m.$

2. **Feasible solution :** The values of $x_1, x_2 \ldots\ldots x_n, s_1, \ldots\ldots s_m$ is known as feasible solution if these values satisfy the equations (2) and (3).

3. **Basic solution :** By making any n variables out of $n + m$ equal to zero, the values of the remaining m variable is called the basic solution, if the determinant of the coefficients of these slack variables is non negative.

4. **Basic variables :** The variables of the basic solution are called basic variables (Some of them may be zero). The other n variables are called non-basic variables.

5. **Basic feasible solution :** The basic solution of non-negative variable is called basic feasible solution.

6. **Non- degenerate basic feasible solution :** If all m basic feasible variables non-negative, then the set of these variables known as non degenerate basic feasible solution.

7. **Degenerate basic feasible solution :** If one or more basic feasible variable is zero, then the solution known as degenerate basic feasible solution.

8. **Optimal basic feasible solution :** It contains basic feasible solution that optimize the objective function.

9. **Sets of points :** Linear equation in two variables represents a line and a linear equation in three variables represents a plane. Both of them are considered as set of points.

CHAPTER 17: TRANSPORTATION PROBLEM
(N-W corner Method, Least cost Method, Vogel's Approximation method, u-v method)

Definitions

1. **Rim condition.**
   Total quantity of the item available at different sources = Total requirement at different destinations.

2. **Feasible solution**
   A feasible solution to a transportation problem is a set of non negative allocations, x_{ij} that satisfy rim (row and column) restriction.

3. **Basic feasible solution.**
   If it contains no more than $m + n - 1$ non negative allocations where m is rows and n is the number of columns of T.P.

4. **Optimal solution.**
   A feasible solution that minimize the transportation cost is called an optimal solution.

5. **Non-degenerate basic feasible solution**
   A basic feasible solution to a $(m \times n)$ transportation problem is said to be non degerate if:
   (a) the total number of non-negative allocation is exactly $m + n - 1$
   (b) these $m + n - 1$ allocations are in independent position.

6. **Degenerate basic feasible solution.**
   A basic feasible solution in which the total number of non-negative allocation is less than m + n - 1 is called degenerate basic feasible solution.

North West Corner Rule

North West Corner Method (NWCM)

The method start at the north west corner cell of the table.

Step 1. Allocate as much as possible to the north west corner and adjust associated amount of (supply – demand)

D_1	$S_1 - D_1$		S_1
			S_2
			S_3
D_1	D_2	D_3	

(a) $D_1 < S_1$

North west cell is to be filled by min $(D_1, S_1) = D_1$ then $x_{11} = D_1$.

The balance supply $(S_1 - D_1)$ is to be filled in cell $(1, 2)$

D_1			S_1
			S_2
			S_3
D_1	D_2	D_3	

(b) If $D_1 = S_1$, set $x_{11} = D_1$

S_1			S_1
$D_1 - S_1$			S_2
			S_3
D_1	D_2	D_3	

(c) If $D_1 > S_1$, then $x_{11} = $ min $(D_1, S_1) = S_1$ the balance $(D_1 - S_1)$ is to be filled to cell $(2, 1)$. Proceed vertically.

Step 2. Continue in this manner, step by step away from North west corner until final value is reached in the south east corner.

Least cost method

The least cost method depends upon the concentration of cheapest routes. The method assigns as much as possible to the cell with the smallest cost unit. The satisfied row or column is cross out and the amount of supply and demand are adjusted accordingly.

Vogel's Approximation Method (VAM) or penalty method or Regret method

Vogel's method is preferred to the methods described earlier. The solution found by VAM is very close if not equal to the optimal solution.

Working Rule

Step 1. Write the difference between the smallest and second smallest element in each column below the corresponding column and the difference between the smallest and second smallest element in each row on the right of the row.

Step 2. Identify the row or column with the greatest difference (penalty) and allocate min (D_1, S_1) in the lowest cost cell.

If there is a tie in the minimum cost, then select that cell which will have maximum assignment.

Step 3. If whole D_1 is completed in lowest cost cell, then the remaining elements of the column are crossed. If the whole S_1 is written in lowest cost cell then the remaining elements of the row are crossed.

Step 4. In the new table replace the original supply/demand by the remaining supply/demand.

Step 5. Recompute the row and column difference in the reduced transportation table until crossed elements of rows or columns and identify the maximum difference (penalty).

Step 6. Repeat the above procedure unit all the rim conditions are satisfied.

CHAPTER 18: ASSIGNMENT PROBLEM

Working Rule (Hungarian's Method)

Step 1. (a) Subtract the minimum cost of each row from all the elements of the respective row.

(b) From the resulting matrix, subtract the minimum cost of each column from all the elements of the respective column.

Step 2. (a) Row assignment

Choose the row containing single zero and draw square around it and draw a line in the column containing squared zero.

(b) Column Assignment

Choose the column containing single zero and make a square around it and draw a line in the row containing squared zero.

Step 3. Choose the smallest uncrossed element and subtract from all remaining uncrossed elements and add this smallest element to all the elements at the cross-section of two lines. Repeat the above process.

Step 4. If the number of lines drawn is not equal to number of rows repeat the above method. If the number of lines is equal to number of rows stop and get a optimal solution.

CHAPTER 19: NETWORK ANALYSIS IN PROJECT PLANNING

Drawing of a network

1. Each activity is represented by one and only one arrow.
2. Time flows from left to right.
3. Arrows should be in straight line not in a curve.
4. Angles between arrows should be obtuse as large as possible.
5. Arrow should not cross each other.
6. There must be no loop.
7. Each activity must have a tail and a head event.
8. Two or more activity should not have the same tail and head event.
9. In a network there must be only one initial event and only one end event.
10. If necessary dummy activity can be introduce.

Numbering the event

When a network is drawn in a logical sequence and the events are written as 1,2,3 inside the node circle . The number sequence of the events should be such as to reflect the flow of the network. The rule is framed by Dr. Fulkerson is written by the following steps.

Step 1. The initial event which has all outgoing arrows with no incoming arrow is numbered as (1).

Step 2. Delete all the arrows coming out from node (i) In this way some nodes will become initial events. Number these events 2, 3.......

Step 3. Delete all the arrows coming out from these numbered arrows (2) and (3) by deleting the arrows coming out from (2) and (3) some more events will be the initial events. These events numbered as 4, 5.........

Step 4. Continue until the terminal node which has all arrows incoming but no arrow going out is numbered.

Critical Path Method (CPM)

The critical path of a network gives the shortest time in which the whole project can be completed. Any activity is said to be critical if a delay in its start will cause a further delay in the completion of the project. And thus the target time will not be maintained in a network analysis.

Critical path method (CPM) can be defined as the identification of the sequence of the activities which will acquire greatest normal time to accomplish the period.

Working steps for critical path

Step 1. Write down the activity time just above the arrow representing that activity.

Step 2. Calculate earliest starting time (ES), earliest finishing time (EF), and latest finishing time (LF) for each activity.

Step 3. A table is prepared in which

    (1)   normal activity time

    (2)   earliest finish time

    (3)   latest finish time for each activity are written.

Step 4. The slack time is calculated by the formula LF_j-EF_i for each activity.

Step 5. The activities with zero slack are the critical activities. All critical activities are connected from beginning to end node by double arrows. This is the critical path.

PERT (Project evaluation and Review technique)

Time estimation in CPM are taken as deterministic. But in practical problem project time completion of a project can not be definite with certainty. But there are certain project in which research work is involve and there is no experience to handle such projects in such cases an uncertainty remains regarding the duration of various activity. So in such cases we need probabilistic approach to handle such problems. In PERT network time estimate are based on probabilistic approach. The time duration in PERT analysis is random variable characterize by some probability distribution usually a β-distribution. To estimate the parameters of the β-distribution (mean and variance). The PERT system is based on three time estimate of the performance time of an activity as follows:

1. The optimistic time estimate:

The shortest possible time required for the completion of an activity in normal circumstances.

  2.   **The pessimistic time estimate.** The maximum possible time of the activity will take, if every thing goes bad. However earth quick, flood, storm and labour trouble are not taken into account while estimating this time.

CHAPTER 20: SIMULATION

Simulation

Simulation is the imitation of reality which may be in the physical form or in the form of mathematical equations. For example; testing of an air craft model in a wind tunnel from which we determine the performance of the actual aircraft under real operating conditions.

Simulation can be defined as a representation of reality through the use of a model or other devices which will react in the same manner as reality under given set of conditions.

Advantages of Simulation technique

  1.   Some important managerial decision problems are solved by mathematical programming and experimentation with the actual system.

  2.   By simulation management we can foresee the difficulty and bottlenecks which may come due to new technique, equipment or process.

  3.   Operating personnels and non technical managers can understand easily the proposed plans.

  4.   It is flexible and can be modify according to changing conditions.

  5.   Computers simulation can express the performance of a system into few minutes.

  6.   It is easier than mathematical models.

  7.   Simulation has been used for training the managers.

CHAPTER 1

Approximation and Errors
(Truncation & Round off)

1.1 INTRODUCTION

Generally, we come across with two types of numbers, one exact and the other approximate.

For example; Exact numbers are 1, 3, 5, 7, 10, $\frac{5}{2}$, 6.23.

Approximate value of $\pi = 3.1416$,

Approximate value of $\frac{4}{3} = 1.3333$

Approximate value of $\sqrt{2} = 1.4142$

True value + Error = Measured value/observed value.

Correction. The error with sign changed is called correction.

Measured value + correction = True value.

1.2 SIGNIFICANT FIGURES

The digits used to express a number are called significant digits (figures).

8123, 3.187, 0.8725, contains 4 significant figures. While the numbers 0.0163, 0.00127, 0.000365 and 0.0000345 contain only three significant figures (digits).

Since zeroes before decimal and after decimal only helps to fix the position of decimal point.

Similarly, the numbers 52000 and 870.00 have two significant figures only.

1.3 TYPES OF ERRORS

(1) Inherent errors

Errors which are already present in the statement of problem before solution are called inherent errors.

Input data being approximated due to the limitations of computer, calculator, mathematical tables etc. It arises inherent error.

(2) Rounding off errors

During computation we round off the numbers due to this rounding off errors arise.

For example; $\frac{22}{7} = 3.14285143$

In practice it is convenient to limit such number as 3.14 or 3.143.

The dropping of the digits is called rounding off.

Rule: (1) To round of a number to *n* significant numbers discard all the digits to the right of *n*th digit if the discarded digit is.

   (*a*) Less than 5, leave the *n*th digit unchanged.

   (*b*) Greater than 5 increase the *n*th digit by one.

   (*c*) Exactly 5 (*i*) In case when *n*th number is odd increase the odd number by one.

                   (*ii*) In case when *n*th number is even, then *n*th number should not be changed.

Reason: Even numbers are more exactly divisible by many more numbers than odd numbers and so there will be fewer left over error in computation when the rounded numbers are left even.

The following numbers are rounded off to three significant figures:

5.783 to 5.78	7.767 to 7.77
15.876 to 15.9	95767 to 95800
8.4355 to 8.44	87.656 to 87.6

Also the numbers 7.284359, 15.864651, 9.464762 rounded off to four places of decimals at 7.2844, 15.8646, 9.4648 respectively.

Example 1. *The value of rounding off number 0.0021365 to four significant figure is :*

   (*a*) 0.0021   (*b*) 0.002137   (*c*) 0.002136   (*d*) *None of these*

                                                                    (*Bihar SBTE, 2011*)

Solution. Here, to the right of fourth significant figure is 5 and the fourth significant figure is 6. i.e.; even. So fourth significant figure remains unchanged.

Hence (*c*) is correct answer.   **Ans.**

Example 2. *The value of rounding off a number 2.32653 to three decimal place is:*

   (*a*) 2.326   (*b*) 2.327   (*c*) 2.32   (*d*) *None of these*

                                                        (*Bihar SBTE 2011*) **Ans.** (*b*)

Solution. To the right of third decimal place is 5.

In third decimal place the digit (6) is even so the third decimal digit remains unchanged.

Hence (*a*) is correct answer.   **Ans.**

Example 3. *Round off the number 0.046025 to four significant figures after decimal place.*

                                                                    (*Bihar SBTE 2009*)

Solution. To the right of fourth significant figure is 5 and fourth significant figure is 2 i.e.; even. Thus 2 remains unchanged.

Hence, answer is 0.04602.   **Ans.**

Example 4. *The value of rounding off number 56.4235 to three decimal place is:*

   (*a*) 56.424   (*b*) 56.423   (*c*) 56.42   (*d*) *None of these*

                                                                    (*Bihar SBTE 2009*)

Solution. Here 5 is to the right of third decimal place and the third decimal place digit is 3 i.e. odd.

Hence, the digit 3 is increased by 1 i.e.; 4.

Hence (*a*) is correct answer.   **Ans.**

Example 5. *The value of rounding off number 53.3651 to four significant figure is:*

    (*a*) *53.36*   (*b*) *53.37*   (*c*) *53.365*   (*d*) *None of these*

                                                                             (Bihar SBTE 2009)

Solution.   Here 5 is to the right of fourth significant figure and the fourth significant figure is 6 i.e.; even.

                  So, the digit 6 remains unchanged.

                  Hence (*a*) is correct answer.   **Ans.**

Example 6. *Rounding off the number 0.00022175 to four significant figures, the new number is* ....

                                                           (Bihar SBTE, 2008)

Solution.   Here 5 is to the right of fourth significant figure and the fourth significant figure is 7 i.e.; odd.

                  So, the digit 7 is increased by 1 i.e; 8.

                  Hence, answer is 0.0002218.   **Ans.**

Example 7. *Rounding off the number 37.46251 to three decimal significant figure, the new number is* ....   *(Bihar SBTE, 2004)*

Solution.   Here 5 is to the right of third decimal significant figure and third decimal significant digit is 2 i.e.; even. So, 2 remaining unchange.

                  Hence the answer is 37.462.   **Ans.**

(3) Truncation Error

Truncation error arises by droping some numbers of a decimal fraction or on replacing an infinite series by a finite one. If a decimal number is required of the length of five digits by dropping more than five digits. For example 12.36789 gives 12.367 on truncation but rounding off 12.36789 gives 12.368.

On truncation the remaining digits do not change but on rounding off the last digit changes according to rounding off rules. As we have seen above.

    For example ;   If $e^x = 1 + x + \dfrac{x^2}{2!} + \dfrac{x^3}{3!} + \dfrac{x^4}{4!} + ... \infty = X$ (say)

    is replaced by $1 + x + \dfrac{x^2}{2!} + \dfrac{x^3}{3!} + \dfrac{x^4}{4!} = X'$ (say)

    then the truncation error $= X - X'$

(4) Absolute Error

If X is the true value and X' is approximate value then $|X - X'|$ is called the absolute error.

Example 8.   *If a = 10.00 ± 0.05,   b = 15300 ± 100,   d = 62000 ± 500*

              Find maximum value of absolute error in a + 2b – d.   *(Bihar SBTE, 2009, 2008)*

Solution.   Maximum value of absolute error $|E|$ in $(a + 2b - d)$ is

                $|E| = a + 2b - d = 0.05 + 2 \times 100 + 500 = 0.05 + 200 + 500$

                $|E| = 700.05$ **Ans.**

(5) Relative Error

        Relative error $= \left| \dfrac{X - X'}{X} \right|$

Example 9. *The number 0.0205 is rounded off to three significant figures, the relative error is* ....

(Bihar SBTE, 2009, 2008)

Solution.   In the given number there are three significant figures and rounding off to three significant figure is meaning less.

Error = True value – Approximate value = 0.0205 – 0.0205 = 0

$$\text{Relative error} = \frac{\text{Error}}{\text{True value}} = \frac{0}{0.0205} = 0 \qquad\qquad\qquad\qquad \textbf{Ans.}$$

(6) Percentage error:

If X is the true value of a quantity and X' is its approximate value then $|X - X'|$ is called the absolute error E_a.

The relative error is defined by $E_r = \left| \dfrac{X - X'}{X} \right|$ and the percentage error is $E_p = 100\ E_r$

$$\text{Percentage error} = \frac{100\,|X - X'|}{X}$$

Example 10. *Define percentage error and calculate percentage error in* $\dfrac{a}{b}$ *if:*

$a = 620 \pm 5$ *and* $b = 10 \pm 0.05$ *(Bihar SBTE 2010)*

Solution.   % error of a:

$a_1 = 620 + 5 = 625$ $\qquad\qquad\qquad\qquad\qquad\qquad$ $a_2 = 620 - 5 = 615$

$$\therefore \quad \%\ a = \frac{625 - 615}{625} \times 100 = 1.6\%$$

% error of b:

$b_1 = 10 + 0.05 = 10.05$ $\qquad\qquad\qquad\qquad$ $b_2 = 10 - 0.05 = 9.95$

$$\therefore \quad \%\ b = \frac{10.05 - 9.95}{10.05} \times 100 = 0.995\%$$

$$\therefore \quad \%\ \text{error in}\ \frac{a}{b} = \frac{1.6}{0.995} = 1.60 \qquad\qquad\qquad\qquad \textbf{Ans.}$$

Example 11. *Round off the number 857623, to four significant digits and determine absolute, relative and percentage error.*

Solution.   857623, is rounded off to four significant digits is 857600.

Absolute error = Original number – Rounded off number

Absolute error = 857623 – 857600 = 23

$$\text{Relative error}\ =\ \frac{\text{Absolute error}}{\text{Original number}}\ =\ \frac{23}{857623}\ =\ 2.68 \times 10^{-5}$$

Percentage error = 100 × Relative error = 2.68 × 10⁻³ $\qquad\qquad$ **Ans.**

Example 12. *Round off the numbers 754126 and 16.73117 to four significant figures. Compute absolute error relative error and percentile error.*

Solution.   (*i*) Number rounded off to 4 significant figure equal to 754100

Absolute error = $|X - X'| = |\ 754126 - 754100\ | = |\ 26\ | = 26$

$$\text{Relative error} = \left| \frac{X - X'}{X} \right| = \frac{26}{754126} = 3.45 \times 10^{-5}$$

$$\text{Percentile error} = \frac{|X - X'|}{X} \times 100 = 3.45 \times 10^{-5} \times 100 = 3.45 \times 10^{-3}$$

(ii)   Number rounded off to four significant figure is 16.73

$$\text{Absolute error} = |X - X'| = |16.73117 - 16.73| = 0.00117$$

$$\text{Relative error} = \left|\frac{X - X'}{X}\right| = \frac{0.00117}{16.73117} = 6.99 \times 10^{-5}$$

$$\text{Percentile error} = \left|\frac{X - X'}{X}\right| \times 100 = 6.99 \times 10^{-5} \times 100 = 6.99 \times 10^{-3} \qquad \textbf{Ans.}$$

Example 13. *If 0.333 is the approximate value of 1/3, find the absolute and relative errors.*

(R.G.P.V., Bhopal III Semester, June 2007)

Solution. Here we have,

$$\text{Exact number} = \frac{1}{3}$$

$$\text{Approximate number} = 0.333 = \frac{333}{1000}$$

$$\text{Absolute error} = \text{Exact number} - \text{Approximate number} = \left|\frac{1}{3} - \frac{333}{1000}\right| = \frac{1}{3000}$$

$$\text{Relative Error} = \left|\frac{\text{Exact Number} - \text{Approximate number}}{\text{Exact number}}\right| = \frac{\frac{1}{3000}}{\frac{1}{3}} = \frac{1}{1000} = 0.001 \qquad \textbf{Ans.}$$

1.4 ERROR IN APPROXIMATE NUMBER

If a approximate number is correct to n decimal places, then the error in the approximate number is $\frac{1}{2} \times 10^{-n}$.

For example,

(i)   The approximate number of $\sqrt{2} = 1.141$ is correct to three decimal places, then the error in the

approximate number $= \frac{1}{2} \times 10^{-3} = \frac{1}{2} \times 0.001 = 0.0005$

(ii)   If a number is rounded to three places then the error $= \frac{1}{2} \times 10^{-3}$.

If approximate number is 0.879, correct to three decimal places then the error $= \frac{1}{2} \times 10^{-3}$.

Theorem. If the first significant figure of a number is k and the number is correct to n significant figures,

then the relative error is less than $\dfrac{1}{k \times 10^{n-1}}$.

Example 14. *Verify the formula by finding the relative error in the number 453.64 correct to 5 significant figures.*

Solution. Here, $n = 5$

$$\text{Absolute error} \quad > \quad \frac{1}{2} \times 10^{-2} = 0.01 \times \frac{1}{2} = 0.005$$

$$\text{Relative error} \ = \ \frac{0.005}{453.64} = \frac{5}{453640} = \frac{1}{2 \times 45364} < \frac{1}{2 \times 40000} < \frac{1}{4 \times 10^4} \quad \textbf{Verified.}$$

EXERCISE 1.1

1. Round off the following numbers correct to three significant figures :

   (a) 0.0031614   **Ans.** 0.00316   (b) 16.132102   **Ans.** 46.1

   (c) 0.30617   **Ans.** 0.306   (d) 2945567   **Ans.** 2940000

   (e) 45.56735   **Ans.** 45.6   (f) 5.26521   **Ans.** 5.26

2. Round off the following numbers to four significant figures:

   (a) 4.35921   **Ans.** 4.359   (b) 84.46825   **Ans.** 84.47

   (c) 2387871   **Ans.** 2388000   (d) 0.60046   **Ans.** 0.6005

   (e) 0.00046819   **Ans.** 0.000468   (f) 28.370101   **Ans.** 28.37

3. Find the percentage error if 887.492 is approximated to three significant figures.   **Ans.** 0.0554

4. $\sqrt{29}$ = 5.385 and $\sqrt{\pi}$ = 1.772 are correct to four significant figures. Find the relative error in their sum and difference.   **Ans.** 1.4×10^{-4}, 2.8×10^{-4}

5. What are different types of errors and safeguards against them ?

6. Explain significant figures, Truncation errors, Absolute error, Relative error, Percentage error.

   (R.G.P.V., Bhopal, III Semester, Dec. 2003)

7. Find the relative error if $\dfrac{1}{3}$ is approximated to 0.334.   **Ans.** 0.002

8. Find the percentage error if 625.483 is approximated to three significant figures.   **Ans.** 0.077

9. $\sqrt{29}$ = 5.385 and π = 3.317 correct to four significant figures. Find the relative errors in their sum and difference.

   Ans. 1.149×10^{-4}, 4.836×10^{-4}

10. Round off 14.64499 to four significant digits. Find out absolute, relative and percentage error.

    Ans. 0.00499, 3.41×10^{-4}, 3.41×10^{-2}

11. Define absolute error and relative error. Also find relation between them.   *(Bihar SBTE 2009)*

Objective Type Questions

Select the most appropriate answer from the alternatives given under:

12. The value of rounding off number 53.3651 to four signficiant figure is:

    (a) 53.36   (b) 53.37   (c) 53.365   (d) none of thes

    (Bihar SBTE, 2009) **Ans.** (b)

13. The value of rounding off number 56.4235 to three decimal place is

    (a) 56.424   (b) 56.423   (c) 56.42   (d) none of thes

    (Bihar SBTE, 2009) **Ans.** (a)

14. The value of rounding off number 0.0021365 to four signficiant figure is:

    (a) 0.0021   (b) 0.002137   (c) 0.002136   (d) none of these   **Ans.** (c)

15. The value of rounding off a number 2.32653 to three decimal place is:

    (a) 2.326   (b) 2.327   (c) 2.32   (d) none of the   **Ans.** (a)

16. Select the most suitable answer from the alternatives

    (i) Round off the number 6.0009 correct upto 4 signficiant figures is ....

    (Bihar SBTE, 2009) **Ans.** 6.00

17. Floating point representation of 5834.876 is:

(a) 583.4876×10 (b) 58.34876×10^2 (c) 5.834876×10^3 (d) 0.5834876×10^4

(Bihar SBTE, 2010) **Ans.** *(c)*

18. The error in the common logarithm of the number is $1'$.

19. Match the following:

(i) $(83.7)^{\frac{3}{4}}$ (a) 0.475   **Ans.** (i) — (c)

(ii) $\dfrac{1}{2.1}$ (b) 2.13   **Ans.** (ii) — (a)

(iii) $\sqrt{4.05}$ (c) 28.56   *(Bihar SBTE, 2014)* **Ans.** (iii) — (b)

20. If x be true value of a quantity and x_1 be the approximate value then the percentage error is..........

(a) $\dfrac{x - x_1}{x} \times 100$ (b) $\dfrac{x - x_1}{x} - x\,100$ (c) $\dfrac{x_1 - x}{x}$ (d) $\dfrac{x_1 - x}{x}$ **Ans.** (a)

(Bihar SBTE, 2010)

21. If 0.333 is the approximate value of $\dfrac{1}{3}$, then the relative error is

(a) 0.003   (b) 0.02   (c) 0.001   (d) none of these

(Bihar SBTE, 2014) **Ans.** *(d)*

22. Round off the number 0.046025 to four significant figures after decimal place.   **Ans.** (0.04602)

23. The number 0.0205 is rounded off to three significant figures, the relative error is..............   **Ans.** (zero)

(Bihar SBTE, 2008)

Choose the correct alternative :

24. If X and X' are exact and approximate values respectively of a quantity, then relative error in X' is :

(a) $\left| \dfrac{X - X'}{X} \right|$ (b) $\dfrac{X - X'}{X}$ (c) $|X - X'|$ (d) $X - X'$ **Ans.** (a)

(R.G.P.V., Bhopal, III Semester, June 2007)

Fill up the blanks:

25. The numerical difference between the exact value and the approximate value of a quantity is ...... error.

(Absolute/Relative)   **Ans.** Absolute

26. If two numbers are added or subtracted, then the magnitude of the absolute error in the result is the .... of the magnitude of the absolute errors is the two numbers.

(Prdoduct/sum)   **Ans.** Sum

27. If the number x is rounded to n decimal places, then the maximum absolute error is ....   **Ans.** $\dfrac{1}{2} \times 10^{-n}$

1.5 ERROR DUE TO APPROXIMATION OF THE FUNCTION

Let $z = f(x, y)$ be a function of two variables x and y.

If δx, δy be the errors in x and y, then the error in z is given by $z + \delta z = f(x + \delta x, y + \delta y)$.

Expanding $f(x, y)$ by Taylor's series, we get

$$z + \delta z = f(x, y) + \left(\frac{\partial f}{\partial x} \delta x + \frac{\partial f}{\partial y} \delta y \right) + \text{terms involving higher powers of } \delta x \text{ and } \delta y. \quad \ldots(1)$$

If δx and δy be so small that their squares and higher powers can be neglected, then (1) can be written as

$$\delta z = \frac{\partial z}{\partial x} \delta x + \frac{\partial z}{\partial y} \delta y \qquad \text{(app.)}$$

In general, if $z = f(x_1, x_2, \ldots x_n)$ and there are errors in $x_1, x_2 \ldots x_n$, then

$$\delta z = \frac{\partial z}{\partial x_1} dx_1 + \frac{\partial z}{\partial x_2} dx_2 + \frac{\partial z}{\partial x_3} dx_3 + \ldots + \frac{\partial z}{\partial x_n} dx_n.$$

Example 15. *If* $u = \dfrac{5x^3 y^4}{z^5}$ *and errors in x, y, z be 0.001, and compute the relative maximum error when*

$x = 1, y = 1, z = 1.$

Solution. $u = \dfrac{5x^3 y^4}{z^5}$ $\ldots(1)$

$$\delta x = \delta y = \delta z = 0.001$$

and $x = y = z = 1$

Differentiating (1) partially, with respect to 'x', we get

$$\frac{\delta u}{\delta x} = \frac{15x^2 y^4}{z^5}, \qquad \frac{\delta u}{\delta y} = \frac{20x^3 y^3}{z^5}, \qquad \frac{\delta u}{\delta z} = -\frac{25x^3 y^4}{z^6}$$

Now, we know that

$$\delta u = \frac{\partial u}{\partial x} \delta x + \frac{\partial u}{\partial y} \delta y + \frac{\partial u}{\partial z} \delta z = \frac{15x^2 y^4}{z^5} \delta x + \frac{20x^3 y^3}{z^5} \delta y - \frac{25x^3 y^4}{z^6} \delta z$$

The error being maximum

$$(\delta u)_{max} = \left| \frac{15x^2 y^4}{z^5} \delta x \right| + \left| \frac{20x^3 y^3}{z^5} \delta y \right| + \left| \frac{25x^3 y^4}{z^6} \delta z \right|$$

$$= \left| \frac{15(1)(1)}{(1)} (0.001) \right| + \left| \frac{20(1)(1)}{(1)} (0.001) \right| + \left| \frac{25(1)(1)}{(1)} (0.001) \right|$$

$$= 0.015 + 0.020 + 0.025 = 0.06$$

Relative error $= \dfrac{(\delta u)_{max}}{u} = \dfrac{0.06}{5} = 0.012$ **Ans.**

EXERCISE 1.2

1. The fractional error in the measurement of x is 0.001. What is the corresponding error in expansion of e^x.

2. If $R = \dfrac{4xy^2}{z^3}$ and errors in x, y, z be 0.001, show that the maximum relative error at $x = y = z = 1$ is 0.006.

3. Find the absolute error if $f(x, y) = 4x^3 y^6$ at $x = 1, y = 2$ and the errors in x and y be 0.01 and 0.02 respectively.

                                                         Ans. 23.04

4. If $f(x, y) = \dfrac{1}{2}\left(\dfrac{x^2}{y} + y\right)$, then find relative error in $f(x, y)$ when $x = 2$, $y = 3$ and errors in x and y are 0.01 and 0.03 respectively.

Ans. 6.923×10^{-3}

1.6 ERROR IN A SERIES

By Taylor's series $f(x)$ at $x = a$

$$f(x) = f(a + \overline{x-a}) = f(a) + (x-a)\, f'(a) + \frac{1}{2!}(x-a)^2\, f''(a) + ... + \frac{1}{(n-1)!}(x-a)^{n-1}\, f^{n-1}(a) + R_n(x)$$

Where $\quad R_n(x) = \dfrac{(x-a)^n}{n!}\, f^n(\theta), \qquad a < \theta < x$

Maximum error $= \dfrac{(x-a)^n}{n!}\, f^n(x)$.

Example 16. *Find the number of terms of the exponential series such that their sum gives the value of e^x correct to five decimal places at $x = 1$.*

Solution. We have, by Taylor's Series

$$e^x = 1 + \frac{x}{1!} + \frac{x^2}{2!} + \frac{x^3}{3!} + ... + \frac{x^{n-1}}{(n-1)!} + R_n(x)$$

Where $\qquad R_n(x) = \dfrac{x^n}{n!} e^\theta, \qquad\qquad 0 < \theta < x$

Maximum absolute error (at $\theta = x$) $= \dfrac{x^n}{n!}(e^x)$

At $\qquad\qquad x = 1$

Maximum error $= \dfrac{1^n}{n!} e$

Maximum relative error $= \dfrac{\text{Maximum error}}{\text{Value of function}} = \dfrac{1.e}{n!e} = \dfrac{1}{n!}$

For six decimal accuracy at $x = 1$, we get

$$\frac{1}{n!} < \frac{1}{2} \times 10^{-5}$$

i.e. $\qquad\qquad n! > 2 \times 10^5 \qquad \Rightarrow \qquad n = 9$

Hence, the series should contain 9 terms in order that the sum is correct to five decimal places. **Ans.**

Example 17. *Expand $\sin x$ by Taylor's series about $x = 0$ in the interval $[0, 2\pi]$ upto four accurate decimal places. How many terms are needed.*

Solution. Taylor's expansion of $\sin x$ about $x = 0$ is

$$\sin x = x - \frac{x^3}{3!} + \frac{x^5}{5!} - .... + \frac{(-1)^{n-1} . x^{2n-1}}{(2n-1)!} + R_n(x)$$

$$R_n(x) = \frac{(-1)^n (x)^{2n+1}}{(2n+1)!} M$$

$$M = \sin x, \qquad \text{Maximum } |M| = 1, \qquad 0 \le x \le 2\pi$$

$$\left| \frac{(-1)^n x^{2n+1}}{(2n+1)!} M \right| \le \frac{1}{2} \times 10^{-\frac{1}{4}}$$

$$\frac{(2\pi)^{2n+1}}{(2n+1)!} \le 0.00005$$

Which gives $n = 12.$ **Ans.**

Example 18. *Obtain the linearised form $f(x, y)$ of the function $f(x, y) = x^2 - xy + \dfrac{1}{2} y^2 + 3$ at the point*

(3, 2) using Taylor's Series expansion. Find the maximum error and relative error in magnitude in the approximation $f(x, y) = t(x, y)$ over the rectangle are : $|x - 3| < 0.1, |y - 2| < 0.1.$

Solution.   Here, we have

$$x = 3, \qquad\qquad y = 2, \qquad \delta x = 0.1, \qquad\qquad \delta y = 0.1$$

$$f(x, y) = x^2 - xy + \frac{1}{2} y^2 + 3 \qquad \Rightarrow \qquad f(3, 2) = 9 - 6 + 2 + 3 = 8$$

$$\frac{\partial f}{\partial x} = 2x - y \qquad \Rightarrow \qquad \left(\frac{\partial f}{\partial x} \right)_{(3,2)} = 2(3) - 2 = 4$$

$$\frac{\partial f}{\partial y} = -x + y \qquad \Rightarrow \qquad \left(\frac{\partial f}{\partial y} \right)_{(3,2)} = -3 + 2 = -1$$

$$\Rightarrow \qquad f(x + \delta x, \ y + \delta y) = f(x, y) + \frac{\partial f}{\partial x} \delta x + \frac{\partial f}{\partial y} \delta y$$

$$\Rightarrow \qquad\qquad f(3.1, 2.1) = f(3, 2) + (4)(0.1) + (-1)(0.1)$$

$$f(3.1, 2.1) - f(3, 2) = 0.4 + (-0.1)$$

Maximum $\delta f = |0.4| + |-0.1| = 0.4 + 0.1 = 0.5$

$$\text{Maximum relative error} = \frac{\text{Absolute error}}{\text{Value of the function}} = \frac{0.5}{8} = 0.0625 \qquad\qquad\qquad \textbf{Ans.}$$

EXERCISE 1.3

1.   The function $f(x) = \tan^{-1} x$ can be expanded as

$$\tan^{-1} x = x - \frac{x^3}{3} + \frac{x^5}{5} - \ldots + (-1)^{n-1} \frac{x^{2n-1}}{(2n-1)} + \ldots$$

Find n such that the series determines $\tan^{-1} 1$ correct to two significant digits.   **Ans.** $n = 100$

2.   Derive the series

$$\log_e \frac{1+x}{1-x} = 2 \left[x + \frac{x^3}{3} + \frac{x^5}{5} + \ldots \right]$$

and use it to compute the value of $\log_e (1.2)$ correct to seven decimal places. Determine the number of terms required if the series for $\log_e (1 + x)$ were used instead.

Ans. $\log_e 1.2 = 0.1823215, n = 9.$

3.   Find the number of terms of the exponential series such that their sum gives the value of e^x correct to five decimal places for all values of x in the range $0 \le x \le 1.$ **Ans.** 9

4. Determine the number of terms required in the expansion of e^x such that their sum is correct to six decimal places at $x = 1$.

Ans. 10

5. Obtain a second degree polynomial approximation to $f(x) = [1 + x]^{\frac{1}{2}}$ over $[0, 1]$ by means of the Taylor expansion about $x = 0$. Use first three terms of expansion to approximate $f(0.05)$. Obtain a bound of the error in the interval $[0, 1]$.

Ans. 0.0625

6. How many terms of the series $\log (1 + x)$ are required to get five decimal places of accuracy for computation at $x = 0.9$.

Ans. 75

7. Find the number of terms of $\log (1 + x)$ such that $\log (1.2)$ correct to six decimal places.

Ans. 10

8. Define percentage error of number.

(Bihar SBTE, 2012)

Answer any two of the following:

If $a = 10.00 \pm 0.05$

$b = 15300 \pm 100$

$d = 62000 \pm 500$

Find maximum value of absolute error in $a + 2b - d$.

(Bihar SBTE, 2008) **Ans.** 700.05

CHAPTER 2
Numerical Solution of Non-linear and Transcendental Equations
(Bisection Method, Regula Falsi, Newton-Raphson Method, Iteration Method)

2.1 METHODS OF SOLVING EQUATIONS

The polynomial equations and the transcendental equations can be solved by the following method:

(*i*)   Bisection method   (*ii*) Newton-Raphson method

(*iii*)   False position method   (*iv*) Iterative method

2.2 BISECTION METHOD (BOLZANO METHOD)

If $f(x)$ is continuous in the interval (a, b) such that $f(a)$ and $f(b)$ are of opposite signs, then

$$f(a).f(b) < 0$$

The curve crosses the *x*-axis between *a* and *b* then the first approximation to the root is

$$x_1 = \frac{1}{2}(a+b)$$

Now there are three cases, if

1. $f(x_1) = 0$, then x_1 is the root of $f(x) = 0$

2. $f(x_1) < 0$, then root lies between x_1 and b.   [If $f(a) = -$ve, $f(b) = +$ve]

3. $f(x_1) > 0$, then root lies between a and x_1.

Suppose $f(x_1) > 0$, then $a < \text{Root} < x_1$

Second approximation to the root is $x_2 = \frac{1}{2}(a+x_1)$

Again there are three cases, if

1. $f(x_2) = 0$, then x_2 is the root of $f(x) = 0$.

2. $f(x_2) < 0$, then root lies between x_2 and x_1.

3. $f(x_2) > 0$, then root lies between a and x_2.

Suppose $f(x_2) > 0$.

Third approximation $x_3 = \frac{1}{2}(a+x_2)$

If **1.** $f(x_3) = 0$, then x_3 is the root of $f(x) = 0$

2. $f(x_3) > 0$, then root lies between a and x_3

and so on.

2.3 CONVERGENCE OF BISECTION METHOD

The successive approximations x_n of a root $x = \alpha$ of the equation $f(x) = 0$ is said to converge to $x = \alpha$ with order $q \geq 1$.

If $|x_{n+1} - \alpha| \leq C|x_n - \alpha|$

Here, q, $n > 0$, n and C is some constant greater than 0.

when $q = 1$ and $0 < C < 1$, then the convergence is said to be of first order and C is called the rate of convergence.

$$|x_{n+1} - \alpha| \leq C|x_n - \alpha|$$

Example 1. *If the roots of the equation $3x^3 + 5x - 40 = 0$ lies in the interval (2, 3), find the first approximate root of the equation using bisection method.* *(Bihar SBTE, 2010)*

Solution. $f(x) = 3x^3 + 5x - 40 = 0$

$f(2) = 24 + 10 - 40 = -6$

$f(3) = 81 + 15 - 40 = +56$

$f(2) \cdot f(3) < 0$

The first approximate root lies between 2 and 3.

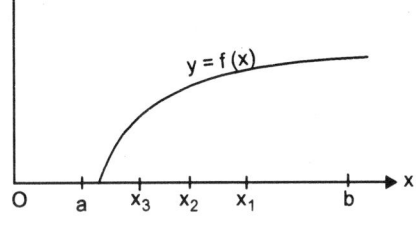

$$x_1 = \frac{2+3}{2} = \frac{5}{2} = 2.5$$

$f(2.5) = 3(2.5)^3 + 5(2.5) - 40 = 19.375$

$f(2) \cdot f(2.5) < 0$

The second approximation root lies between 2 and 2.5. **Ans.**

Example 2. *Perform five iterations of the bisection method to obtain the smallest positive root of the equation $f(x) = x^3 - 5x + 1 = 0$* *(U.P. Jan. 2011)*

Solution. We have, $f(x) = x^3 - 5x + 1 = 0$

$f(0) = +1$

$f(1) = 1 - 5 + 1 = -3$

As $f(0) \cdot f(1) < 0$,

The first approximate root lies between 0 and 1.

$$x_1 = \frac{0+1}{2} = \frac{1}{2}$$

$$f\left(\frac{1}{2}\right) = \frac{1}{8} - \frac{5}{2} + 1 = -1.375$$

As $f(0) \, f\left(\frac{1}{2}\right) < 0$

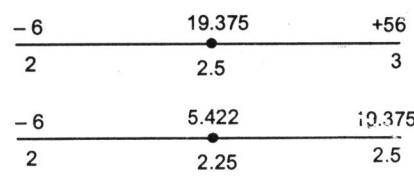

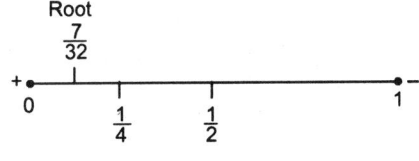

The second approximate root lies between 0 and $\frac{1}{2}$.

$$x_2 = \frac{0 + \frac{1}{2}}{2} = \frac{1}{4}$$

$$f\left(\frac{1}{4}\right) = \frac{1}{64} - \frac{5}{4} + 1 = -0.234375$$

As $f(0).f\left(\dfrac{1}{4}\right) < 0$

The third approximate root lies between 0 and $\dfrac{1}{4}$.

$$x_3 = \dfrac{0 + \dfrac{1}{4}}{2} = \dfrac{1}{8}$$

$$f\left(\dfrac{1}{8}\right) = \dfrac{1}{512} - \dfrac{5}{8} + 1 = +0.37695$$

As $f\left(\dfrac{1}{8}\right).f\left(\dfrac{1}{4}\right) < 0$

The fourth approximate root lies between $\dfrac{1}{8}$ and $\dfrac{1}{4}$.

$$x_4 = \dfrac{\dfrac{1}{8} + \dfrac{1}{4}}{2} = \dfrac{3}{16}$$

$$f\left(\dfrac{3}{16}\right) = \left(\dfrac{3}{16}\right)^3 - \dfrac{15}{16} + 1 = +0.06909$$

As $f\left(\dfrac{3}{16}\right).f\left(\dfrac{1}{4}\right) < 0$

Then $\dfrac{3}{16} <$ root $< \dfrac{1}{4}$.

$$x_5 = \dfrac{\dfrac{3}{16} + \dfrac{1}{4}}{2} = \dfrac{7}{32}, \qquad \text{Root} = \dfrac{7}{32}$$ **Ans.**

Example 3. *Find the real root of the equation $x^3 + x - 1 = 0$ by bisection method (to find third approximation of the root).*
                        (Bihar SBTE, 201, 2009)

Solution. Let $f(x) = x^3 + x - 1$

Here $f(0) = -1 = $ –ve and $f(1) = 1 = +$ ve.

Hence a root lies between 0 and 1.

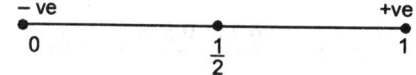

$f(0.5) = (0.5)^3 + (0.5) - 1 = 0.125 + 0.5 - 1 = -0.375$

and $f(1) = 1 = +$ ve i.e. $f(1).f(0.5) < 0$

Hence, the root lies between 0.5 and 1.

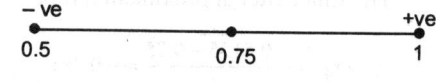

Let $x_1 = \dfrac{0.5 + 1}{2} = 0.75$

$f(0.75) = (0.75)^3 + (0.75) - 1 = 0.421 + 0.75 - 1 = 0.1718 = +$ ve

Hence the root lies between 0.5 to 0.75.

Now, $x_2 = \dfrac{0.5 + 0.75}{2} = 0.625$

$f(0.625) = (0.625)^3 + (0.625) - 1 = 0.2441 + 0.625 - 1 = -0.1308 = -$ ve
Hence the root lies between 0.625 and 0.75.

Now $x_3 = \dfrac{0.625 + 0.75}{2} = 0.687$

$f(0.687) = (0.687)^3 + 0.687 - 1 = 0.324 + 0.687 - 1 = 0.0119 = 0$ nearly

The root of the given equation is 0.687 correct upto 1 decimal place. **Ans.**

Example 4. *Discuss Bolzano's bisection method for solving algebraic and transcedental equations and hence find a root of the equation $x^3 - x^2 + 1 = 0$.* *(Bihar SBTE, 2008)*

Solution. $f(x) = x^3 - x^2 + 1 = 0$

Now, $f(0) = 1$

$f(-1) = -1$

Since $f(0) . f(-1) < 0$.

The first better approximation root in the interval $(-1, 0)$.

$\therefore \quad x_0 = \dfrac{-1+0}{2} = -0.5$

$f(-0.5) = (-0.5)^3 - (-0.5)^2 + 1 = 0.625$

Since, $f(-0.5) . f(-1) < 0$

The second better approximation root in the interval $(-1, -0.5)$.

$\therefore \quad x_1 = \dfrac{-1-0.5}{2} = -0.75$

$f(-0.75) = (-0.75)^3 - (-0.75)^2 + 1 = 0.0156$

Since, $f(-0.75) . f(-1) < 0$.

The third better approximation root in the interval $(-1, -0.75)$.

$\therefore \quad x_2 = \dfrac{-1-0.75}{2} = -0.875$

$f(-0.875) = (-0.875)^3 - (-0.875)^2 + 1 = -0.436$

Since, $f(-0.875) . f(-0.75) < 0$

The fourth better approximation root in the interval $(-0.75, -0.875)$.

$\therefore \quad x_3 = \dfrac{-0.75 - 0.875}{2} = -0.8125$

$f(-0.8125) = (-0.8125)^3 - (-0.8125)^2 + 1 = -0.1965$

The fifth better approximation root in the interval $(-0.8125, -0.75)$.

$\therefore \quad x_4 = \dfrac{-0.8125 - 0.75}{2} = -0.781$

$f(-0.781) = -0.0863 = 0$ nearly

So -0.781 is the correct root upto one decimal place. **Ans.**

Example 5. *Find the 1st positive root of $f(x) = 2x^3 - 3x - 6 = 90$ by Newton. Rapsion method upto third approximation of roat.* *(Bihar SBTE, 2012)*

Solution. $f(x) = 2x^3 - 3x - 6$

$f'(x) = 6x^2 - 3$ and $f''(x) = 12x$

Here $f(1) = 2 - 3 - 6 = -7$ (−ve)

$f(2) = 16 - 6 - 6 = 4$ (+ve)

also $f(1)$ & $f(2)$ both are of opposite signs.

So the root of $f(x) = 0$ can lie between 1 & 2

By Newton Method

The approximate root $b = x - \dfrac{f(x)}{f'(x)} = 2 - \dfrac{2(2)}{f'(2)} = 2 - \dfrac{4}{21} = 1.81$

$\gamma = 1.81 - \dfrac{f(1.81)}{f'(1.81)} = 1.81 - \dfrac{0.429}{16.657} = 1.785$

$\gamma = 1.785 - \dfrac{f(1.785)}{f'(1.785)} = 1.785 - \dfrac{0.20}{16.117} = 1.784$ **Ans.**

EXERCISE 2.1

1. Solve $x^3 - 9x + 1 = 0$ for the root between $x = 2$ and $x = 4$ by the method of Bolzano Method. **Ans.** 2.9375
2. Find the positive root of $x^3 - x = 1$ correct to four decimal places by Bisection method.   **Ans.** 1.3248
3. Find a real root of the equation $x^3 + x^2 - 1 = 0$ using Bisection method.   (*Bihar SBTE, 2009*) **Ans.** 0.75486
4. Find a root of the equation $x^3 - 3x - 5 = 0$ by Bisection method.   **Ans.** 2.280
5. Solve $x^3 - 4x - 9 = 0$ by Bolzano method.   **Ans.** 2.706
6. Find the 1st positive root of the question $5x^3 + 2x^2 + x - 9 = 0$ by Bisection method up to third approximation of the root.   (*Bihar SBTE, 2014*)

Objective Type Questions

Choose the correct alternative:

7. Ist positive root of the equation $x^4 + 3x^3 + 5x^2 + 7x - 6 = 0$ lies between ....
   (a) (0, 1)   (b) (1, 2)   (c) (2, 3)   (d) (3, 4)
   (*Bihar SBTE, 2014*) **Ans.** (*a*)
8. A first negative real root of the equation $x^3 + 3x^2 - 3 = 0$ lies between ....
   (a) (0, – 1)   (b) (–1, –2)   (c) (–2, –3)   (d) (–3, –4)
   (*Bihar SBTE, 2011*) **Ans.** (b)
9. The integral part of a root of the equation $x^2 - 3x + 1 = 0$ is ....
   (a) 0   (b) 1   (c) 2   (d) 3
   (*Bihar SBTE, 2008*) **Ans.** (a)
10. The function $f(x) = 2x^3 - x - 6 = 0$ has a real root lying between ....
    (a) (0, 1)   (b) (1, 2)   (c) (2, 3)   (d) (3, 4)   **Ans.** (b)
11. If the roots of the equation $3x^3 + 5x - 40 = 0$ lies in the interval (2, 3), find the first approximate root of the equation using bisection method.   (*Bihar SBTE, 2010*) **Ans.** 2.5
12. Approximate root of the equation $f(x) = 0$ is obtained by ....
    (c) Simplex method   (b) Gauss-Jordan Method
    (c) Graphical Method   (d) None of these   (*Bihar SBTE, 2014*) **And.** (d)
13. If $f(x_1)$ and $f(x_2)$ of $f(x) = 0$ are of opposite sign then between x_1 and x_2 there exists
    (*a*) Almost one root   (*b*) Only one root
    (*c*) More than one root   (*d*) At least one root   **Ans.** (*d*)
14. Bisection method is a slow process but after each iteration it gives
    (*a*) Some time worse   (*b*) Some time better approximation
    (*c*) Always better approximation   (*d*) None of them   **Ans.** (*c*)

15. If x_1 and x_2 are approximate root of $f(x) = 0$ then $\dfrac{x_1 + x_2}{2}$ is also an approximate root if

(a) $f(x_1) \cdot f(x_2) > 0$ (b) $f(x_1) \cdot f(x_2) = 0$

(c) $f(x_1) \cdot f(x_2) < 0$ (d) None of them   **Ans.** (c)

2.4 NEWTON – RAPSON METHOD   (Bihar, SBTE, 2003, 2009, U.P., III Semester Dec. 2009)

Let x_0 be an approximate root of $f(x) = 0$ and let $x_1 = r_0 + h$ be the correct root so that $f(x_0 + h) = 0$

To find h, we expand $f(x_0 + h)$ by Taylor's Series

$$f(x_0 + h) = f(x_0) + hf'(x_0) + \frac{h^2}{2!} f''(x_0) + \dots \qquad [f(x_0 + h) = 0]$$

$$0 = f(x_0) + hf'(x_0) \qquad \text{[Neglecting the second and higher order derivative]}$$

$$h = -\frac{f(x_0)}{f'(x_0)}$$

But $\qquad x_1 = x_0 + h$

Putting the value of h, we get $\qquad \Rightarrow \qquad x_1 = x_0 - \dfrac{f(x_0)}{f'(x_0)}$

x_1 is better approximation than x_0. $\qquad\qquad x_2 = x_1 - \dfrac{f(x_1)}{f'(x_1)}$

x_2 is better approximation than x_1.

Successive approximations are $x_3, x_4, \dots\dots x_{n+1}$.

$$x_{n+1} = x_n - \frac{f(x_n)}{f'(x_n)}$$

Which is the Newton - Raphson formula.

Note 1. Newton method is the best known procedure for finding the roots of an equation. It is applicable to the solution of all types of equations *i.e.*, algebraic and transcendental.

    2. This method is useful in case of large value of $f'(x)$. For large $f'(x)$, h will be small.

    3. This formula converges rapidly. If the initial approximation x_0 is taken very close to the root α. Thus proper choice of x_0 is very important for the success of this method.

2.5 GEOMETRICAL INTERPRETATION

Let $P_0 P$ be a curve $y = f(x)$.

Slope of the tangent $P_0 B$ to the curve at the point $P_0 (x_0, y_0) = f'(x_0)$.

Tangent $P_0 B$ cuts the x-axis at B i.e. $(x_1, 0)$.

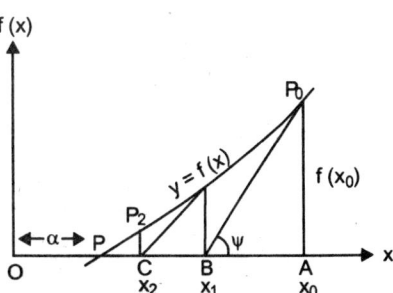

$$\begin{aligned}
x_1 &= OB \\
&= OA - AB \\
&= x_0 - P_0 A \cot \Psi \\
&= x_0 - \frac{P_0 A}{\tan \psi} \qquad \left[AN = f(x_0), \frac{BA}{AN} = \cot \psi \right] \\
&= x_0 - \frac{f(x_0)}{f'(x_0)} \qquad \text{(First approximation)}
\end{aligned}$$

The tangent to the curve at P_1 (corresponding to x_1) cuts the axis at C $(x_2, 0)$.

Using x_1 as the starting point, then

Similarly $\quad x_2 = x_1 - \dfrac{f(x_1)}{f'(x_1)}$

Now x_2 is nearer to α than x_2 (second approximation).

The process can be repeated and the root α is approached very fast.

$$x_{n+1} = x_n - \dfrac{f(x_n)}{f'(x_n)}$$

2.6 WORKING RULE TO SOLVE $f(x) = 0$ BY NEWTON - RAPHSON METHOD

Step 1. Choose two close numbers b and c such that $f(b)$ and $f(c)$ are of opposite signs. then the root α lies between b and c.

Step 2. Out of $f(b)$ and $f(c)$ choose which is nearer to zero. If $f(b)$ is nearer to zero then b is an initial approximate root (x_0) of the given equation.

Step 3. Apply Newton-Raphson formula to find out better approximate root x_1.

$$x_1 = x_0 - \dfrac{f(x_0)}{f'(x_0)}$$

Repeat the process to get successive approximation.

Step 4. Stop when two approximate roots are equal.

Example 6. *Find the smallest positive root of the equation* $x^3 - 2x + 0.5 = 0$

Solution. Here $\qquad f(x) = x^3 - 2x + 0.5$

$\qquad\qquad f(0) = 0.5$

$\qquad\qquad f(0.1) = 0.001 - 0.2 + 0.5 = 0.3001\,(+ve)$

$\qquad\qquad f(0.3) = 0.027 - 0.6 + .5 = -0.073\,(-ve)$

Hence, the root lies between 0.1 and 0.3.

$f(0.3)$ is near to zero.

So, 0.3 is first approximation, we have a better approximation as $0.3 + h$.

By Newton's method

$$f'(x) = 3x^2 - 2, \quad f'(0.3) = 3(0.3)^2 - 2 = -1.73$$

Second approximate root, $\quad x_2 = 0.3 - \dfrac{f(0.3)}{f'(0.3)} = 0.3 - \dfrac{-0.073}{-1.73} = 0.3 - \dfrac{73}{1730} = 0.3 - 0.0422$

$$= 0.2578 \qquad\qquad\qquad\qquad\textbf{Ans.}$$

Example 7. *Starting with* $x_0 = 3$, *find a root of* $x^3 - 3x - 5 = 0$. *Use four iterations of Newton-Raphson Method.* (Bihar SBTE, 2011, 2010, 2004, 2002)

Solution. $\quad f(x) = x^3 - 3x - 5 = 0, \quad f'(x) = 3x^2 - 3$

$\qquad\qquad f(3) = 27 - 9 - 5 = 13, \quad f'(3) = 27 - 3 = 24$

$$x_1 = x_0 - \dfrac{f(x_0)}{f'(x_0)} = 3 - \dfrac{f(3)}{f'(3)} = 3 - \dfrac{13}{24} = 3 - 0.5417 = 2.4583$$

$$f(2.4583) = (2.4583)^3 - 3(2.4583) - 5 = 2.4812$$
$$f'(2.4583) = 3(2.4583)^2 - 3 = 15.1297$$

$$x_2 = 2.4583 - \frac{f(2.4583)}{f'(2.4583)} = 2.4583 - \frac{2.4812}{15.1297} = 2.4583 - 0.1640 = 2.2943$$

$$f(2.2943) = (2.2943)^3 - 3(2.2943) - 5 = 0.1939$$
$$f'(2.2943) = 3(2.2943)^2 - 3 = 12.7914$$

$$x_3 = 2.2943 - \frac{f(2.2943)}{f'(2.2943)} = 2.2943 - \frac{0.1939}{12.7914} = 2.2943 - 0.0152 = 2.2791$$

$$f(2.2791) = (2.2791)^3 - 3(2.2791) - 5 = 0.0010219$$
$$f'(2.2791) = 3(2.2791)^2 - 3 = 12.5829$$

$$x_4 = 2.2791 - \frac{f(2.2791)}{f'(2.2791)} = 2.2791 - \frac{0.0010219}{12.5829} = 2.2791 - 8.1213 \times 10^{-5} = 2.2790$$

Hence, the root of the given equation is 2.2790 upto three decimal place.   **Ans.**

Example 8. *Find the real root of the following equation, correct to three decimal places, using Newton-Raphson Method.*   (R.G.P.V., III Semester, June 2007)

$$x^3 - 2x - 5 = 0$$

Solution. $x^3 - 2x - 5 = 0$ ... (1)

Let $f(x) = x^3 - 2x - 5$
$$f(2) = 8 - 4 - 5 = -1$$
$$f(2.5) = (2.5)^3 - 2(2.5) - 5 = +5.625$$

Since $f(2)$ and $f(2.5)$ are, of opposite signs, the root of (1) lies between 2 and 2.5; $f(2)$ is near to zero than $f(2.5)$, So 2 is better appropriate root than 2.5.

$$f'(x) = 3x^2 - 2 \Rightarrow f'(2) = 12 - 2 = 10$$

Let 2 be an approximate root of (1). By Newton-Raphson method

$$x_1 = x_0 - \frac{f(x_0)}{f'(x_0)} = 2 - \frac{f(2)}{f'(2)} = 2 - \frac{-1}{10} = 2.1$$

$$f(2.1) = (2.1)^3 - 2(2.1) - 5 = 9.261 - 4.2 - 5 = 0.061$$
$$f'(2.1) = 3(2.1)^2 - 2 = 11.23$$

$$x_2 = 2.1 - \frac{f(2.1)}{f'(2.1)} = 2.1 - \frac{0.061}{11.23} = 2.1 - 0.00543 = 2.09457$$

$$f(2.09457) = (2.09457)^3 - 2(2.09457) - 5$$
$$= 9.1893 - 4.18914 - 5 = 0.00016$$
$$f'(2.09457) = 3(2.09457)^2 - 2 = 13.16167 - 2 = 11.16167$$

$$x_3 = 2.09457 - \frac{f(2.09457)}{f'(2.09457)} = 2.09457 - \frac{0.00016}{11.16167} = 2.09457 - 0.000014 = 2.09456$$

As $x_3 = x_2$ correct upto four places of decimal.

Hence, the root of the given equation is 2.0945 correct upto four places of decimal.

Example 9. *Derive the Newton-Raphson formula for finding a root of a non-linear equation. Find a root of $f(x) = x^3 + 2x^2 + 10x - 20 = 0$ up to 10 iterations.*

(U.P., III Semester, Dec. 2009)

Solution. For derivation of the formula see Art. 2.4 on page 15.

Newton-Raphson Formula $x_{n+1} = x_n - \dfrac{f(x_n)}{f'(x_n)}$

$f(x) = x^3 + 2x^2 + 10x - 20$ ... (1) $\Rightarrow f'(x) = 3x^2 + 4x + 10$

$f(1) = 1^3 + 2(1)^2 + 10(1) - 20 = 1 + 2 + 10 - 20 = -7$

$f(2) = 2^3 + 2(2)^2 + 10(2) - 20 = 8 + 8 + 20 - 20 = 16$

Since $f(1)$ and $f(2)$ are of opposite signs, so the root of (1) lies between 1 and 2.

As $f(1)$ is near to zero than $f(2)$. So, 1 is better approximate root than 2.

$f'(1) = 3(1)^2 + 4(1) + 10 = 3 + 4 + 10 = 17$

$$x_1 = x_0 - \frac{f(x_0)}{f'(x_0)} \qquad \text{(Newton-Raphson formula)}$$

First Iteration

$$x_1 = x_0 - \frac{f(1)}{f'(1)} = 1 - \frac{-7}{17} = \frac{24}{17} \qquad f\left(\frac{24}{17}\right) = 0.9175656$$

Second Iteration

$$x_2 = x_1 - \frac{f(x_1)}{f'(x_1)} = \frac{24}{17} - \frac{f\left(\dfrac{24}{17}\right)}{f'\left(\dfrac{24}{17}\right)} = 1.4117647 - \frac{0.9175656}{21.6262976} \qquad f'\left(\frac{24}{17}\right) = 21.6262976$$

$$= 1.4117647 - 0.04242823 = 1.36933647$$

Third Iteration

$$x_3 = x_2 - \frac{f(x_2)}{f'(x_2)} = 1.36933647 - \frac{f(1.36933647)}{f'(1.36933647)}$$

$$= 1.36933647 - \frac{0.01114811}{21.1025930}$$

$$= 1.36933647 - 0.00052828 = 1.36880819$$

Fourth Iteration

$$x_4 = x_3 - \frac{f(x_3)}{f'(x_3)} = 1.36880819 - \frac{f(1.36880819)}{f'(1.36880819)}$$

$$= 1.36880819 - \frac{0.00000173}{21.09614034} = 1.36880819 - 0.00000008$$

$$= 1.36880811$$

Fifth Iteration

$$x_5 = x_4 - \frac{f(x_4)}{f'(x_4)} = 1.36880811 - \frac{f(1.36880811)}{f'(1.36880811)}$$

$$= 1.36880811 - \frac{0.00000005}{21.09618937} = 1.36880811 - 0.00000000$$

$$= 1.36880811$$

The root of the given equation is 1.36880811 correct upto eighth decimal place after fifth iteration. For the accuracy more than 8 decimal places, we can iterate further. **Ans.**

Example 10. *Find the real root of the equation $x^4 - x - 9 = 0$ by Newton-Raphson Method, correct to three places of decimal.* *(R.G.P.V., Bhopal, III Semester, June 2006)*

Solution. $f(x) = x^4 - x - 9 = 0$, $\qquad f'(x) = 4x^3 - 1$

$f(0) = -9$

$f(1) = 1^4 - 1 - 9 = 1 - 1 - 9 = -9, f(2) = 2^4 - 2 - 9 = 16 - 2 - 9 = 5$

As $f(2)$ is nearer to zero we take 2 as an approximate root of $f(x)$.

$f'(2) = 4(2)^3 - 1 = 4 \times 8 - 1 = 32 - 1 = 31$

By Newton-Raphson method

$$x_1 = x_0 - \frac{f(x_0)}{f'(x_0)}$$

$$x_1 = 2 - \frac{f(2)}{f'(2)} = 2 - \frac{5}{31} = 1.8387$$

$f(1.8387) = (1.8387)^4 - 1.8387 - 9 = 11.4299 - 1.8387 - 9 = 0.5912$

$f'(1.8387) = 4(1.8387)^3 - 1 = 23.8652$

$$x_2 = 1.8387 - \frac{f(1.8387)}{f'(1.8387)} = 1.8387 - \frac{0.5912}{23.8652} = 1.8387 - 0.02477 = 1.8139$$

$f(1.8139) = (1.8139)^4 - 1.8139 - 9 = 0.01173$

$f'(1.8139) = 4(1.8139)^3 - 1 = 22.8726$

$$x_3 = 1.8139 - \frac{f(1.8139)}{f'(1.8139)} = 1.8139 - \frac{0.01173}{22.8726}$$

$$= 1.8139 - 0.0005 = 1.8134$$

$f(1.8134) = (1.8134)^4 - 1.8134 - 9 = 10.8137 - 1.8134 - 9 = 0.0003$

$f'(1.8134) = 4(1.8134)^3 - 1 = 22.8529$

$$x_4 = 1.8134 - \frac{f(1.8134)}{f'(1.8134)} = 1.8134 - \frac{0.0003}{22.8529} = 1.8134 - 0.000013$$

$$x_4 = 1.8134$$

As x_3 and x_4 are equal, so root = 1.8134

Hence, real root of the given equation is 1.8134 **Ans.**

Example 11. *Find a positive root of the equation $2x^3 - 3x^2 - 9 = 0$ by Newton-Rapson Method upto third approximation of the root.* *(Bihar SBTE, 2011)*

Solution. $f(x) = 2x^3 - 3x^2 - 9 = 0$

$f'(x) = 6x^2 - 6x$

$f(0) = -9$

$f(1) = 2 \times 1^3 - 3 \times 1^2 - 9 = 2 - 3 - 9 = -10$

$f(2) = 2(2)^3 - 3(2)^2 - 9 = 16 - 12 - 9 = -5$

$f(3) = 2\ (3)^3 - 3\ (3)^2 - 9 = 54 - 27 - 9 = 18$

Since $f(2)$ and $f(3)$ are of opposite signs, so the root of eq. lies between 2 and 3. As $f(2)$ is near to zero than f (3), so 2 is better approximation root than 3.

$f^1(2) = 6\ (2)^2 - 6\ (2) = 24 - 12 = 12$

By Newton rapshon formula

$$x_1 = x_0 - \frac{f(x_0)}{f^1(x_0)}$$

$$x_1 = 2 - \frac{f(2)}{f'(2)} = 2 - \frac{(-5)}{12} = 2 + \frac{5}{12} = 2.4166$$

$f(2.4166) = 2\ (2.4166)^3 - 3\ (2.4166)^2 - 9 = 1.7058$

$f'(2.4166) = 6\ (2.4166)^2 - 6\ (2.4166) = 20.5401$

$$x_2 = 2.4166 - \frac{f(2.4166)}{f'(2.4166)} = 2.4166 - \frac{1.7058}{20.5401} = 2.3335$$

$$x_3 = 2.3335 - \frac{f(2.3335)}{f'(2.3335)} = 2.3335 - \frac{0.0791}{18.6725} = 2.3293$$

$$x_4 = 2.3293 - \frac{f(2.3293)}{f'(2.3293)} = 2.3293 = \frac{-0.0010}{18.5780} = 2.3293$$

As x_3 and x_4 are equal 00 root = 2.3293.

Hence real root of the given equation 2.3293.   **Ans.**

Example 12. *By using Newton-Raphson's Method, find the root of $x^4 - x - 10 = 0$, which is near to $x = 2$ correct to three places of decimai.*

(GDTU, 2012, R.G.P.V., Bhopal, III Semester, June 2008, Dec. 2003)

Solution. $f(x) = x^4 - x - 10,$ $f'(x) = 4x^3 - 1$

$f(2) = 16 - 2 - 10 = 4$

$f'(2) = 32 - 1 = 31$

By Newton-Raphson's method

$$x_1 = x_0 - \frac{f(x_0)}{f'(x_r)} = 2 - \frac{f(2)}{f'(2)} = 2 - \frac{4}{31} = 2 - 0.129 = 1.871$$

$f(1.871) = (1.871)^4 - 1.871 - 10 = 12.2545 - 1.871 - 10 = 0.3835$

$f'(1.871) = 4\ (1.871)^3 - 1 = 4 \times 6.5497 - 1 = 25.1988$

$$x_2 = 1.871 - \frac{f(1.871)}{f'(1.871)} = 1.871 - \frac{0.3835}{25.1988} = 1.871 - 0.0152 = 1.8558$$

$f(1.8558) = (1.8558)^4 - (1.8558) - 10 = 11.8611 - 11.8558 = 0.0053$

$f'(1.8558) = 4\ (1.8558)^3 - 1 = 4 \times 6.3914 - 1 = 24.5656$

$$x_3 = 1.8558 - \frac{f(1.8558)}{f'(1.8558)} = 1.8558 - \frac{0.0053}{24.5656} = 1.8558 - 0.00022 = 1.85558$$

As $x_2 = x_3$ correct upto three places of decimal, so the correct root of the given equation is 1.856   **Ans.**

Example 13. *Using Newton – Raphson Method find an iterative scheme to compute the reciprocal of a positive number.*

Solution.   Let x be the reciprocal of a given positive number N.

$$x = \frac{1}{N} \text{ or } N - \frac{1}{x} = 0$$

$$f(x) = N - \frac{1}{x} = 0, \quad f'(x) = \frac{1}{x^2}$$

By Newton - Raphson Method

$$x_{n+1} = x_n - \frac{f(x_n)}{f'(x_n)} = x_n - \frac{N - \frac{1}{x_n}}{\frac{1}{x_n^2}} = x_n - x_n^2 \left(N - \frac{1}{x_n} \right)$$

$$= x_n - N x_n^2 + x_n = 2x_n - N x_n^2 = x_n \left[2 - N x_n \right] \qquad \textbf{Ans}$$

Example 14. *Find the value of $\frac{1}{18}$ by Newton - Raphson Method.*

Solution. Let $\quad x = \frac{1}{18}$

Let $\quad\quad\quad f(x) = \frac{1}{x} - 18 = 0$ $\qquad\qquad\qquad\qquad\qquad\qquad$ ... (1)

$\Rightarrow \quad\quad\quad f'(x) = -\frac{1}{x^2}$

Let the approximate root of equation (1) be 0.05

By Newton - Raphson Method $\quad\quad x_n + 1 = x_n (2 - N x_n)$

$$x_1 = x_0 (2 - 18 x_0) = 0.05 (2 - 18 \times 0.05) = 0.05 (2 - 0.90) = 0.05 \times 1.10 = 0.055$$
$$x_2 = x_1 (2 - 18 x_1) = 0.055 (2 - 18 \times 0.055) = 0.055 (2 - 0.99) = 0.055 \times 1.01$$
$$= 0.05555$$
$$x_3 = x_2 (2 - 18 x_2) = 0.05555 (2 - 18 \times 0.05555) = 0.05555 (1.0001) = 0.05556$$

Hence, $\frac{1}{18} = 0.0556$ approximately. $\qquad\qquad\qquad\qquad\qquad\qquad\qquad$ **Ans.**

Example 15. *Using Newton-Raphson method find an iterative scheme to compute the cube root of a positive number.*

Solution.   Let x be the cube root of a given positive number N.

$$x = (N)^{\frac{1}{3}} \quad \text{or} \quad x^3 = N \text{ or } x^3 - N = 0$$

Let $\quad\quad\quad f(x) = x^3 - N = 0 \quad\quad \Rightarrow f'(x) = 3x^2$

By Newton-Raphson Method

$$x_{n+1} = x_n - \frac{f(x_n)}{f'(x_n)} = x_n - \frac{x_n^3 - N}{3x_n^2} = \frac{3x_n^3 - x_n^3 + N}{3x_n^2} = \frac{2x_n^3 + N}{3x_n^2} \text{ for } n = 0, 1, 2.... \textbf{ Ans.}$$

Example 16. *Find an iterative formula to find $\sqrt{N}$ (where N is a positive number) and hence find*
$\sqrt{24}$.   *(A.M.I.E., Summer 2001, 2000)*

Solution. Let x be the square root of a given positive number N.

$$x = (N)^{\frac{1}{2}} \quad\Rightarrow\quad x^2 = N \;\Rightarrow\; x^2 - N = 0$$

Let $\quad f(x) = x^2 - N = 0 \quad\Rightarrow\quad f'(x) = 2x$

By Newton-Raphson Method

$$x_{n+1} = x_n - \frac{f(x_n)}{f'(x_n)}$$

$$x_{n+1} = x_n - \frac{x_n^2 - N}{2x_n} = \frac{2x_n^2 - x_n^2 + N}{2x_n} = \frac{x_n^2 + N}{2x_n}$$

$$x_{n+1} = \frac{x_n^2 + N}{2x_n}; n = 0,.... \text{ This is the iterative formula to find } \sqrt{N}.$$

$$N = 24, \quad x_0 = \text{app. root of } 24 = 5$$

$$x_1 = \frac{(5)^2 + 24}{2\times5} = 4.9, \quad x_2 = \frac{(4.9)^2 + 24}{2\times4.9} = 4.899$$

$$x_3 = \frac{(4.899)^2 + 24}{2\times4.899} = 4.89898$$ **Ans.**

Example 17. *By iteration method find the value of $(48)^{1/3}$, correct to three decimal places.*
                                                     (R.G.P.V., Bhopal, III Semester Dec. 2003)

Solution. Let $\quad x = \sqrt[3]{N} \Rightarrow x^3 = N \Rightarrow x^3 \quad N = 0$

Let $\quad f(x) = x^3 - N = 0 \quad\Rightarrow f'(x) = 3x^2$

By Newton-Raphson Method

$$x_{n+1} = x_n - \frac{f(x_n)}{f'(x_n)}$$

$$x_{n+1} = x_n - \frac{x_n^3 - N}{3x_n^2}, n = 0, 1, 2.....$$

$\Rightarrow \qquad x_{n+1} = \frac{3x_n^3 - x_n^3 + N}{3x_n^2} \qquad\Rightarrow\qquad x_{n+1} = \frac{2x_n^3 + N}{3x_n^2}, n = 0, 1, 2, ...$ ...(1)

Putting $\qquad N = 48, \quad x_0 = \text{Approximate root of } 48 = 3.5$ in (1), we get

$$x_1 = \frac{2(3.5)^3 + 48}{3(3.5)^2} = \frac{133.75}{36.75} = 3.6395$$

$$x_2 = \frac{2(3.6395)^3 + 48}{3(3.6395)^2} = \frac{144.4173}{39.7379} = 3.6342$$

$$x_3 = \frac{2(3.6342)^3 + 48}{3(3.6342)^2} = \frac{143.9967}{39.6222} = 3.6342$$

As $x_2 = x_3$, so $(48)^{\frac{1}{3}} = 3.6342$ is correct upto 4 decimal places.

Hence, $(48)^{\frac{1}{3}} = 3.6342$

Ans.

EXERCISE 2.2

1. Using Newton-Raphson Method, find the root that lies in (0,1) of the equation $x^3 - 6x + 4 = 0$, correct to 4 decimal places.

   [**Hint:** Take $\alpha_0 = 1$]

   Ans. 0.73205

2. Discuss Newton-Raphson method for numerical solution of equation and hence find the smallest positive root of the equation $3x^3 - 9x^2 + 8 = 0$. *(Bihar SBTE, 2009)* Ans. 1.2

3. A root of the equation $x^3 - x^2 - x + 1 = 0$ is to be determined by the Newton-Raphson Method. The initial approximation to the root is given as 0.9. Find the root of the equation. Ans. 1.0001

4. Find an interval of length 1, in which the root of $f(x) = 3x^3 - 4x^2 - 4x - 7 = 0$ lies. Take the middle point of this interval as the starting approximation and iterate two times, using the Newton-Raphson Method. Ans. 2.334

5. Find the number of real roots of the equation

   $$3x^3 - 8x^2 + 10x - 4 = 0$$

   Find an interval of unit length in which the smallest positive root of the equation lies. Taking mid-point of this interval as initial approximation, perform two iterations of Newton-Raphson method. Ans. 0.6664277

6. Solve: $x^3 - 3x + 1 = 0$ *(R.G.P.V., Bhopal, III Semester, Dec. 2007)* Ans. 1.532

7. Find the real root to four decimals of the equation $x^6 - x^4 - x^3 - 1 = 0$ lying between (1, 2). Ans. 1.4036

8. Using Newton-Raphson Method, find one negative root of $3x^3 + 8x^2 + 8x + 5 = 0$

   [**Hint:** Root of $f(-x)$]

   Ans. -1.67

9. Find the 1st negative root of the equation $f(x) = x^3 + 2x + 10 = 0$ up to third approximation of the root by Newton-Rapson method.

   (Bihar. SBTE, 2014)

10. Evaluate $\sqrt{41}$ by Newton-Raphson method. Ans. 6.4031

11. Find cube root of 10 to four places of decimal by Newton-Raphson method. Ans. 2.1547

Choose the correct alternative :

12. If $f(x) = 0$ is an algebraic equation then Newton-Raphson method is given by :

    $$x_{n+1} = x_n - \frac{f(x_n)}{?}$$

    (R.G.P.V., Bhopal, III Semester, June 2007)

    (a) $f(x_{n-1})$ (b) $f'(x_{n-1})$ (c) $f'(x_n)$ (d) $f''(x_n)$ Ans. (c)

13. The order of convergence in Newton-Raphson method is :

    (a) 2   (b) 3   (c) 0   (d) 1   Ans. (a)

    (R.G.P.V., Bhopal, III Semester, Dec. 2006)

14. Let $p(x)$ be a polynomial satisfying $p(0) = 0$, $p(1) = 1$ and $p(2) = 2$; then

    (a) $p(x)$ is unique   (b) $p(x)$ is of degree 1

    (c) $p(x)$ is of degree 2   (d) $p(x)$ is not unique   Ans. (a)

15. Given that the coefficient in the equation $ax^3 + bx^2 + cx + d = 0$ are integers and that one of its root is $a + b\sqrt{2}$, then

    (a) The other two roots are complex

(b) One of the other two roots is real and other is complex

(c) The other two roots are real

(d) One of the other two root is real and other is $a - b\sqrt{2}$ **Ans. (d)**

16. Newton's iterative formula to find the value of $\sqrt{N}$ is

(a) $x_{n+1} = \frac{1}{2}\left(x_n + \frac{1}{N x_n}\right)$ (b) $x_{n+1} = x_n(2 - N x_n)$ **Ans. (c)**

(c) $x_{n+1} = \frac{1}{2}\left(x_n + \frac{N}{x_n}\right)$ (d) $x_{n+1} = \frac{x_n^2 + N}{x_n}$ *(R.G.P.V., Bhopal, III Semester, Dec. 2007)*

17. Newton's method approximate the curve of $f(x)$ by

(a) Tangents   (b) Chords   (c) Arbitrary   (d) All of these   **Ans. (a)**

18. Newton's method fails when at or near a root of $f(x) = 0$, the derivative $f'(x)$ is

(a) 1   (b) 0   (c) ∞ (d) None of them   **Ans. (b)**

19. For any approximate root Newton's method yields

(a) Always better approximation   (b) Some time better approximation

(c) Always worse approximation   (d) All of them   **Ans. (b)**

20. If the first approximate root of the equation $f(x) = 0$ is 2 such that $f(2) = 0$, then the root is

(a) less than 3   (b) more than 3   (c) 3   (d) None of them   **Ans. (c)**

2.7 REGULA - FALSI METHOD OR FALSE POSITION METHOD   *(Bihar, SBTE, 2011, 2011)*

The oldest method for computing the real root of a numerical equation is the method of false position, or "(Regula falsi)".

Let the root lie between a and b. These numbers a and b should be as close - together as possible. Since the root lies between a and b, the graph of $y = f(x)$ must cross the x- axis between $x = a$ and $x = b$, and $f(a)$ and $f(b)$ must have opposite signs.

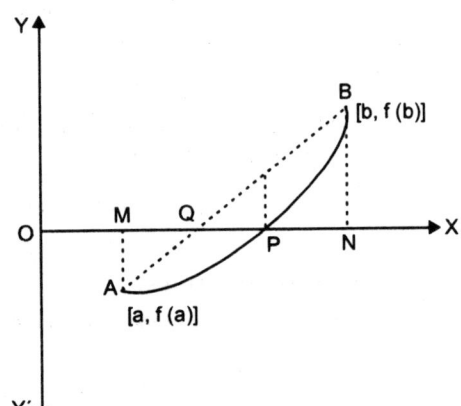

Now since any portion of a smooth curve is practically straight line for a short distance, it is legitimate to assume that change in $f(x)$ is proportional to change in x over a short interval. The method of False position is based on this principle, for it assumes that the graph of $y = f(x)$ is a straight line between the points (x_1, y_1) and (x_2, y_2), these points being on opposite sides of x-axis.

2.8 RULE OF FALSE POSITION (REGULA FALSI)

Let $f(x) = 0$

Let $y = f(x)$ be represented by the curve AB.

The curve AB cuts the x-axis at P.

The real root of (1) is OP.

The false position of the curve AB is taken as the chord AB. The chord AB cuts the x-axis at Q. The approximate   root   of $f(x) = 0$ is OQ.

By this method, we find OQ.

Let $[a, f(a)]$, B $[b, f(b)]$ be the extremities of the chord AB.

The equation of the chord AB is

$$y - f(a) = \frac{f(b) - f(a)}{b - a}(x - a) \qquad \text{(Two points form)}$$

To find OQ, put $y = 0$

$$0 - f(a) = \frac{f(b) - f(a)}{b - a}(x - a)$$

$$(x - a) = \frac{-(b - a) f(a)}{f(b) - f(a)}$$

$$\Rightarrow \quad x = a + \frac{(a - b) f(a)}{f(b) - f(a)} = \frac{a f(b) - a f(a) + a f(a) - b f(a)}{f(b) - f(a)}$$

$$\boxed{x = \frac{a f(b) - b f(a)}{f(b) - f(a)}}$$

Repeat the above rule.

Example 18. *If a function $f(x)$ is such that $f(2) = -1$ and $f(3) = 16$ find the first approximate root of the equation $f(x) = 0$ lying between 2 and 3 using method of false position.*

(Bihar SBTE, 2009)

Solution.   Given that, $f(2) = -1$ and $f(3) = 16$

By using method of false position.

The first approximate root of $f(x)$ lying between 2 and 3.

Here, $a = 2$, $b = 3$, $f(a) = -1$, $f(b) = 16$

$$x = \frac{a \; f(b) - b \; f(a)}{f(b) - f(a)} = \frac{2(16) - 3(-1)}{16 - (-1)} = \frac{35}{17} = 2.05855235 \qquad \textbf{Ans.}$$

Example 19. *Find a first positive root of the equation $x^3 - 2x - 5 = 0$ by the method of a false position correct to three decimal places.*   *(Bihar SBTE, 2005, 2009)*

Solution.   Given that $x^3 - 2x - 5 = 0$

Let $f(x) = x^3 - 2x - 5$

$f(0) = -5$

$f(1) = 1^3 - 2(1) - 5 = 1 - 2 - 5 = -6$

$f(2) = 2^3 - 2(2) - 5 = 8 - 4 - 5 = -1$

$f(3) = 3^3 - 3.2 - 5 = 27 - 6 - 5 = 16$

Clearly $f(2) \cdot f(3) < 0$

The equation $f(x) = 0$ has at least one root in $(2, 3)$.

Now by using method of a false position, we have

$$f(2) = -1, f(3) = 16$$

The first approximation root is given by

$$x_2 = \frac{x_0 \; f(x_1) - x_1 \; f(x_0)}{f(x_1) - f(x_0)} = \frac{2 \, f(3) - 3 \, f(2)}{f(3) - f(2)} = \frac{2 \times 6 - 3(-1)}{16 - (-1)} = \frac{32 + 3}{16 + 1} = \frac{35}{17} = 2.058823529$$

$$\therefore x_2 = 2.058823529$$

$$f(x_2) = f(2.058823529) = -39079916$$

Clearly $f(2.058823529) \cdot f(3) < 0$

$$x_3 = \frac{2.058823529 \, f \, (3) - 3f \, (2.058823529)}{f \, (3) - f \, (2.058823529)} = \frac{2.058823529 \times 16 - 3 \times (-0.39079916)}{16 - (-0.39079916)}$$

$$x_3 = 2.081263676$$

$$f(x_3) = f(2.081263676) = 0.147203881$$

Clearly $f(2.081263676) \cdot f(3) < 0$

$$\therefore x_4 = \frac{2.081263676 \, f \, (3) - 3f \, (-2.081263676)}{f \, (3) - f \, (2.081263676)} = \frac{2.081263676 \times 16 - 3 \times (0.147203881)}{16 - (-0.147203881)}$$

$$\therefore x_4 = 2.089639216$$

$$f(x_4) = f(2.089639216) = -0.054676437 = 0 \text{ nearly}$$ **Ans.**

Example 20. *Find an approximate value of the root of the equation $x^3 + x - 1 = 0$ near $x = 1$, using the method of false position (regula falsi) two times.*

Solution. $\quad f(x) = x^3 + x - 1 = 0$

$$f(1) = 1 + 1 - 1 = +1$$

$$f(0.5) = (0.5)^3 + (0.5) - 1 = -0.375, \qquad f(1) \cdot f(0.5) < 0$$

The root lies between 0.5 and 1.

Let $\qquad\qquad a = 0.5$ and $\qquad b = 1$

$$x_1 = \frac{a \, f \, (b) - b \, f \, (a)}{f \, (b) - f \, (a)} \Rightarrow x_1 = \frac{0.5 \, f \, (1) - 1 \, f \, (0.5)}{f \, (1) - f \, (0.5)} = \frac{0.5 \, (1) - 1 \, (-0.375)}{1 + 0.375} = 0.6364$$

Now $\qquad f(0.6364) = (0.6364)^3 + 0.6364 - 1 = -0.1059$

$\qquad\qquad$ and $f(1) = 1$

$\therefore$ Root lies between 0.6364 and 1. $\qquad\qquad f(0.6364) \cdot f(1) < 0$

$$a = 0.6364, \qquad b = 1$$

$$x_2 = \frac{0.6364 \, f \, (1) - 1 \, f \, (0.6364)}{f \, (1) - f \, (0.6364)} = \frac{0.6364 - 1(-0.1059)}{1 + 0.1059} = 0.6712$$

Now, $\qquad f(0.6712) = -0.0264$ and $f(1) = 1$

$$a = 0.6712 \text{ and } b = 1 \qquad\qquad\qquad [f(0.6712) \cdot f(1) < 0]$$

$$x_3 = \frac{0.6712 \, f \, (1) - 1 \, f \, (0.6712)}{f \, (1) - f \, (0.6712)} = \frac{0.6712 - (-0.0264)}{1 - (-0.0264)} = 0.6797$$ **Ans.**

Example 21. *Regula - Falsi method is used to obtain the smallest positive root of $x^4 - 3x - 8 = 0$. The root lies in the interval (1.5, 2). The first iterate is 1.8940. Find the next iterate.*

Solution. Let $\qquad f(x) = x^4 - 3x - 8 = 0$

$$a = 1.5, \quad b = 2$$

$$x_1 = 1.8940, \qquad f(1.5) = (1.5)^4 - 3(1.5) - 8 = -7.4375$$

$$f(1.8940) = (1.8940)^4 - 3(1.8940) - 8 = -0.8137379$$

$$f(2) = (2)^4 - 3(2) - 8 = +2$$

Since $f(2)$ and $f(1.8940)$ are of opposite signs, so root lies between 1.8940 and 2.

By Regula falsi method

$$x_2 = \frac{a f(b) - b f(a)}{f(b) - f(a)} = \frac{2 f(1.8940) - 1.8940 f(2)}{f(1.8940) - f(2)}$$

$$= \frac{2(-0.8137379) - 1.8940(2)}{-0.8137379 - 2} = 1.9246554$$

$$= 1.9247 \qquad \text{(App.)} \qquad \qquad \textbf{Ans.}$$

Example 22. *Find a positive root of $x^3 - 4x + 1 = 0$ by the method of false position.*

(R.G.P.V., Bhopal, III Semester, June 2007)

Solution. Let

$$f(x) = x^3 - 4x + 1 = 0$$

$$f(0) = 1$$

$$f(1) = 1 - 4 + 1 = -2$$

Since $f(0)$ and $f(1)$ are of opposite signs, so the root lies between 0 and 1.

By False position method

$$x = \frac{a f(b) - b f(a)}{f(b) - f(a)}$$

$$x_1 = \frac{0 f(1) - 1 f(0)}{f(1) - f(0)} = \frac{-1}{-2 - (1)} = \frac{1}{3}$$

$$f\left(\frac{1}{3}\right) = \left(\frac{1}{3}\right)^3 - 4\left(\frac{1}{3}\right) + 1 = \frac{1}{27} - \frac{4}{3} + 1 = \frac{-8}{27}$$

Since $f\left(\frac{1}{3}\right)$ and $f(0)$ are of opposite signs, so the root lies between $\frac{1}{3}$ and 0.

$$x_2 = \frac{\frac{1}{3} f(0) - 0 f\left(\frac{1}{3}\right)}{f(0) - f\left(\frac{1}{3}\right)} = \frac{\frac{1}{3}(1) - 0}{1 - \left(\frac{-8}{27}\right)} = \frac{9}{35}$$

$$f\left(\frac{9}{35}\right) = \left(\frac{9}{35}\right)^3 - 4\left(\frac{9}{35}\right) + 1 = \frac{729}{42875} - \frac{36}{35} + 1 = -\frac{496}{42875}$$

Since $f\left(\frac{9}{35}\right)$ and $f(0)$ are of opposite signs, so the root lies between $\frac{9}{35}$ and 0.

$$x_3 = \frac{\frac{9}{35} f(0) - 0 f\left(\frac{9}{35}\right)}{f(0) - f\left(\frac{9}{35}\right)} = \frac{\frac{9}{35}(1) - 0}{1 - \left(-\frac{496}{42875}\right)} = \frac{1225}{4819}$$

Since $\left(x_2 = \frac{9}{35} = 0.2571\right) = \left(x_3 = \frac{1225}{4819} = 0.2542\right)$ correct upto two decimal places,

so the root of the given equation is 0.25

$$\qquad \qquad \textbf{Ans.}$$

Example 23. *Find the root of the equation $x^3 - 5x - 7 = 0$ which lies between 2 and 3 by the method of false position.* (R.G.P.V., Bhopal, III Semester, June 2005)

Solution.   Here, we have

Let
$$f(x) = x^3 - 5x - 7 = 0$$
$$f(2) = 8 - 10 - 7 = -9$$
$$f(3) = 27 - 15 - 7 = +5$$

As $f(2)$ and $f(3)$ are of opposite signs, so the root lies between 2 and 3.

$$x_1 = \frac{a f(b) - b f(a)}{f(b) - f(a)}$$

$$x_1 = \frac{2 f(3) - 3 f(2)}{f(3) - f(2)} = \frac{2(5) - 3(-9)}{5 - (-9)} = \frac{37}{14} = 2.6429$$

$$f(2.6429) = (2.6429)^3 - 5(2.6429) - 7 = -1.7541$$

Now, $f(2.6429) = -1.7541$ and $f(3) = 5$

∴ Root lies between 2.6429 and 3.

$$a = 2.6429, b = 3$$

$$x_2 = \frac{2.6429 f(3) - 3 f(2.6429)}{f(3) - f(2.6429)} = \frac{2.6429(5) - 3(-1.7541)}{5 - (-1.7541)} = \frac{18.4768}{6.7541} = 2.7356$$

Now,
$$f(2.7356) = (2.7356)^3 - 5(2.7356) - 7 = -0.2061$$
and
$$f(3) = 5$$

∴ Root lies between 2.7356 and 3.

$$a = 2.7356 \qquad \text{and } b = 3$$

$$x_3 = \frac{2.7356 f(3) - 3 f(2.7356)}{f(3) - f(2.7356)} = \frac{2.7356(5) - 3(-0.2061)}{5 - (-0.2061)} = \frac{14.2963}{5.2061} = 2.7461$$

$$f(2.7461) = (2.7461)^3 - 5(2.7461) - 7 = -0.02198$$

Now,
$$f(2.7461) = -0.02198 \text{ and } f(3) = +5$$

∴ Root lies between 2.7461 and 3.

$$a = 2.7461 \text{ and } b = 3$$

$$x_4 = \frac{2.7461 f(3) - 3 f(2.7461)}{f(3) - f(2.7461)} = \frac{2.7461(5) - 3(-0.02198)}{5 - (-0.02198)} = \frac{13.79644}{5.02198} = 2.7472$$

Hence, the root of the given equation is 2.7472.

Ans.

Example 24. *Find by the method of Regula falsi a root of the equation $x^3 + x^2 - 3x - 3 = 0$ lying between 1 and 2.* (R.G.P.V., Bhopal, III Semester, June 2005)

Solution.
$$f(x) = x^3 + x^2 - 3x - 3 = 0$$
$$f(1) = 1 + 1 - 3 - 3 = -4 = -ve$$
$$f(2) = 8 + 4 - 6 - 3 = +3 = +ve$$

The root lies between 1 and 2 as $f(1)$ is –ve and $f(2)$ is +ve.

By Regula Falsi method

$$x_1 = \frac{1 f(2) - 2 f(1)}{f(2) - f(1)} = \frac{1 \times 3 - 2 \times (-4)}{3 - (-4)} = \frac{11}{7} = 1.571$$

$$f(1.571) = (1.571)^3 + (1.571)^2 - 3(1.571) - 3 = 3.877 + 2.468 - 4.713 - 3 = -1.368 = -ve$$

The root lies between 1.571 and 2 as $f(1.571)$ is $-ve$ and $f(2)$ is $+ve$.

$$x_2 = \frac{1.571 f(2) - 2 f(1.571)}{f(2) - f(1.571)} = \frac{1.571 \times 3 - 2 \times (-1.368)}{3 - (-1.368)} = \frac{4.713 + 2.736}{4.368} = 1.705$$

$$f(1.705) = (1.705)^3 + (1.705)^2 - 3(1.705) - 3 = 4.956 + 2.907 - 5.115 - 3 = -0.252 = -ve.$$

The root lies between 1.705 and 2 as $f(1.705)$ is $-ve$ and $f(2)$ is $+ve$.

$$x_3 = \frac{1.705 f(2) - 2 f(1.705)}{f(2) - f(1.705)} = \frac{1.705 \times 3 - 2 \times (-0.252)}{3 - (-0.252)} = 1.728 \qquad \textbf{Ans.}$$

Example 25. *Find a real root of the $2x^3 + 5x - 9 = 0$ by Regula-Falsi Method upto third approximation of root.* (Bihar SBTE, 2011)

Solution.   Given that $2x^3 + 5x - 9 = 0$

Let $f(x) = 2x^3 + 5x - 9$

$f(0) = -9$

$f(1) = -2$

$f(2) = 16 + 10 - 9 = 17$

As f (1) and f (2) are of opposite signs so root lie between 1 and 2

$$x_1 = \frac{af(b) - bf(a)}{f(b) - f(a)}$$

$$x_1 = \frac{1 f(2) - f(1)}{f(2) - f(1)} = \frac{1 \times 17 - 2 \times (-2)}{17 + 2} = \frac{17 + 4}{19} = \frac{21}{19} = 1.105$$

$\therefore$ Root lies between 1.105 and 2, $a = 1.105$ $b = 2$

Now, $f(1.105) = -0.7765$ and $f(2) = 17$

$$x_2 = \frac{1.105 f(2) - 2 f(1.105)}{f(2) - f(1.105)}$$

$$x_2 = \frac{1.105 \times 17 - (-0.7765) \times 2}{17 + 0.7765} = 1.144$$

$f(1.144) = -0.2856$ and f (2) = 17

Now, $f(1.144) = 2(1.44)^3 + 5(1.44) - 9 = -0.2856$

$$x_3 = \frac{1.144 \times f(2) - 2 f(1.144)}{f(2) - f(1.144)} = \frac{1.144 \times 17 - 2x(-0.2856)}{17 + 0.2856} = 1.158$$

$f(1.158) = -0.1043$ and $f(2) = 17$

Now, $f(1.158) = 2(1.158)^3 + 5.(1.158) - 9 = -0.1043$

So, root lies b/w 1.158 and 2

$a = 1.158$ and $b = 2$

$$x_4 = \frac{1.158 f(2) - 2 f(1.158)}{f(2) - f(1.158)} = \frac{1.158 \times 17 - 2 \times (20.1043)}{(7 + 0.1043)} = 1.163 \qquad \textbf{Ans.}$$

EXERCISE 2.3

Solve the following equations by Regula falsi method :

1. $x^3 - 10x^2 + 40x - 35 = 0$ **Ans.** 1.1875
2. $x^3 + x^2 + 3x + 4 = 0$ **Ans.** -1.22248
3. $x^6 - x^4 - x^3 - 1 = 0$ **Ans.** 1.4036
4. $x^3 - 9x + 1 = 0$ (Root between 2 & 3) **Ans.** 2.9428
5. $x^3 - x - 1 = 0$ **Ans.** 1.315
6. $3x^3 + 8x^2 + 8x + 5 = 0$ **Ans.** -1.65
7. $x^3 - 6x + 4 = 0$ **Ans.** 0.732

Choose the correct alternative :

8. In finding the root of an equation $f(x) = 0$, the following values were obtained.

   $f(1.8) = -1.3024, f(2.0) = 4.0$

   Regula-falsi method gives the next approximation of the root as

   (a) 1.8761 (b) 1.8553 (c) 1.8226 (d) 1.8491 **Ans.** (d)

9. Given that the equation $x^3 + 3x - 7 = 0$ has a root in the interval $(1, 2)$ a single application of the false position method gives the following approximation for the root.

   (a) 1.2 (b) 1.3 (c) 1.4 (d) 1.406 **Ans.** (b)

10. If $a < b$ and $f(a)$ and $f(b)$ be of opposite signs for the equation $f(x) = 0$, then by Regula falsi method, the 1st approximation of the root of the equation is ....

    (a) $\dfrac{af(b) - bf(a)}{n - a}$ (b) $\dfrac{af(b) - bf(a)}{f(a) - f(b)}$ (c) $\dfrac{af(b) - bf(a)}{f(b) - f(a)}$ (d) none of the *(Bihar SBTE, 2011)* **Ans.** (c)

2.9 ITERATION METHOD

Consider the equation $f(x) = 0$ ... (1)

We rewrite the equation in the form

$\quad x = \phi(x)$ .... (2)

Let us draw two curves

$\quad y = x$ and $y = \phi(x)$

The point of intersection of two curves is the root of (1).

Let $x = x_0$ be an initial approximate root, then first approximation x_1 is found by

$\quad x_1 = \phi(x_0)$

Now taking x_1 as initial value, x_2 second approximation is given by

$\quad x_2 = \phi(x_1)$ and so on.

$\quad x_{n+1} = \phi(x_n)$

This is also known as successive approximation method.

Remember. The equation $f(x) = 0$ is written as $x = \phi(x)$.

This form $x = \phi(x)$ can be choosen in many ways. We have to choose $\phi(x)$ in such a way that initial approximation x_0 should satisfy the condition $| \phi'(x_0) | < 1$.

Then $x_0, x_1, x_2 \ldots\ldots x_n$ converge to the root α of the equation $f(x) = 0$.

Note. 1. The rate of convergence is more if the value of $\phi'(x)$ is smaller.

   2. For real roots, the method is very useful.

Example 26. *Apply the iterative method to find the real roots of $x^3 + x^2 - 1 = 0$, assuming the initial approximation is as $x_0 = 0.8$.*   *(Bihar SBTE, 2011, 2010)*

Solution.   Let $f(x) = x^3 + x^2 - 1 = 0$ ... (1)

Equation (1) can be written as $x = \phi(x)$ in many ways

   i.e. (i) $x^3 = 1 - x^2 \implies x = (1 - x^2)^{\frac{1}{3}}$, Here $\phi(x) = (1 - x^2)^{\frac{1}{3}}$

$\implies \qquad \phi'(x) = \frac{1}{3}(1 - x^2)^{-\frac{2}{3}}(-2x) \implies |\phi'(0.8)| = 1.05 > 1,$

Which does not satisfy the condition.

(ii) $\qquad x^2 = 1 - x^3 \implies x = (1 - x^3)^{\frac{1}{2}}$.

Here $\phi(x) = (1 - x^3)^{\frac{1}{2}}$

$\implies \phi'(x) = \frac{1}{2}(1 - x^3)^{-\frac{1}{2}}(-3x^2) \qquad \implies |\phi'(0.8)| = 1.37 > 1$

Again it also does not satisfy the condition.

(iii) $x^3 + x^2 - 1 = 0 \implies x^2(x+1) = 1 \implies x^2 = \frac{1}{1+x} \implies x = \frac{1}{\sqrt{1+x}}$

Here, $\phi(x) = \frac{1}{\sqrt{1+x}}$

$\implies \phi'(x) = -\frac{1}{2}(1+x)^{-\frac{3}{2}} \qquad \implies |\phi'(0.8)| = 0.2 < 1$

Which satisfies the condition.

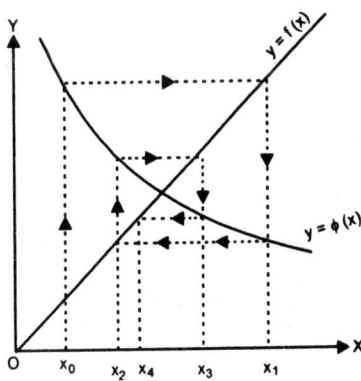

By Iterative method

$$x = \phi(x) \qquad\qquad\qquad \implies x = \frac{1}{\sqrt{1+x}} \qquad\qquad\qquad ... (2)$$

On putting the initial approximation $x_0 = 0.8$ in (2), we get

$$x_1 = \frac{1}{\sqrt{1+0.8}} = 0.7454$$

Again putting $x_1 = 0.7454$ in (2), we have

$$x_2 = \frac{1}{\sqrt{1+0.7454}} = 0.7569$$

Similarly, putting the successive values of x (approximate root) in (2), we obtain

$$x_3 = \frac{1}{\sqrt{1+0.7569}} = 0.7544$$

$$x_4 = \frac{1}{\sqrt{1+0.7544}} = 0.7550$$

$$x_5 = \frac{1}{\sqrt{1+0.7550}} = 0.7549$$

$$x_6 = \frac{1}{\sqrt{1+0.7549}} = 0.7549$$

$$x_7 = \frac{1}{\sqrt{1+0.7549}} = 0.7549$$

Since $x_6 = x_7$, so the correct root of the given equation is 0.7549

Ans.

Example 27. *Find the root of the equation $x^3 - 5x - 11 = 0$ by the method of iteration correct to three decimal places.*

Solution. $f(x) = x^3 - 5x - 11$

$f(2) = 8 - 10 - 11 = -13$

$f(3) = 27 - 15 - 11 = +1$

Since $f(2)$ and $f(3)$ are of opposite sign, the root lies between 2 and 3.

$f(3)$ is nearer to zero than $f(2)$. Hence we take $x_0 = 3$.

The equation can be written as

$$x^3 = 5x + 11 \qquad \Rightarrow \qquad x = (5x+11)^{\frac{1}{3}} \qquad \Rightarrow x = \phi(x) \qquad \qquad \ldots (1)$$

$$\phi(x) = (5x+11)^{\frac{1}{3}}$$

$$\phi'(x) = \frac{5}{3}(5x+11)^{-\frac{2}{3}}$$

$$|\phi'(3)| = \frac{5}{3}(5 \times 3 + 11)^{-\frac{2}{3}} = \frac{5}{3}\frac{1}{(26)^{\frac{2}{3}}} = 0.1899 < 1$$

In (1) we substitute 3 for x and we get

$$x_1 = (15+11)^{\frac{1}{3}} = (26)^{\frac{1}{3}} = 2.9625$$

Putting 2.9625 for x in (1), we get

$$x_2 = (5 \times 2.9625 + 11)^{\frac{1}{3}} = (14.8125 + 11)^{\frac{1}{3}} = (25.8125)^{\frac{1}{3}} = 2.9554$$

Similarly, putting the successive values of x (approximate root) in (1), we get

$$x_3 = (5 \times 2.9554 + 11)^{\frac{1}{3}} = (14.777 + 11)^{\frac{1}{3}} = (25.777)^{\frac{1}{3}} = 2.9540$$

$$x_4 = (5 \times 2.9540 + 11)^{\frac{1}{3}} = (14.77 + 11)^{\frac{1}{3}} = (25.77)^{\frac{1}{3}} = 2.9537$$

$$x_5 = (5 \times 2.9537 + 11)^{\frac{1}{3}} = (14.7685 + 11)^{\frac{1}{3}} = (25.7685)^{\frac{1}{3}} = 2.9537$$

$$\begin{cases} \phi'(x) = -e^{-x} \\ \phi'(0.5) = -e^{-0.5} \\ \qquad = -0.6065 \\ |\phi'(0.5)| < 1 \end{cases}$$

Hence we take $x = 2.953$ correct to three decimal places.

Ans.

Example 28. *Find a solution of $x^3 + x - 1 = 0$ by iteration.* *(Bihar SBTE, 2010)*

Solution. $f(x) = x^3 + x - 1 = 0$

$$\Rightarrow \quad x(x^2 + 1) = 1 \quad \Rightarrow \quad x = \frac{1}{1+x^2}$$

The approximate root of the given equation is 1 as shown by rough sketch. We can write the equation in the form

$$x = \frac{1}{1+x^2} \qquad \qquad \qquad \dots(1)$$

Let $\qquad \phi(x) = \frac{1}{1+x^2}, \ |\phi'(x)| = \left| \frac{-2x}{(1+x^2)^2} \right| \quad \Rightarrow \quad |\phi'(1)| = \frac{1}{2} < 1$

Putting $\qquad x = 1$ in (1), we get

$$x_1 = \frac{1}{1+1^2} = 0.5$$

Again putting $\quad x = 0.5$ in (1), we get

$$x_2 = \frac{1}{1+(0.5)^2} = 0.800$$

Similarly, putting the successive values of x in (1), we get

$$x_3 = \frac{1}{1+(0.8)^2} = 0.610$$

$$x_4 = \frac{1}{1+(0.610)^2} = 0.729$$

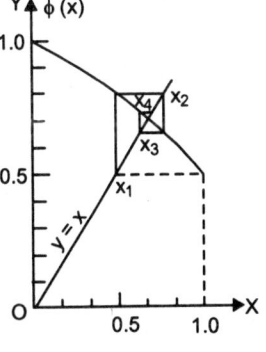

Similarly, $\qquad x_5 = 0.653, \ x_6 = 0.701$

The exat root is 0.682328. **Ans.**

Example 29. *Find a real root of $2x - \log_{10} x = 7$ correct to three decimal places using iteration method.*
(R.G.P.V., Bhopal, III Semester, Dec. 2006)

Solution. The given equation is

$$2x - \log_{10} x = 7$$

which can be written as

$$x = \frac{1}{2}\left(\log_{10} x + 7\right)$$

Here, $\qquad \phi(x) = \frac{1}{2}\left(\log_{10} x + 7\right) \qquad \qquad \dots(1)$

Let $\qquad x_0 = 3.8$

Putting the value of $x = 3.8$ in (1), we get

$$x_1 = \frac{1}{2}\left(\log_{10} 3.8 + 7\right) = 3.79$$

On putting $\qquad x = 3.79$ in (1), we get

$$x_2 = \frac{1}{2}\left(\log_{10} 3.79 + 7\right) = 3.7893$$

Again putting $\qquad x = 3.7893$ in (1), we get

$$x_3 = \frac{1}{2} (\log_{10} 3.7893 + 7) = 3.7893$$

Since $x_2 = x_3$ the root of the given equation is 3.7893. **Ans.**

Example 30. *Apply iterative scheme method to find the real root of $x\,e^x = 1$, correct to three decimals, assuming the initial approximation as $x_0 = 0.5$.*

Solution. The condition for the convergence of the iterative scheme is

$$|\phi'(x_k)| < 1.$$

Here $\qquad x e^x = 1 \qquad \Rightarrow x = e^{-x}$ ... (1)

$\Rightarrow \qquad \phi(x) = e^{-x}$

Putting $\qquad x = 0.5$ in (1), we get

$$x_1 = e^{-0.5} = 0.6065$$

Again putting $\qquad x = 0.6065$ in (1), we have

$$x_2 = e^{-0.6065} = 0.5453$$

Similarly putting the successive values of x in (1), we get

$$x_3 = e^{-.5453} = 0.5797$$
$$x_4 = e^{-.5797} = 0.5601$$
$$x_5 = e^{-.5601} = 0.5712$$
$$x_6 = e^{-.5712} = 0.5648$$
$$x_7 = e^{-.5648} = 0.5685$$
$$x_8 = e^{-.5685} = 0.5664$$
$$x_9 = e^{-.5664} = 0.5676$$
$$x_{10} = e^{-.5676} = 0.5669$$
$$x_{11} = e^{-.5669} = 0.5673$$
$$x_{12} = e^{-0.5673} = 0.5671$$
$$x_{13} = e^{-0.5671} = 0.5672$$

The above iterates are written to show the convergence of the iterates.

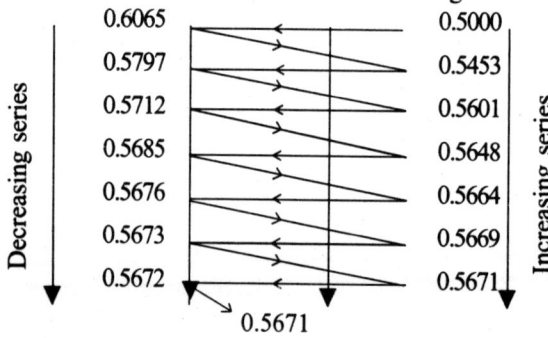

Ans.

EXERCISE 2.4

Solve the following equations by Iteration method :

1. $x^3 + x^2 - 100 = 0$ **Ans.** 4.3311

2. $x^3 - 5x - 11 = 0$, between 2 and 3.   **Ans.** 2.95

3. Use the method of iteration to find a root, near 2 of the equation $x^3 = x^2 + x + 1$, carry out 5 iterations.

                                                                                           Ans. 0.8408

4. $1 + \log x = \dfrac{x}{2}$ **Ans.** 5.36

5. $\sin x = \dfrac{x+1}{x-1}$ [**Hint:** Approximate root = $- 5.5$]   **Ans.** $- 5.5174$

6. $e^x - \sin x = 0$ **Ans.** 0.61413

Choose the correct alternative :

7. The equation $\sin x = (x-1)^2 + 0.5$ has

   (*a*) No real root   (*b*) One real root   (*c*) Two real roots   (*d*) Infinitely many real roots

   [**Hint.** $x^2 - 2x + 1.5 - \sin x = 0$ has two variations of signs]   **Ans.** (*c*)

8. The number of positive real roots of the equation $e^x = 2 \sin x$ is

   (*a*) zero   (*b*) 1   (*c*) 2   (*d*) infinitely many

   [**Hint.** $e^x - 2 \sin x = 0$ ... (1)

                     (+ve) (–ve)

   On putting positive value of x in (1) we get one change in sign from +ve to –ve, so there is only one real positive root.]   **Ans.** (*b*)

9. The integral part of a root of the equation $x^3 - 3x + 1 = 0$ is ....

   (*a*) 0   (*b*) 1   (*c*) 2   (*d*) 3   (*Bihar SBTE, 2008*) **Ans.** (*a*)

CHAPTER 3

Solution of Linear Simultaneous Equations

(Gaussian Elimination Method, Gauss Jordan Method)

3.1 INTRODUCTION

We have already solved simultaneous equations of two or three unknowns. When the number of unknowns in simultaneous equations is large, then it becomes tedious to solve them by the known methods.

Now, we will use the following methods to solve such simultaneous equations.

    (1) Gaussian elimination method   (2) Gauss-Jordan method

3.2 GAUSSIAN ELIMINATION METHOD

In this method the unknowns of equations below are eliminated and the system is reduced to an upper triangular system. The unknowns are obtained by back substitution.

Let a system of simultaneous equations in n unknowns $x_1, x_2 \ldots \ldots x_n$ be

$$a_{11} x_1 + a_{12} x_2 + \ldots \ldots \ldots + a_{1n} x_n = b_1 \qquad \ldots (1)$$
$$a_{21} x_1 + a_{22} x_2 + \ldots \ldots \ldots + a_{2n} x_n = b_2 \qquad \ldots (2)$$

$$\text{--}$$

$$\text{--}$$

$$a_{n1} x_1 + a_{n2} x_2 + \ldots \ldots \ldots + a_{nn} x_n = b_n \qquad \ldots (n)$$

Method to solve the above equations

Step 1. We eliminate x_1 from 2nd, 3rd ....nth equation with the help of the first equation.

Step 2. We again eliminate x_2 from 3rd, 4th..... nth equation with the help of second equation.

Step 3. Continue this process.

In this way the given system is reduced to the triangular form.

Backward Substitution

We first find out the value of x_n from the last equation, then substitute the value of x_n in the $(n-1)$th equation to get the value of x_{n-1}. Again substitute the value of x_{n-1} in $(n-2)$th equation to get the value of x_{n-2}. By this backward substitution we can find the values of all the unknowns.

Example 1. *Solve the following equations by using Gaussian-elimination method :*

$$2x_1 + 4x_2 + x_3 = 3$$
$$3x_1 + 2x_2 - 2x_3 = -2$$
$$x_1 - x_2 + x_3 = 6 \qquad \text{(R.G.P.V. Bhopal, III Semester, June 2006)}$$

Solution. Third equation is written as first equation, the system becomes as

$$x_1 - x_2 + x_3 = 6 \qquad \ldots (1)$$
$$2x_1 + 4x_2 + x_3 = 3 \qquad \ldots (2)$$
$$3x_1 + 2x_2 - 2x_3 = -2 \qquad \ldots (3)$$

Step 1. Subtracting 2 (1) from (2), and 3 (1) from (3), we get

$$x_1 - x_2 + x_3 = 6$$
$$6x_2 - x_3 = -9 \qquad \qquad \text{...(4)}$$
$$5x_2 - 5x_3 = -20 \qquad \qquad \text{...(5)}$$

Step 2. Operate $\dfrac{6}{5}$ (5) − (4)

$$x_1 - x_2 + x_3 = 6 \qquad \qquad \text{...(1)}$$
$$6x_2 - x_3 = -9 \qquad \qquad \text{...(6)}$$
$$-5x_3 = -15 \qquad \qquad \text{...(7)}$$

Step 3. Backward substitution

From (7), $x_3 = \dfrac{-15}{-5} = 3$

From (6), $6x_2 - 3 = -9 \qquad \Rightarrow 6x_2 = -6 \qquad \Rightarrow \qquad x_2 = -1$

From (1), $x_1 - (-1) + 3 = 6 \qquad \Rightarrow x_1 = 6 - 3 - 1 = 2$

Hence, $x_1 = 2, \ x_2 = -1, \ x_3 = 3$ **Ans.**

Example 2. *Solve the following equations:*

$$3x + y - z = 3$$
$$2x - 8y + z = -5$$
$$x - 2y + 9z = 8$$

Using Gaussian elimination method.   (*Bihar SBTE, 2005*)

Solution. The following equations are written in matrix form as

$$\begin{bmatrix} 3 & 1 & -1 \\ 2 & -8 & 1 \\ 1 & -2 & 9 \end{bmatrix} \begin{bmatrix} x \\ y \\ z \end{bmatrix} = \begin{bmatrix} 3 \\ -5 \\ 8 \end{bmatrix}$$

By Gaussian elimination method

$$\begin{bmatrix} 3 & 1 & -1 \\ 0 & -26/3 & 5/3 \\ 0 & -7/3 & 28/3 \end{bmatrix} \begin{bmatrix} x \\ y \\ z \end{bmatrix} = \begin{bmatrix} 3 \\ -7 \\ 7 \end{bmatrix} \quad \begin{array}{l} R_2 \to R_2 - \dfrac{2}{3} R_1 \\[2mm] R_3 \to R_3 - \dfrac{1}{3} R_1 \end{array}$$

$$\begin{bmatrix} 3 & 1 & -1 \\ 0 & -\dfrac{26}{3} & \dfrac{5}{3} \\ 0 & 0 & \dfrac{693}{3 \times 26} \end{bmatrix} \begin{bmatrix} x \\ y \\ z \end{bmatrix} = \begin{bmatrix} 3 \\ -7 \\ \dfrac{231}{26} \end{bmatrix} \quad R_3 \to R_3 - \dfrac{7}{26} R_2$$

$$3x + y - z = 3 \qquad \qquad \text{...(1)}$$

$$-\dfrac{26}{3}y + \dfrac{5}{3}z = -7 \qquad \qquad \text{...(2)}$$

$$\dfrac{693}{3 \times 26}z = \dfrac{231}{26} \qquad \Rightarrow \quad z = \dfrac{231}{26} \times \dfrac{3 \times 26}{693} \qquad \Rightarrow \ z = 1 \quad \text{...(3)}$$

Backward Substitution.

Putting the value of z in (2), we get

$$-\frac{26}{3} y + \frac{5}{3}(1) = -7 \qquad \Rightarrow y = 1 \qquad \textbf{Ans.}$$

Putting the values of y and z in (1), we get

$$3x + 1 - 1 = 3 \qquad \Rightarrow x = 1 \qquad \textbf{Ans.}$$

Example 3. *Solve the following equations:*

$$x_1 + x_2 + x_3 + x_4 = 2$$
$$x_1 + x_2 + 3x_3 - 2x_4 = -6$$
$$2x_1 + 3x_2 - x_3 + 2x_4 = 7$$
$$x_1 + 2x_2 + x_3 - x_4 = -2$$

By Gaussian elimination method. (*Bihar SBGTE, 2003*)

Solution. The given system of linear equations can be written in the matrix form as.

$$\begin{bmatrix} 1 & 1 & 1 & 1 \\ 1 & 1 & 3 & -2 \\ 2 & 3 & -1 & 2 \\ 1 & 2 & 1 & -1 \end{bmatrix} \begin{bmatrix} x_1 \\ x_2 \\ x_3 \\ x_4 \end{bmatrix} = \begin{bmatrix} 2 \\ -6 \\ 7 \\ -2 \end{bmatrix}$$

By using Gaussian elimination method, we have

$$\begin{bmatrix} 1 & 1 & 1 & 1 \\ 0 & 0 & 2 & -3 \\ 0 & 1 & -3 & 0 \\ 0 & 1 & 0 & -2 \end{bmatrix} \begin{bmatrix} x_1 \\ x_2 \\ x_3 \\ x_4 \end{bmatrix} = \begin{bmatrix} 2 \\ -8 \\ 3 \\ -4 \end{bmatrix} \begin{matrix} \\ R_2 \to R_2 - R_1 \\ R_3 \to R_3 - 2R_1 \\ R_4 \to R_4 - R_1 \end{matrix}$$

$$\begin{bmatrix} 1 & 1 & 1 & 1 \\ 0 & 1 & 0 & -2 \\ 0 & 1 & -3 & 0 \\ 0 & 0 & 2 & -3 \end{bmatrix} \begin{bmatrix} x_1 \\ x_2 \\ x_3 \\ x_4 \end{bmatrix} = \begin{bmatrix} 2 \\ -4 \\ 3 \\ -8 \end{bmatrix} \begin{matrix} \\ R_2 \leftrightarrow R_4 \\ \\ R_4 \leftrightarrow R_2 \end{matrix}$$

$$\begin{bmatrix} 1 & 1 & 1 & 1 \\ 0 & 1 & 0 & -2 \\ 0 & 0 & -3 & 2 \\ 0 & 0 & 2 & -3 \end{bmatrix} \begin{bmatrix} x_1 \\ x_2 \\ x_3 \\ x_4 \end{bmatrix} = \begin{bmatrix} 2 \\ -4 \\ 7 \\ -8 \end{bmatrix} \begin{matrix} \\ \\ R_3 \to R_3 - R_2 \\ \end{matrix}$$

$$\begin{bmatrix} 1 & 1 & 1 & 1 \\ 0 & 1 & 0 & -2 \\ 0 & 0 & -3 & 2 \\ 0 & 0 & 0 & -\dfrac{5}{3} \end{bmatrix} \begin{bmatrix} x_1 \\ x_2 \\ x_3 \\ x_4 \end{bmatrix} = \begin{bmatrix} 2 \\ -4 \\ 7 \\ -\dfrac{10}{3} \end{bmatrix} \begin{matrix} \\ \\ \\ R_4 \to R_4 + \dfrac{2}{3} R_3 \end{matrix}$$

$$x_1 + x_2 + x_3 + x_4 = 2 \qquad \qquad …(1)$$
$$x_2 - 2x_4 = -4 \qquad \qquad …(2)$$
$$-3x_3 + 2x_4 = 7 \qquad \qquad …(3)$$
$$-\frac{5}{3} x_4 = -\frac{10}{3} \qquad \Rightarrow x_4 = 2 \qquad …(4)$$

Backward Substitution.

Putting the value of x_4 in (3), we get

$$-3x_3 + 4 = -7 \qquad \Rightarrow \quad x_3 = -1$$

Putting the value of x_4 in (2), we get

$$x_2 - 4 = -4 \qquad \Rightarrow \quad x_2 = 0$$

Putting the values of x_2, x_3, x_4 in (1), we get

$$x_1 + 0 - 1 + 2 = 2 \qquad \Rightarrow \quad x_1 = 1$$

Hence, $\quad x_1 = 1, \; x_2 = 0, \; x_3 = -1$ and $x_4 = 2$ **Ans.**

EXERCISE 3.1

Solve the following system of the equations by Gaussian elimination method:

1. $x - y + z = 1, \; -3x + 2y - 3z = -6, \; 2x - 5y + 4z = 5$ **Ans.** $x = -2, \; y = 3, \; z = 6$

2. $x + 10y + z = 12, \; x + y + 10z = 12, \; 10x + y + z = 12$ **Ans.** $x = 1, \; y = 1, \; z = 1$

3. $x + 3y + 10z = 23.89, \; 2x + 17y + 4z = 34.84, \; 28x + 4y - z = 31.88$ **Ans.** $x = 0.99, \; y = 1.50, \; z = 1.84$

4. $2x + 6y - z = -12, \; 5x - y + z = 11, \; 4x - y + 3z = 10$ **Ans.** $x = 1.64, \; y = -2.49, \; z = 0.32$

5. $5x_1 - x_2 = 9, \quad -x_1 + 5x_2 - x_3 = 4, \qquad -x_2 + 5x_3 = -6$ **Ans.** $x_1 = 1.842, \; x_2 = 0.21, x_3 = -1.15$

                                               (Bihar SBTE, 2010)

6. $2x + 2y - z + u = 4, \; 4x + 3y - z + 2u = 6, \; 8x + 5y - 3z + 4u = 12, \; 3x + 3y - 2z + 2u = 6$

                                       Ans. $x = 1, \; y = 1, \; z = -1, u = -1.$

7. $x_1 + 2x_2 - 12x_3 + 8x_4 = 27, \; 5x_1 + 4x_2 + 7x_3 - 2x_4 = 4, \; 6x_1 - 12x_2 - 8x_3 + 3x_4 = 49,$

    $3x_1 - 7x_2 - 9x_3 - 5x_4 = -11$ **Ans.** $x_1 = 3, \; x_2 = -2, \; x_3 = 1, x_4 = 5$

8. $x_1 + 2x_2 + 3x_3 + 4x_4 = 32, \; 2x_1 - x_2 + 2x_3 - x_4 = 3, \; 3x_1 + 2x_2 + 4x_3 - x_4 = 17.7,$

    $6x_1 + 7x_2 + 8x_3 - 5x_4 = 30$ **Ans.** $x_1 = 1.1, \; x_2 = 2.2, \; x_3 = 3.5, x_4 = 4$

3.3 GAUSS- JORDAN METHOD

                                           (R.G.P.V. Bhopal, III Semester, Dec. 2007)

This is modification of the Gaussian elimination method.

By this method we eliminate unknowns not only from the equations below but also from the equations above. In this way the system is reduced to a diagonal matrix.

Finally each equation consists of only one unknown and thus, we get the solution. Here, the labour of backward substitution for finding the unknowns is saved.

Gauss-Jordan method is modification of Gaussian elimination method.

Example 4. *Apply Gauss-Jordan method to solve the equations :*

$$x + y + z = 9$$
$$2x - 3y + 4z = 13$$
$$3x + 4y + 5z = 40 \quad \textit{(Bihar SBTE, 2011, 2010, R.G.P.V. Bhopal, III Semester, Dec. 2007)}$$

Solution. The following system of linear equation can be written in matrix form

$$AX = B$$

where, $A = \begin{bmatrix} 1 & 1 & 1 \\ 2 & -3 & 4 \\ 3 & 4 & 5 \end{bmatrix}$, $X = \begin{bmatrix} x \\ y \\ z \end{bmatrix}$, $B = \begin{bmatrix} 9 \\ 13 \\ 40 \end{bmatrix}$

By using Gauss-Jordon method ,we have

$$\begin{bmatrix} 1 & 1 & 1 \\ 2 & -3 & 4 \\ 3 & 4 & 5 \end{bmatrix} \begin{bmatrix} x \\ y \\ z \end{bmatrix} = \begin{bmatrix} 9 \\ 13 \\ 40 \end{bmatrix}$$

$$\Rightarrow \begin{bmatrix} 1 & 1 & 1 \\ 0 & -5 & 2 \\ 0 & 1 & 2 \end{bmatrix} \begin{bmatrix} x \\ y \\ z \end{bmatrix} = \begin{bmatrix} 9 \\ -5 \\ 13 \end{bmatrix} \begin{matrix} \\ R_2 \to R_2 - 2R_1 \\ R_3 \to R_3 - 3R_1 \end{matrix}$$

$$\Rightarrow \begin{bmatrix} 1 & 1 & 1 \\ 0 & -5 & 2 \\ 0 & 0 & \dfrac{12}{5} \end{bmatrix} \begin{bmatrix} x \\ y \\ z \end{bmatrix} = \begin{bmatrix} 9 \\ -5 \\ 12 \end{bmatrix} \begin{matrix} \\ \\ R_3 \to R_3 + \dfrac{1}{5} R_2 \end{matrix}$$

$$\Rightarrow \begin{bmatrix} 1 & 0 & \dfrac{7}{5} \\ 0 & -5 & 2 \\ 0 & 0 & \dfrac{12}{5} \end{bmatrix} \begin{bmatrix} x \\ y \\ z \end{bmatrix} = \begin{bmatrix} 8 \\ -5 \\ 12 \end{bmatrix} \begin{matrix} R_1 \to R_1 + \dfrac{1}{5} R_2 \\ \\ \end{matrix}$$

$$\Rightarrow \begin{bmatrix} 1 & 0 & 0 \\ 0 & -5 & 0 \\ 0 & 0 & \dfrac{12}{5} \end{bmatrix} \begin{bmatrix} x \\ y \\ z \end{bmatrix} = \begin{bmatrix} 1 \\ -15 \\ 12 \end{bmatrix} \begin{matrix} R_1 \to R_1 - \dfrac{7}{12} R_3 \\ R_2 \to R_2 - \dfrac{5}{6} R_3 \\ \end{matrix}$$

$$\Rightarrow \begin{bmatrix} 1 & 0 & 0 \\ 0 & 1 & 0 \\ 0 & 0 & 1 \end{bmatrix} \begin{bmatrix} x \\ y \\ z \end{bmatrix} = \begin{bmatrix} 1 \\ 3 \\ 5 \end{bmatrix} \begin{matrix} \\ R_2 \to -\dfrac{1}{5} R_2 \\ R_3 \to \dfrac{5}{12} R_3 \end{matrix}$$

Thus, we have

$x = 1$, $y = 3$, $z = 5$ **Ans.**

Example 5. *Solve the following system of equations by Gauss-Jordan elimination method:*

$$5x_1 - x_2 = 9$$
$$-x_1 + 5x_2 - x_3 = 4$$
$$x_2 + 5x_3 = 6$$

(Bihar SBTE, 2009)

Solution. The given system of linear equations can be written in matrix form $AX = B$

where, $A = \begin{bmatrix} 5 & -1 & 0 \\ -1 & 5 & -1 \\ 0 & 1 & 5 \end{bmatrix}$, $X = \begin{bmatrix} x_1 \\ x_2 \\ x_3 \end{bmatrix}$, $B = \begin{bmatrix} 9 \\ 4 \\ 6 \end{bmatrix}$

By using Gauss-Jordan elimination method, we have

$$\begin{bmatrix} 5 & -1 & 0 \\ -1 & 5 & -1 \\ 0 & 1 & 5 \end{bmatrix} \begin{bmatrix} x_1 \\ x_2 \\ x_3 \end{bmatrix} = \begin{bmatrix} 9 \\ 4 \\ 6 \end{bmatrix}$$

$$R_2 \to R_2 + \frac{1}{5} R_1$$

$$\Rightarrow \begin{bmatrix} 5 & -1 & 0 \\ 0 & \dfrac{24}{5} & -1 \\ 0 & 1 & 5 \end{bmatrix} \begin{bmatrix} x_1 \\ x_2 \\ x_3 \end{bmatrix} = \begin{bmatrix} 9 \\ \dfrac{2\ 9}{5} \\ 6 \end{bmatrix}$$

$$R_3 \to R_3 - \frac{5}{24} R_2$$

$$\begin{bmatrix} 5 & -1 & 0 \\ 0 & \dfrac{24}{5} & -0 \\ 0 & 0 & \dfrac{125}{24} \end{bmatrix} \begin{bmatrix} x_1 \\ x_2 \\ x_3 \end{bmatrix} = \begin{bmatrix} 9 \\ \dfrac{29}{5} \\ \dfrac{115}{24} \end{bmatrix}$$

$$R_2 \to R_2 + \frac{24}{125} R_3$$

$$\begin{bmatrix} 5 & -1 & 0 \\ 0 & \dfrac{24}{5} & -1 \\ 0 & 0 & \dfrac{125}{24} \end{bmatrix} \begin{bmatrix} x_1 \\ x_2 \\ x_3 \end{bmatrix} = \begin{bmatrix} 9 \\ \dfrac{168}{25} \\ \dfrac{115}{24} \end{bmatrix}$$

$$R_1 \to R_1 + \frac{5}{24} R_2$$

$$\begin{bmatrix} 5 & 0 & 0 \\ 0 & \dfrac{24}{5} & 0 \\ 0 & 0 & \dfrac{125}{24} \end{bmatrix} \begin{bmatrix} x_1 \\ x_2 \\ x_3 \end{bmatrix} = \begin{bmatrix} \dfrac{52}{5} \\ \dfrac{168}{25} \\ \dfrac{115}{24} \end{bmatrix}$$

$$R_1 \to \frac{1}{5} R_1, \ R_2 \to \frac{5}{24} R_2, \ R_3 \to \frac{24}{125} R_3$$

$$\begin{bmatrix} 1 & 0 & 0 \\ 0 & 1 & 0 \\ 0 & 0 & 1 \end{bmatrix} \begin{bmatrix} x_1 \\ x_2 \\ x_3 \end{bmatrix} = \begin{bmatrix} \dfrac{52}{25} \\ \dfrac{7}{5} \\ \dfrac{23}{25} \end{bmatrix}$$

Thus, $x_1 = \dfrac{52}{25}$, $x_2 = \dfrac{7}{5}$, $x_3 = \dfrac{23}{25}$ **Ans.**

Example 6. *Solve the system of equations by Gauss-Jordan method*

$$x + y + z + u = 2$$
$$2x - y + 2z - u = -5$$
$$3x + 2y + 3z + 4u = 7$$
$$x - 2y - 3z + 2u = 5$$

Solution. The following system of equations is written in matrix form:

$$\begin{bmatrix} 1 & 1 & 1 & 1 \\ 2 & -1 & 2 & -1 \\ 3 & 2 & 3 & 4 \\ 1 & -2 & -3 & 2 \end{bmatrix} \begin{bmatrix} x \\ y \\ z \\ u \end{bmatrix} = \begin{bmatrix} 2 \\ -5 \\ 7 \\ 5 \end{bmatrix}$$

Using Gauss-Jordan method, we have

$$R_2 \to R_2 - 2R_1, \ R_3 \to R_3 - 3R_1, \ R_4 \to R_4 - R_1$$

$$\begin{bmatrix} 1 & 1 & 1 & 1 \\ 0 & -3 & 0 & -3 \\ 0 & -1 & 0 & 1 \\ 0 & -3 & -4 & 1 \end{bmatrix} \begin{bmatrix} x \\ y \\ z \\ u \end{bmatrix} = \begin{bmatrix} 2 \\ -9 \\ 1 \\ 3 \end{bmatrix}$$

$$R_2 \leftrightarrow R_3$$

$$\begin{bmatrix} 1 & 1 & 1 & 1 \\ 0 & -1 & 0 & 1 \\ 0 & -3 & 0 & -3 \\ 0 & -3 & -4 & 1 \end{bmatrix} \begin{bmatrix} x \\ y \\ z \\ u \end{bmatrix} = \begin{bmatrix} 2 \\ 1 \\ -9 \\ 3 \end{bmatrix}$$

$$R_3 \to R_3 - 3R_2$$
$$R_4 \to R_4 - 3R_2$$

$$\begin{bmatrix} 1 & 1 & 1 & 1 \\ 0 & -1 & 0 & 1 \\ 0 & 0 & 0 & -6 \\ 0 & 0 & -4 & -2 \end{bmatrix} \begin{bmatrix} x \\ y \\ z \\ u \end{bmatrix} = \begin{bmatrix} 2 \\ 1 \\ -12 \\ 0 \end{bmatrix}$$

$$R_3 \leftrightarrow R_4$$

$$\begin{bmatrix} 1 & 1 & 1 & 1 \\ 0 & -1 & 0 & 1 \\ 0 & 0 & -4 & -2 \\ 0 & 0 & 0 & -6 \end{bmatrix} \begin{bmatrix} x \\ y \\ z \\ u \end{bmatrix} = \begin{bmatrix} 2 \\ 1 \\ 0 \\ -12 \end{bmatrix}$$

$$R_2 \to -R_2,\ R_3 \to -\frac{1}{4}R_3,\ R_4 \to -\frac{1}{6}R_4$$

$$\begin{bmatrix} 1 & 1 & 1 & 1 \\ 0 & 1 & 0 & -1 \\ 0 & 0 & 1 & \dfrac{1}{2} \\ 0 & 0 & 0 & 1 \end{bmatrix} \begin{bmatrix} x \\ y \\ z \\ u \end{bmatrix} = \begin{bmatrix} 2 \\ -1 \\ 0 \\ 2 \end{bmatrix}$$

$$R_1 \to R_1 - R_4,\ R_2 \to R_2 + R_4,\ R_3 \to R_3 - \frac{1}{2}R_4$$

$$\begin{bmatrix} 1 & 1 & 1 & 0 \\ 0 & 1 & 0 & 0 \\ 0 & 0 & 1 & 0 \\ 0 & 0 & 0 & 1 \end{bmatrix} \begin{bmatrix} x \\ y \\ z \\ u \end{bmatrix} = \begin{bmatrix} 0 \\ 1 \\ -1 \\ -2 \end{bmatrix}$$

$$R_1 \to R_1 - R_3$$

$$\begin{bmatrix} 1 & 1 & 0 & 0 \\ 0 & 1 & 0 & 0 \\ 0 & 0 & 1 & 0 \\ 0 & 0 & 0 & 1 \end{bmatrix} \begin{bmatrix} x \\ y \\ z \\ u \end{bmatrix} = \begin{bmatrix} 1 \\ 1 \\ -1 \\ 2 \end{bmatrix}$$

$$R_1 \to R_1 - R_2$$

$$\begin{bmatrix} 1 & 0 & 0 & 0 \\ 0 & 1 & 0 & 0 \\ 0 & 0 & 1 & 0 \\ 0 & 0 & 0 & 1 \end{bmatrix} \begin{bmatrix} x \\ y \\ z \\ u \end{bmatrix} = \begin{bmatrix} 0 \\ 1 \\ -1 \\ 2 \end{bmatrix}$$

Hence, $x = 0,\ y = 1,\ z = -1,\ u = 2$ **Ans.**

EXERCISE 3.2

Solve the following system by Gauss-Jordan method.

1. $2x - 6y + 8z = 24,\ 5x + 4y - 3z = 2,\ 3x + y + 2z = 16$ **Ans.** $x = 1,\ y = 3,\ z = 5$

2. $x + 2y + z = 8,\ 2x + 3y + 4z = 20,\ 4x + 3y + 2z = 16$ **Ans.** $x = 1,\ y = 2,\ z = 3$

3. $^*3x + 4y + 5z = 18,\ 2x - y + 8z = 13,\ 5x - 2y + 7z = 20$ **Ans.** $x = 3,\ y = 1,\ z = 1$

4. $2x - y + 3z = 9,\ x + y + z = 6,\ x - y + z = 2$ **Ans.** $x = 1,\ y = 2,\ z = 3$

5. $10x + y + 2z = 13, \ 3x + 10y + z = 14, \ 2x + 3y + 10 z = 15$ **Ans.** $x = 1, \ y = 1, \ z = 1$

$(R.G.P.V., Bhopal, III Semester, June 2002)$

6. $2x_1 + 2x_2 + x_3 = 6, \ 4x_1 + 2x_2 + 3x_3 = 4, \ x_1 + x_2 + x_3 = 0$

$(R.G.P.V., Bhopal, M.C.A. June 2001)$ **Ans.** $x_1 = 5, x_2 = 1, x_3 = -6$

7. $x + 3y + 6z = 2, \ 3x - y + 4z = 9, x - 4y + 2z = 7$ **Ans.** $x = 2, \ y = -1, \ z = \dfrac{1}{2}$

8. $x + y + z = 6.6, \ x - y + z = 2.2, \ x + 2y + 3z = 15.2$ **Ans.** $x = 1.2, \ y = 2.2, \ z = 3.2$

9. $9x - 2y + z = 50, \ x + 5y - 3z = 18, \ -2x + 2y + 7z = 19$ **Ans.** $x = 6.13, y = 4.31, z = 3.23$

10. $2x_1 - x_2 - x_3 = -1, \ 2x_2 - x_3 + x_4 = 1, \ x_1 + 2x_3 - x_4 = -1, \ x_1 + x_2 + 3x_4 = 5$

Ans. $x_1 = -1, x_2 = 0, x_3 = 1, x_4 = 2$

11. $3x + 4y + 5z = 12, x + 2y + 3z = 10, 2x + 3y + 2z = 8$ $(Bihar SBTE, 2014)$ **Ans.** $x = \dfrac{-13}{2}, \ y = 6, \ z = \dfrac{3}{2}$

Ans. $x_1 = 2, x_2 = \dfrac{1}{5}, x_3 = 0, x_4 = \dfrac{4}{5}$

12. $10x + y + z = 12, 2x + 10y + z = 13, x + y + 5z = 7$ $(Bihar SBTE, 2009)$ **Ans.** $x = 1, y = 1, z = 1$

13. $2x - 8y + z = -5, x - 2y + 9z = 8, 3x + y - z = 3$ $(Bihar SBTE, 2011)$ **Ans.** $x = 1, y = 1, z = 1$

14. $2x_1 - 3x_2 + x_3 + x_4 = -1, \ x_1 + 4x_2 + 5x_3 + 2x_4 = 25, \ 3x_1 - 4x_2 + x_3 + 3x_4 = 2,$
$x_1 + 2x_2 + 3x_3 + 4x_4 = 16.2$ **Ans.** $x_1 = 8.7, x_2 = 5.7, x_3 = -1.3, x_4 = 0$

15. Simultaneous linear albergraic may be solved by ....

(a) Gauss-Jordan method

(b) Newton-Rapson method

(c) Regula-Falsi method

(d) Hungarian method

$(Bihar STBE, 2011)$ **Ans.** (a)

CHAPTER 4

Finite Differences

4.1 INTRODUCTION

In a table the values of x and corresponding values of y are given

x	x_1	x_2	x_3	x_4	x_5	x_6
y	y_1	y_2	y_3	y_4	y_5	y_6

Now, we want to express y as a function of x.

In this chapter we will try to establish relation between x and y and calculate the value of y at any value of x. This method is known as interpolation.

4.2 FINITE DIFFERENCES

Consider the function $y = f(x)$, where x is known as *argument* and y is called *entry*. Here, the values of the argument are at equal intervals.

$$a, \quad a + h, \quad a + 2h, \quad a + 3h, \quad .... \; a + nh.$$

The corresponding values of y are :

$$f(a), \; f(a + h), \; f(a + 2h), \; f(a + 3h), \; ...., f(a + nh)$$

The following differences are called finite differences

$$f(a + h) - f(a)$$
$$f(a + 2h) - f(a + h)$$
$$f(a + 3h) - f(a + 2h)$$

....

....

$$f(a + n\,h) - f[a + (n - 1)\,h]$$

4.3 FORWARD DIFFERENCE

(U.P., III Semester, Dec. 2009)

If the above differences are denoted by the forward operator Δ then these differences are known as forward differences.

$$\Delta f(a) = f(a + h) - f(a)$$
$$\Delta f(a + h) = f(a + 2h) - f(a + h)$$
$$\Delta f(a + 2h) = f(a + 3h) - f(a + 2h)$$

....

....

$$\Delta f(a + \overline{n - 1}h) = f(a + nh) - f(a + \overline{n - 1}h)$$

In general

$$\boxed{\Delta f(x) = f(x + h) - f(x)}$$

Δ is an operator and is called a *forward difference operator* and h is known as the *interval of differences*.

47

First Forward Difference

$$\Delta f(a) = f(a + h) - f(a) \qquad \qquad ... (1)$$

Second Forward difference

$$\Delta^2 f(a) = \Delta [\Delta f(a)] = \Delta [f(a + h) - f(a)] = \Delta f(a + h) - \Delta f(a)$$
$$= [f(a + 2h) - f(a + h)] - [f(a + h) - f(a)] = f(a + 2h) - 2f(a + h) + f(a) \quad ... (2)$$

Third Forward difference

$$\Delta^3 f(a) = \Delta [\Delta^2 f(a)] \qquad \qquad ... (3)$$

Putting the value of $\Delta^2 f(a)$ from (2) in (3), we get

$$\Delta^3 f(a) = \Delta [f(a + 2h) - 2f(a + h) + f(a)] = \Delta f(a + 2h) - 2\Delta f(a + h) + \Delta f(a)$$
$$= f(a + 3h) - f(a + 2h) - 2[f(a + 2h) - f(a + h)] + f(a + h) - f(a)$$
$$= f(a + 3h) - 3f(a + 2h) + 3f(a + h) - f(a)$$

Similarly, $\Delta^n f(a) = \Delta^{n-1} . \Delta f(a) = \Delta^{n-1} [f(a+h) - f(a)]$

$$\boxed{\Delta^n f(a) = \Delta^{n-1} f(a+h) - \Delta^{n-1} f(a)}$$

Note. 1. $\Delta f(a)$ means $f(a)$ is to be subtracted from next entry.

2. The difference $f(a + h) - f(a)$ is denoted by placing Δ before the second entry.

3. Δ^2 is not the square of the operator Δ but Δ^2 means Δ operated by Δ.

Example 1. *Construct a forward difference table and find $\Delta^4 f(1)$, if*
$$f(1) = 1, f(2) = 3, \quad f(3) = 8, f(4) = 15, f(5) = 25.$$

Solution.

x	$f(x)$	$\Delta f(x)$	$\Delta^2 f(x)$	$\Delta^3 f(x)$	$\Delta^4 f(x)$
1	1				
		2			
			3		
2	3			−1	
		5			2
			2	1	
3	8				
		7			
4	15		3		
		10			
5	25				

From the table, we have $\Delta^4 f(1) = 2$ **Ans.**

Example 2. *Evaluate:* $\Delta (5x^3 - 2x^2 + 3x + 9)$ *(Bihar SBTE, 2014)*

Solution. $\Delta (5x^3 - 2x^2 + 3x + 9) = 5\Delta x^3 - 2\Delta x^2 + 3\Delta x + \Delta 9$

$$= 5 [(x + h)^3 - x^3] - 2 [(x + h)^2 - x^2] + 3 [(x + h) - x] \qquad (\Delta 9 = 0)$$
$$= 5 [x^3 + 3x^2 h + 3xh^2 + h^3 - x^3] - 2 [x^2 + 2xh + h^2 - x^2] + 3 [x + h - x]$$
$$= 5 [3x^2 h + 3xh^2 + h^3] - 2 (2xh + h^2) + 3h$$
$$= 15x^2 h + 15xh^2 + 5h^3 - 4xh - 2h^2 + 3h$$
$$= 15x^2 + 15x + 5 - 4x - 2 + 3 \qquad (\text{If } h = 1)$$
$$= 15x^2 + 11x + 6 \qquad \qquad \textbf{Ans.}$$

Table of Forward differences

Argument x	Entry $f(x)$	First Difference $\Delta f(x)$	Second Difference $\Delta^2 f(x)$	Third Difference $\Delta^3 f(x)$	Fourth Difference $\Delta^4 f(x)$
a	$f(a)$				
		$f(a+h) - f(a) = \Delta f(a)$			
$a+h$	$f(a+h)$		$\Delta f(a+h) - \Delta f(a) = \Delta^2 f(a)$		
		$f(a+2h) - f(a+h) = \Delta f(a+h)$		$\Delta^2 f(a+h) - \Delta^2 f(a) = \Delta^3 f(a)$	
$a+2h$	$f(a+2h)$		$\Delta f(a+2h) - \Delta f(a+h)$ $= \Delta^2 f(a+h)$		$\Delta^3 f(a+h) - \Delta^3 f(a) = \Delta^4 f(a)$
		$f(a+3h) - f(a+2h) =$ $\Delta f(a+2h)$		$\Delta^2 f(a+2h) - \Delta^2 f(a+h)$ $= \Delta^3 f(a+h)$	
$a+3h$	$f(a+3h)$		$\Delta f(a+3h) - \Delta f(a+2h)$ $= \Delta^2 f(a+2h)$		
		$f(a+4h) - f(a+3h) =$ $\Delta f(a+3h)$			
$a+4h$	$f(a+4h)$				

4.4 BACKWARD DIFFERENCE OPERATOR

If the difference $f(x) - f(x - h)$ is denoted by the backward difference operator ∇ then the difference $\nabla f(x) = f(x) - f(x - h)$ is called *backward difference*.

$$\boxed{\nabla f(x) = f(x) - f(x - h)}$$

The operator ∇ is called as **backward difference** operator. It is, to be noted that it is only the notation which changes and not the difference $y_1 - y_0 = \Delta y_0 = \nabla y_1$.

Second backward difference is denoted by $\nabla^2 f(x)$ and

$$\begin{aligned}
\nabla^2 f(x) = \nabla [\nabla f(x)] &= \nabla [f(x) - f(x - h)] \\
&= \nabla f(x) - \nabla f(x - h) \\
&= [f(x) - f(x - h)] - [f(x - h) - f(x - 2h)] \\
&= f(x) - 2f(x - h) + f(x - 2h)
\end{aligned}$$

Third backward difference.

$$\begin{aligned}
\nabla^3 f(x) &= \nabla^2 [\nabla f(x)] = \nabla^2 [f(x) - f(x - h)] \\
&= \nabla^2 f(x) - \nabla^2 f(x - h) \\
&= f(x) - 2f(x - h) + f(x - 2h) - [f(x - h) - 2f(x - 2h) + f(x - 3h)] \\
&= f(x) - 3f(x - h) + 3f(x - 2h) - f(x - 3h)
\end{aligned}$$

In General

$$\begin{aligned}
\nabla^n f(x) &= \nabla^{n-1} [\nabla f(x)] = \nabla^{n-1} [f(x) - f(x - h)] \\
&= \nabla^{n-1} f(x) - \nabla^{n-1} f(x - h)
\end{aligned}$$

Note:

1. The backward difference $f(a) - f(a - h)$ is denoted by placing backward difference operator ∇ before the first entry.

Example 3. *Construct a backward difference table, $f(1) = 4, f(2) = 8, f(3) = 12, f(4) = 18, f(5) = 36$. Find $\nabla^4 f(5)$.*

Solution.

x	$f(x)$	$\nabla f(x)$	$\nabla^2 f(x)$	$\nabla^3 f(x)$	$\nabla^4 f(x)$
1	4				
		4			
2	8		0		
		4		−2	
3	12		2		8
		6		10	
4	18		12		
		18			
5	36				

From the table, we have

$$\nabla^4 f(5) = 8 \qquad\qquad \textbf{Ans.}$$

Table of backward differences

Argument x	Entry $f(x)$	First Difference $\nabla f(x)$	Second Difference $\nabla^2 f(x)$	Third Difference $\nabla^3 f(x)$	Fourth Difference $\nabla^4 f(x)$
a	$f(a)$				
		$f(a+h)-f(a) = \nabla f(a+h)$			
$a+h$	$f(a+h)$		$\nabla f(a+2h) - \nabla f(a+h)$ $= \nabla^2 f(a+2h)$		
		$f(a+2h)-f(a+h) = \nabla f(a+2h)$		$\nabla^2 f(a+3h) - \nabla^2 f(a+2h)$ $= \nabla^3 f(a+3h)$	
$a+2h$	$f(a+2h)$		$\nabla f(a+3h) - \nabla f(a+2h)$ $= \nabla^2 f(a+3h)$		$\nabla^3 f(a+4h) - \nabla^3 f(a+3h)$ $= \nabla^4 f(a+4h)$
		$f(a+3h)-f(a+2h)$ $= \nabla f(a+3h)$		$\nabla^2 f(a+4h) - \nabla^2 f(a+3h)$ $= \nabla^3 f(a+4h)$	
$a+3h$	$f(a+3h)$		$\nabla f(a+4h) - \nabla f(a+3h)$ $= \nabla^2 f(a+4h)$		
		$f(a+4h)-f(a+3h)$ $= \nabla f(a+4h)$			
$a+4h$	$f(a+4h)$				

Example 4. *Given that*

x	1	2	3	4	5
y	2	5	10	17	26

Find the value of $\nabla^2 y_5$.

(R.G.P.V., Bhopal, III Semester, June 2006)

Solution. The difference table is as under :

x	y	∇y	$\nabla^2 y$	$\nabla^3 y$
1	2			
		3		
2	5		2	
		5		0
3	10		2	
		7		$0 = \nabla^3 y_5$
4	17		$2 = \nabla^2 y_5$	
		$9 = \nabla y_5$		
5	$26 = y_5$			

From the table, we get $\nabla^2 y_5 = 2$ **Ans.**

4.5 THE SHIFTING OPERATOR E

(U.P., III Semester, Dec. 2009)

$$E f(x) = f(x + h) \qquad \text{... (1)}$$

The operator E is called the shifting operator.

$$\Delta f(x) = f(x+h) - f(x) \quad \Rightarrow \quad \Delta f(x) = E f(x) - f(x) \qquad \text{[Using (1)]}$$

$$\Rightarrow \quad E f(x) = \Delta f(x) + f(x) \quad \Rightarrow \quad E f(x) = (\Delta + 1) f(x)$$

$$E \equiv \Delta + 1 \quad or \quad \Delta \equiv E - 1 \qquad \text{... (2)}$$

$$\boxed{E = \Delta + 1}$$

(Bihar SBTE, 2014, 2010)

$$\boxed{\Delta = E - 1}$$

Again,

$$E^2 f(x) = EE f(x) = E f(x + h) = f(x + 2h)$$

$$E^3 f(x) = EE^2 f(x) = E f(x + 2h) = f(x + 3h)$$

$$E^4 f(x) = EE^3 f(x) = E f(x + 3h) = f(x + 4h)$$

....

....

$$E^n f(x) = f(x + nh)$$

Similarly, we define $E^{1} f(x) = f(x - h)$

In general, $E^{n} f(x) = f(x - nh)$

Finite Difference Operators

Operator	Definition	Name of the operator
$Ef(x)$	$f(x+h)$	The shift operator
$\Delta f(x)$	$f(x+h)-f(x)$	The forward difference operator
$\nabla f(x)$	$f(x)-f(x-h)$	The backward difference operator

4.6 RELATION BETWEEN ∇ AND E^{-1}

(i) $\quad \nabla = 1 - E^{-1}$ $\qquad$ (ii) $E\nabla = \nabla E = \Delta$

Solution. (*i*) We know that

$$\nabla f(a) = f(a) - f(a-h) \qquad \dots (1)$$

and $\qquad E^{-1} f(a) = f(a-h) \qquad \dots (2)$

Putting the value of $f(a-h)$ from (2) in (1), we get

$$\nabla f(a) = f(a) - E^{-1} f(a)$$

$\Rightarrow \qquad \nabla f(a) = (1 - E^{-1}) f(a)$

$\Rightarrow \qquad \nabla = 1 - E^{-1}$

$\Rightarrow \qquad \boxed{E^{-1} = 1 - \nabla}$ $\qquad\qquad$ (*Bihar SBTE, 2014, 2005*)

This is the relation between ∇ and E^{-1}.

(*ii*) $\qquad E\nabla f(a) = E[f(a) - f(a-h)] = Ef(a) - Ef(a-h)$

$\qquad\qquad\qquad = Ef(a) - f(a) = (E-1) f(a) = \Delta f(a)$

$\therefore \qquad\qquad E\nabla = \Delta \qquad \dots (3)$

$\qquad\qquad \nabla Ef(a) = \nabla f(a+h) = f(a+h) - f(a) = \Delta f(a)$

$\therefore \qquad\qquad \nabla E = \Delta \qquad \dots (4)$

From (3) and (4), we have

$$\boxed{E\nabla = \nabla E = \Delta} \qquad\qquad \textbf{Proved.}$$

4.7 RELATION BETWEEN D AND Δ

$$D \equiv \frac{1}{h}\left[\Delta - \frac{1}{2}\Delta^2 + \frac{1}{3}\Delta^3 \dots\right]$$

Solution. $\qquad Ef(x) = f(x+h)$

$$= \left[f(x) + hf'(x) + \frac{h^2}{2!}f''(x) + \dots\right] = \left[f(x) + hDf(x) + \frac{h^2}{2}D^2 f(x) + \dots\right]$$

$$= \left[1 + hD + \frac{h^2 D^2}{2} + \dots\right] f(x)$$

$$Ef(x) = e^{hD} f(x) \qquad\qquad \left[e^x = 1 + x + \frac{x^2}{2!} + \frac{x^3}{3!} + \dots\right]$$

$$E = e^{hD}$$

$$\log E = \log e^{hD}$$

$$\Rightarrow \quad \log (1 + \Delta) = hD \log_e e \qquad\qquad\qquad [\log_e e = 1]$$

$$\log (1 + \Delta) = hD$$

$$D = \frac{1}{h} \log (1+\Delta) = = \frac{1}{h} \log 3 \qquad\qquad (Bihar\ SBTE,\ 2014,\ 2012)$$

$$\boxed{D = \frac{1}{h}\left[\Delta - \frac{1}{2}\Delta^2 + \frac{1}{3}\Delta^3 + ...\right]}, \quad \left[\log(1+x) = x - \frac{x^2}{2} + \frac{x^3}{3} + ...\right]$$

Table: Relation between E, Δ, ∇

In terms	E	Δ	∇
E	–	$\Delta + 1$	$(1 - \nabla)^{-1}$
Δ	$E - 1$	–	$(1 - \nabla)^{-1} - 1$
∇	$1 - E^{-1}$	$1 - (1 + \nabla)^{-1}$	–

Example 5.   *Evaluate :* $\Delta^n (e^x)$ *where interval of differencing being unity.*   *(Bihar SBTE, 2008)*

Solution. $\Delta\, e^x = e^{x+1} - e^x = (e - 1)\, e^x$

$\Delta^2 e^x = \Delta\,(\Delta e^x) = \Delta\,[(e-1)\,e^x] = (e-1)\,\Delta\,e^x = (e-1)(e-1)\,e^x = (e-1)^2\, e^x$

Similarly

$\Delta^3\, e^x = (e-1)^3\, e^x, \qquad \Delta^4\, e^x = (e-1)^4\, e^x,$

$$\boxed{\Delta^n\, e^x = (e-1)^n\, e^x}$$ **Ans.**

Example 6.   *If* $D = \dfrac{d}{dx}$, *E is shift operator and h is interval of differencing, prove that* $D = \dfrac{1}{h} \log E$.

(Bihar SBTE, 2008)

Solution. $Ef(x) = f(x + h) = f(x) + \dfrac{hf'(x)}{1!} + \dfrac{h^2 f^2(x)}{2!} + ...$

$$Ef(x) = f(x) + \frac{hD f(x)}{1!} + \frac{h^2 D^2 f(x)}{2!} + ...$$

$$Ef(x) = \left[1 + \frac{hD}{1!} + \frac{(hD)^2}{2!} + ...\right] f(x)$$

$$Ef(x) = e^{hD} f(x)$$

$$\therefore \ E = e^{hD}$$

$$\therefore \ hD = \log E$$

$$\boxed{D = \frac{1}{h} \log E}$$ **Proved.**

Example 7.   *Prove with the usual notations that :*

$$\left(E^{1/2} + E^{-1/2}\right)(1+\Delta)^{\frac{1}{2}} = 2 + \Delta \qquad\qquad (R.G.P.V.,\ Bhopal,\ III\ Semester,\ Dec.\ 2005)$$

Solution.   L.H.S. $= \left(E^{\frac{1}{2}} + E^{-\frac{1}{2}}\right)(1+\Delta)^{\frac{1}{2}}$

$= \left(E^{\frac{1}{2}} + E^{-\frac{1}{2}}\right)E^{\frac{1}{2}}$ $\left[\because 1 + \Delta = E\right]$

$= E^{\frac{1}{2}}.E^{\frac{1}{2}} + E^{-\frac{1}{2}}.E^{\frac{1}{2}}$

$= E + 1 = (1 + \Delta) + 1 = 2 + \Delta = \text{R.H.S}$ $\left[\because E = 1 + \Delta\right].$ **Proved.**

Example 8. *If h is interval of differentiating, the value of* $\left(\dfrac{\Delta^2}{E}\right)x^2$ *is*

(a) h   (b) 2h   (c) $2h^2$ (d) $2h^3$ *(Bihar SBTE, 2009)*

Solution. $\left(\dfrac{\Delta^2}{E}\right)x^2 = \left[\dfrac{(E-1)^2}{E}\right]x^2 = \left[\dfrac{E^2 - 2E + 1}{E}\right]x^2 = (E - 2 + E^{-1})x^2$

$= (x + h)^2 - 2x^2 + (x - h)^2 = x^2 + 2hx + h^2 - 2x^2 + x^2 - 2hx + h^2 = 2h^2$ **Ans.** (c)

Example 9. *Find the second difference of the polynomial* $x^4 - 12x^3 + 42x^2 - 30x + 9$ *with interval of differencing h = 2.* *(Bihar SBTE, 2003)*

Solution.   Let $f(x) = x^4 - 12x^3 + 42x^2 - 30x + 9$

$\therefore\quad f(x + 2) = (x + 2)^4 - 12(x + 2)^3 + 42(x + 2)^2 - 30(x + 2) + 9$

$= {}^4C_0\, x^4\, 2^0 + {}^4C_1\, x^3 2^1 + {}^4C_2\, x^2 2^2 + {}^4C_3\, x^1 2^3 + {}^4C_3\, x^0 2^4 - 12(x + 2)^3$
$\quad + 42(x + 2)^2 - 30(x + 2) + 9$

$= x^4 + 4x^3\, 2 + 6x^2.4 + 4x\,.\,8 + 1.1.16 - 12(x^3 + 2^3 + 3.x^2.2 + 3.x.2^2) + 42(x^2 + 2^2 + 2.x.2) - 30x - 60 + 9$

$= x^4 + 8x^3 + 24x^2 + 32x + 16 - 12x^3 - 96 - 72x^2 - 144x + 42x^2 + 168 + 168x - 30x - 60 + 9$

$f(x + 2) = x^4 - 4x^3 - 6x^2 + 26x + 37$

$\Delta f(x) = f(x + 2) - f(x) = (x^4 - 4x^3 - 6x^2 + 26x + 37) - (x^4 - 12x^3 + 42x^2 - 30x + 9)$

$= x^4 - 4x^3 - 6x^2 + 26x + 37 - x^4 + 12x^3 - 42x^2 + 30x - 9$

$\Delta f(x) = 8x^3 - 48x^2 + 56x + 28$

$\Delta^2 f(x) = \Delta f(x + 2) - \Delta f(x)$

$= [\{8(x + 2)^3 - 48(x + 2)^2 + 56(x + 2) + 28\} - \{8x^3 - 48x^2 + 56x + 28\}]$

$= [\{8(x^3 + 2^3 + 3\,.\,x^2.2 + 3.x.2^2) - 48(x^2 + 2^2 + 2.x.2) + 56x + 112 + 28\} - 8x^3 + 48x^2 - 56x - 28$

$= 8x^3 + 64 + 48x^2 + 96x - 48x^2 - 192 - 192x + 56x + 112 + 28 - 8x^3 + 48x^2 - 56x = 28$

$\Delta^2 f(x) = 48x^2 + 288x - 16$ **Ans.**

EXERCISE 4.1

Construct a table of forward differences for the following data :

1.

x	10	20	30	40
y	1.1	2.0	4.4	7.9

2.

x	0	5	10	15	20	25
y	7	11	14	18	24	32

Evaluate $\Delta^3 y\,(0)$.   **Ans. 2**

3. If $u_0 = 3$, $u_1 = 12$, $u_2 = 81$, $u_3 = 2000$, $u_4 = 100$, calculate $\Delta^4 u_0$.   **Ans. – 7459**

4.

x	0	1	2	3	4
y	1.0	1.5	2.2	3.1	4.6

Evaluate $\Delta^3 y$ (2).   **Ans. 0.4**

5.

x	1	2	3	4	5	6	7	8
$f(x)$	0	7	26	63	124	215	342	511

Evaluate $\Delta^3 f$ (4).   **Ans. 6**

Construct a table of backward differences for the following data :

6.

x	0	1	2	3	4	5
y	2	9	28	65	126	217

Find $\nabla^3 y$ (5).   **Ans. 6**

7.

x	0	1	2	3	4	5	6
y	4	10	30	75	160	294	490

Find $\nabla^3 y$ (4).   **Ans. 15**

8.

x	0	1	2	3	4	5	6	7
y	0	0	1	6	24	60	120	210

Find $\nabla^4 y$ (7).   **Ans. 0**

9.   Find the value ∇y_{20} for the given set of values .

x	10	15	20	25	30	35
y	19.97	25.5	22.47	23.52	24.65	25.89

                                                                                                (Bihar SBTE, 2009)

[**Hint :** $\nabla y_{20} = y_{15} - y_{20} = 25.5 - 22.47 = -3.03$ **Ans.**] $\left[\nabla y_k = y_{(k-h)} - y_k\right]$

Choose the correct answer:

10.   Given the set of values :

x	5	6	$\dot{}$	8	9	10
y	10	13	14	16	20	25

                                                                                                (Bihar SBTE, 2008)

the value of Δy_9 is ....

(a) 1   (b) 2   (c) 4   (d) 5   **Ans. (d)**

11.   The values of the independent variables are called:

(a) Argument   (b) Intries   (c) Constant   (d) None of these

                                                                                (Bihar SBTE, 2009) **Ans. (c)**

12.   For a function $y = f(x)$ if $f(1)$ is positive and $f(2)$ is negative, then there exists ..... between the interval (1, 2).

(a) at least one root   (b) two roots

(c) more than one root   (d) none of these

                                                                                (Bihar SBTE, 2005) **Ans. (a)**

13. Which of the following is true?

(a) $E = 1 - \nabla$ (b) $E = 1 - \Delta$ (c) $E = (1 - \nabla)^{-1}$ (d) $E = (1 + \Delta)^{-1}$ (*Bihar SBTE, 2008*) **Ans.** (c)

14. $\Delta^3 x^3$ is a ....

(a) constant   (b) variable   (c) both of these   (d) none of these

(*Bihar SBTE 2003, 2011*) **Ans.** (a)

15. $\Delta [f(x) \cdot g(x)] =$

(a) $\Delta f(x) \cdot \Delta g(x)$ (b) $\Delta f(x) + \Delta g(x)$ (c) $f(x) \cdot \Delta g(x) + g(x) \Delta f(x)$ (d) $\Delta f(x) \cdot g(x)$

(*Bihar SBTE, 2010*)   **Ans.** (c)

16. For a given sequence of numbers $y_1, y_2, ...$ the difference $y_3 - y_2$ is defined as:

(a) Δy_2 (b) ∇y_2 (c) Δy_3 (d) none of these   (*Bihar SBTE, 2004*) **Ans.** (a)

17. The nth difference of a polynomial of nth degree is ....

(a) zero   (b) constant   (c) one   (d) none   (*Bihar SBTE, 2003*)   **Ans.** (b)

18. The difference between two consecutive arguments are called:

(a) finite difference   (b) infinite difference   (c) constant difference   (d) None of these

(*Bihar SBTE, 2011*) **Ans.** (c)

Fill up the blanks:

19. The difference between two consecutive entries are called ....

(*Bihar SBTE, 2005*) **Ans.** Finite difference

4.8 THE FACTORIAL POLYNOMIAL

The factorial polynomial is the continued product of the factors in which the first factor is x and the successive factors decrease by a constant (h) and is denoted by $x^{(n)}$.

$$x^{(n)} = x(x-h)(x-2h)(x-3h)..... [x-(n-1)h]$$

For example; $x^{(1)} = x$

$x^{(2)} = x(x-h)$

$x^{(3)} = x(x-h)(x-2h)$

$x^{(n)} = x(x-h)(x-2h) .... [x-(n-1)h]$

4.9 DIFFERENCES OF A FACTORIAL POLYNOMIAL   (*Nagpur University, Winter 2004*)

$\Delta x^{(n)} = (x+h)^{(n)} - x^{(n)}$

$= [(x+h)x(x-h) .... \{x-(n-2)h\}] - [x(x-h)(x-2h) .... \{x-(n-2)h\} \{x-(n-1)h\}]$

$= [x(x-h)(x-2h) .... \{x-(n-2)h\}] [(x+h) - \{x-(n-1)h\}] = x^{(n-1)} [x+h-x+nh-h]$

$= nh\, x^{(n-1)} = n\, x^{(n-1)},$ ($h = 1$)

$$\boxed{\Delta x^{(n)} = n\, x^{(n-1)}}$$

$\Delta x^{(n)} = n\, x^{(n-1)}$ analogous to $\dfrac{d}{dx} x^n = n x^{n-1}$

Also $\dfrac{1}{\Delta} x^{(n)} = \dfrac{x^{(n+1)}}{n+1}$ is analogous to $\displaystyle\int x^n\, dx = \dfrac{x^{n+1}}{n+1}$

$$\boxed{\dfrac{1}{\Delta} x^{(n)} = \dfrac{x^{(n+1)}}{(n+1)}}$$

Remember

1. The result of the operation of Δ on factorial function $x^{(n)}$ is similar to the result of the operation of $\dfrac{d}{dx}$ on ordinary polynomial x^n *i.e.*

$$\Delta x^{(n)} = n\, x^{(n-1)}, \quad \dfrac{d}{dx}(x^n) = n x^{n-1}$$

2. The result of the operation of $\dfrac{1}{\Delta}$ on factorial function $x^{(n)}$ is similar to the result of the operation of $\int$ on ordinary polynomial x^n i.e.

$$\frac{1}{\Delta} x^{(n)} = \frac{x^{(n+1)}}{n+1}$$

$$\int x^n \, dx = \frac{x^{n+1}}{n+1}$$

4.10 RELATIONS BETWEEN FACTORIAL POLYNOMIAL AND ORDINARY POLYNOMIAL

1. Relation

$$x^{(n)} = n! \, {}^x C_n \qquad\qquad\qquad \text{where interval of difference is 1.}$$

Proof. We have, $h = 1$

$$x^{(n)} = x \, (x-1) \, (x-2) \, (x-3) \, ... \, (x-\overline{n-1})$$

$$= \frac{[x(x-1)(x-2)\,(x-3)\,.......\,(x-\overline{n-1})]\,[(x-n)\,(x-\overline{n+1})\,.....1]}{[(x-n)\,(x-\overline{n+1})\,.......1]} = \frac{x!}{(x-n)!} = \frac{n! \, x!}{n! \, (x-n)!}$$

$$\boxed{x^{(n)} = n! \, {}^x C_n} \qquad\qquad\qquad ... (1) \qquad\qquad \textbf{Proved.}$$

2. Relation

$$\Delta^r \, {}^x C_n = {}^x C_{n-r}$$

Proof. We know that,

$$\Delta^r x^{(n)} = n \, (n-1) \, (n-2) \, ... \, (n-\overline{r-1}) \, x^{(n-r)} \qquad\qquad \left[\Delta x^{(n)} = n\,x^{(n-1)}\right]$$

$$= n(n-1) \, (n-2) \, ... \, (n-\overline{r-1}) \, (n-r)! \; {}^x C_{n-r}$$

$$\text{[On putting } n \to n-r \text{ Relation 1 becomes } x^{(n-r)} = (n-r)! \; {}^x C_{n-r}]$$

$$\Rightarrow \qquad\qquad \Delta^r x^{(n)} = n! \; {}^x C_{n-r} \qquad\qquad\qquad ... (2)$$

On putting the value of $x^{(n)}$ from relation (1) in (2), we get

$$\Delta^r \left[n! \; {}^x C_n \right] = n! \; {}^x C_{n-r} \qquad\qquad \Rightarrow \qquad \boxed{\Delta^r \; {}^x C_n = {}^x C_{n-r}}$$

3. Relation

$$\Delta^n \; {}^x C_n = 1$$

Proof. Putting $r = n$ in Relation (2), we get

$$\Delta^n \; {}^x C_n = {}^x C_{n-n} = {}^x C_0 = 1$$

$$\Rightarrow \qquad \boxed{\Delta^n \; {}^x C_n = 1}$$

Example 10. *Find the value of* $\dfrac{\Delta}{\Delta x}\left[4x^{(5)}\right]$

Solution. Here, we have

$$\frac{\Delta}{\Delta x}\left[4x^{(5)}\right] = 5.4 \, x^{(4)} = 20x^{(4)} \qquad\qquad\qquad \left[\frac{\Delta}{\Delta x} = \frac{d}{dx}\right]$$

$$= 20x \, (x-h) \, (x-2h) \, (x-3h) \qquad\qquad\qquad \textbf{Ans.}$$

Example 11. *Evaluate* $\dfrac{\Delta}{\Delta x}\left[100\,x^{(3)}\right]$

Solution. Here, we have

$$\frac{\Delta}{\Delta x}\left[100\,x^{(3)}\right] \;=\; 100.3x^{(2)} = 300\,x\,(x-h) \qquad\qquad \text{Ans.}$$

Example 12. *Evaluate* $\dfrac{\Delta}{\Delta x}\left[6x^{(5)}+4x^{(4)}+2x^{(3)}+3x^{(2)}-15x-1\right],$ \qquad\qquad *here $h = 1$.*

Solution. Here, we have

$$\frac{\Delta}{\Delta x}\left[6x^{(5)}+4x^{(4)}+2x^{(3)}+3x^{(2)}-15x-1\right]$$

$$= 6.5x^{(4)}+4.4\,.\,x^{(3)}+2.3x^{(2)}+3.2x-15$$

$$= 30x\,(x-1)\,(x-2)\,(x-3)+16x\,(x-1)\,(x-2)+6x\,(x-1)+6x-15$$

$$= 30\,(x^{4}-6x^{3}+11x^{2}-6x)-16\,(x^{3}-3x^{2}+2x)+6x^{2}-6x+6x-15$$

$$= 30x^{4}-180x^{3}+330x^{2}-180x-16x^{3}+48x^{2}-32x+6x^{2}-15$$

$$= 30x^{4}-196x^{3}+384x^{2}-212x-15 \qquad\qquad \text{Ans.}$$

Example 13. *Express $y = 2x^{3} - 3x^{2} + 3x - 10$ in factorial notation and hence show that $\Delta^{3} y = 12$.* \qquad (R.G.P.V., Bhopal, III Semester, June 2007, Dec. 2006)

Solution. By Synthetic Division Method

Let \qquad\qquad $y \;=\; Ax^{(3)}+Bx^{(2)}+Cx^{(1)}+D.$ \qquad\qquad ... (1)

Using the method of synthetic division, we divide by $x, x-1, x-2$, etc. successively. Then

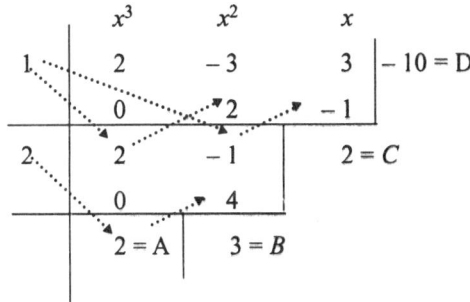

On adding

2	– 3	3
0	2	– 1
2	– 1	2

In arrow-lines the numbers are multiplied as $1 \times 2 = 2$ and it is written in the next column of the above lines.

Similarly, in arrow-line $1 \times (-1) = -1$ and it is also written in the next column of the above line.

On putting the values of A, B, C and D in (1), we get the required factorial polynomial

$$y \;=\; 2x^{(3)}+3x^{(2)}+2x-10$$

$$\Delta y \;=\; 6x^{(2)}+6x+2$$

$$\Delta^{2} y \;=\; 12x+6$$

$$\Delta^{3} y \;=\; 12 \qquad\qquad \text{Ans.}$$

Example 14. *Evaluate* $\Delta^2 (2x^2 + 4x^2 + 7x - 6)$ *(Bihar SBTE, 2012)*

Solution. $\Delta^2 (2x^2 + 4x^2 + 7x - 6) = \Delta^2 f(x)$

$f(x)$ is converted in factorial polynomial by synthetic division.

1	2	4	7	−6
	0	2	6	
2	2	6	13 = x	
	0	4		
3	2	10 = x^{(2)}		
	0			
	2 = x^{(3)}			

$\Delta f(x) = \Delta [2x^{(3)} + 10x^{(2)} + 13^{(1)} - 6] = 6x^{(2)} + 10 \times 2x + 13 = 6x^{(2)} + 20x + 13$

$\Delta^2 f(x) = 12x^{(1)} + 20 = 12x + 20$ **Ans.**

Example 15. *Express* $f(x) = 2x^4 - 9x^3 + 4$ *in factorial notation, the interval of differences being unity and find out* $\Delta^4 f(x)$.

Solution. Direct Method. Here, we have

$$f(x) = 2x^4 - 9x^3 + 4$$

Let $2x^4 - 9x^3 + 4 = ax^{(4)} + bx^{(3)} + cx^{(2)} + dx^{(1)} + e$ ... (1)

$$= ax(x-1)(x-2)(x-3) + bx(x-1)(x-2) + cx(x-1) + dx + e$$

$$= a(x^4 - 6x^3 + 11x^2 - 6x) + b(x^3 - 3x^2 + 2x) + cx^2 - cx + dx + e$$

$$= ax^4 + (-6a + b)x^3 + (11a - 3b + c)x^2 + (-6a + 2b - c + d)x + e$$

Equating the coefficients of like powers of x, we get

$$a = 2$$ ... (2)

$$-6a + b = -9 \quad \Rightarrow \quad -12 + b = -9 \quad \Rightarrow \quad b = 3$$ ... (3)

$$11a - 3b + c = 0 \quad \Rightarrow \quad 22 - 9 + c = 0 \quad \Rightarrow \quad c = -13$$ ... (4)

$$-6a + 2b - c + d = 0 \quad \Rightarrow -12 + 6 + 13 + d = 0 \quad \Rightarrow \quad d = -7$$ ... (5)

$$e = 4$$ ... (6)

Now, putting the values of a, b, c, d, e in (1), we get

$$f(x) = 2x^{(4)} + 3x^{(3)} - 13x^{(2)} - 7x + 4$$

$\Rightarrow \qquad \Delta f(x) = 8x^{(3)} + 9x^{(2)} - 26x^{(1)} - 7$

$\Rightarrow \qquad \Delta^2 f(x) = 24x^{(2)} + 18x - 26$

$\Rightarrow \qquad \Delta^3 f(x) = 48x^{(1)} + 18$

$\Rightarrow \qquad \Delta^4 f(x) = 48$ **Ans.**

Example 16. *Represent the function* $f(x) = x^4 - 12x^3 + 24x^2 - 30x + 9$ *and its successive differences in factorial notation.*   *(Bihar SBTE, 2004, R.G.P.V., Bhopal, III Semester, June 2004)*

Solution. Let $f(x) = x^4 - 12x^3 + 24x^2 - 30x + 9 = Ax^{(4)} + Bx^{(3)} + Cx^{(2)} + Dx^{(1)} + E$ ... (1)

By synthetic division method

	x^4	x^3	x^2	x	
1	1	-12	24	-30	$9 = E$
	0	1	-11	13	
2	1	-11	13	$-17 = D$	
	0	2	-18		
3	1	-9	$-5 = C$		
	0	3			
	$1 = A$	$-6 = B$			

Putting the values of A, B, C, D and E in (1), we get

$$f(x) = x^{(4)} - 6x^{(3)} - 5x^{(2)} - 17x^{(1)} + 9$$

$\Rightarrow \qquad \Delta f(x) = 4x^{(3)} - 18x^{(2)} - 10x^{(1)} - 17$

$\Rightarrow \qquad \Delta^2 f(x) = 12x^{(2)} - 36x^{(1)} - 10 \quad \Rightarrow \quad \Delta^3 f(x) = 24x^{(1)} - 36$

$\Rightarrow \qquad \Delta^4 f(x) = 24$ **Ans.**

Example 17. *Find the function $f(x)$ whose first difference is $9x^2 + 11x + 5$.*

(R.G.P.V., Bhopal, III Semester, Dec. 2004)

Solution. Synthetic Method.

First we shall convert the given first difference function into factorial function.

Let $\quad \Delta f(x) = 9x^2 + 11x + 5 = Ax^{(2)} + Bx^{(1)} + C$ ... (3)

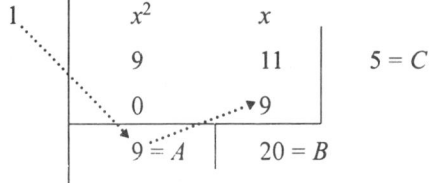

Substituting the values of A, B and C in (3), we get

$$9x^2 + 11x + 5 = 9x^{(2)} + 20x^{(1)} + 5$$

As given $\qquad \Delta f(x) = 9x^{(2)} + 20x^{(1)} + 5$

$\therefore \qquad f(x) = 9\Delta^{-1} x^{(2)} + 20\Delta^{-1} x^{(1)} + \Delta^{-1} (5)$ $\qquad [\Delta^{-1} = f]$

$\Rightarrow \qquad f(x) = \dfrac{9 x^{(3)}}{3} + \dfrac{20 x^{(2)}}{2} + \dfrac{5 x^{(1)}}{1} + d$

$\Rightarrow \qquad f(x) = 3x^{(3)} + 10 x^{(2)} + 5x^{(1)} + d = 3x (x-1)(x-2) + 10 x (x-1) + 5x + d$

$\qquad = 3x^3 - 9x^2 + 6x + 10x^2 - 10x + 5x + d = 3x^3 + x^2 + x + d$ **Ans.**

Example 18. *Obtain the function whose first difference is $x^4 - 12x^3 + 24x^2 - 30x + 9$.*

Solution. Let $\quad f(x)$ be the required function.

$$\Delta f(x) = x^4 - 12x^3 + 24x^2 - 30x + 9 = Ax^{(4)} + Bx^{(3)} + Cx^{(2)} + Dx^{(1)} + E$$

$f(x)$ is converted in factorial polynomial by **synthetic division method.**

	1	-12	24	-30	$9 = E$
1	0	1	-11	13	
2	1	-11	13	$-17 = D$	
	0	2	-18		
3	1	-9	$-5 = C$		
	0	3			
4	1	$-6 = B$			
	0				
	$1 = A$				

Now $\Delta f(x) = Ax^{(4)} + Bx^{(3)} + Cx^{(2)} + Dx^{(1)} + E$ ... (1)

Putting the values of A, B, C, D and E in (1), we get

$$\Delta f(x) = x^{(4)} - 6x^{(3)} - 5x^{(2)} - 17x + 9$$

$$\Rightarrow \quad f(x) = \Delta^{-1} x^{(4)} - 6\Delta^{-1} x^{(3)} - 5\Delta^{-1} x^{(2)} - 17\Delta^{-1} x^{(1)} + \Delta^{-1}$$

$$\Rightarrow \quad f(x) = \frac{x^{(5)}}{5} - \frac{6}{4}x^{(4)} - \frac{5}{3}x^{(3)} - \frac{17}{2}x^{(2)} + 9x + d$$

Hence $f(x) = \frac{x^{(5)}}{5} - \frac{3}{2}x^{(4)} - \frac{5}{3}x^{(3)} - \frac{17}{2}x^{(2)} + 9x^{(1)} + d$ **Ans.**

EXERCISE 4.2

Find the value of :

1. $\dfrac{\Delta}{\Delta x}\left[4x^{(6)}\right]$ **Ans.** $24x^{(5)}$ 2. $\Delta\left[3x^{(-4)} - 3x^{(2)} + 4x^{(-2)}\right]$ **Ans.** $-12x^{(-5)} - 6x^{(1)} - 8x^{(-3)}$

3. $\Delta^2\left[4x^{(5)} + 6x^{(2)}\right]$ **Ans.** $80x^{(3)} + 12$ 4. $\Delta^2\left[x^4 - 2x^2 + 6x - 2\right]$ **Ans.** $12x^{(2)} + 36x^{(1)} + 10$

Express the following function in factorial notations:

5. $x^3 - 2x^2 + x - 1$ **Ans.** $x^{(3)} + x^{(2)} - 1$ (*R.G.P.V., Bhopal, III Semester Dec. 2007*)

6. $3x^4 - 4x^3 + 6x^2 + 2x + 1$. Hence find $\Delta^4 f(x)$.   **Ans.** $3 x^{(4)} + 14 x^{(3)} + 15 x^{(2)} + 7 x^{(1)} + 1, \Delta^4 f(x) = 72$

7. $x^4 + x + 1$ **Ans.** $x^{(4)} + 6 x^{(3)} + 7 x^{(2)} + 2 x^{(1)} + 1$

8. $2x^4 + 5x^2 + 4x + 6$ **Ans.** $2 x^{(4)} + 12 x^{(3)} + 19 x^{(2)} + 11 x^{(1)} + 6$

Obtain the function whose first difference is

9. $2x^{(3)} + 3x^{(2)} - 5x + 4$ **Ans.** $\dfrac{1}{2} x^4 + 3x^3 + 4x + C$ (*R.G.P.V., Bhopal, III Semester Dec. 2007*)

10. $4x^{(3)} - 18x^{(2)} - 10x - 17$ **Ans.** $x^4 - 12x^3 + 24x^2 - 30x + C$

11. $x^{(3)} + 3x^{(2)} + 5x^{(1)} + 12$ **Ans.** $\dfrac{1}{4} x^4 + 2x^3 + \dfrac{9}{2} x^2 + 12x + C$

12. Find $\Delta^{-1} [x (x + 1) (x + 2)]$ **Ans.** $\dfrac{1}{4} (x+2) (x+1) x (x-1)$

13. Find $\Delta^{-1}\left[\dfrac{1}{x(x+1)(x+2)}\right]$ **Ans.** $-\dfrac{1}{2}\left[\dfrac{1}{x(x+1)}\right] + C$

14. A product of form $x(x-1)(x-2)(x-3)$ is called a factorial and is denoted by ....

(a) 4 ! (b) $x^{(4)}$ (c) 4 (d) None

(Bihar SBTE, 2004) **Ans.** (b)

OBJECTIVE TYPE QUESTIONS

Choose the correct alternative

15. $\Delta^2 y_0 = $ ....

(a) $y_1 - y_0$ (b) $y_2 - 2y_1 - y_0$ (c) $y_2 - 2y_1 + y_0$ (d) None

(Bihar SBTE, 2011) **Ans.** (c)

16. If h is interval of differentiating, the value of $\left(\dfrac{\Delta^2}{E}\right)x^2$ is

(a) h (b) $2h$ (c) $2h^2$ (d) $2h^3$ *(Bihar SBTE, 2011)* **Ans.** (c)

17. $\Delta^5 (x^6 + 7x^5 + 9x^4 + 7)$ is a ....

(a) Variable (b) Constant (c) Both of these (d) None of these

(Bihar SBTE, 2012) **Ans.** (a)

18. Which of the following is true?

(a) $E = e^{hD}$ (b) $E = e^{2hD}$ (c) $E^2 = e^{hD}$ (d) None of these

(Bihar SBTE, 2011) **Ans.** (a)

19. A product of the form: $x(x-1)(x-2)(x-3) \dots (x-9)$ is denoted by ....

(a) $[x]^9$ (b) $\lceil x \rceil^{10}$ (c) ??? (d) None of these

(Bihar SBTE, 2011) **Ans.** (b)

CHAPTER

5 Interpolation

5.1 INTERPOLATION

The technique of determining an approximate value of $f(x)$ for a non-tabuler value of x which lies in the interval $[a, b]$ is called interpolation.

Let $y = f(x)$ be a function of x. The corresponding values of y for a set of $a, a + h, a + 2h, \ldots a + nh$ are given

as
$$\begin{aligned} y_0 &= f(a) \\ y_1 &= f(a + h) \\ y_2 &= f(a + 2h) \\ y_n &= f(a + nh) \end{aligned}$$

Interpolation is the process of finding the values of y for any intermediate value of x. between a and $a + nh$.

5.2 EXTRAPOLATION

The technique of determining the value of $f(x)$ for a value of x lying out side the interval $[a, b]$ is called extrapolation.

Extrapolation is the process of obtaining the value of y for a value of x outside the interval a and $a + nh$.

5.3 METHODS OF INTERPOLATION

For equal Interval

    1. Newton – Gregory forward Interpolation method

    2. Newton's Backward Interpolation method

5.4 NEWTON-GREGORY FORWARD INTERPOLATION FORMULA FOR EQUAL INTERVAL

(Bihar SBTE, 2014, 2012, 2010, 2008)

We know that this formula is applied for interpolation near the beginning of the tabulated values.

We know that, $\quad f(a + ph) = E^p f(a)$

$$f(a + ph) = (1 + \Delta)^p f(a)$$

On expanding $(1 + \Delta)^p$ by Binomial theorem, we get

$$f(a + ph) = \left[1 + p\Delta + \frac{p(p-1)}{2} \Delta^2 + \ldots \right] f(a)$$

$$\Rightarrow \qquad f(a + ph) = f(a) + p\Delta f(a) + \frac{p(p-1)}{2} \Delta^2 f(a) + \ldots$$

It is known as the Newton-Gregory formula of interpolation.

This formula can also be written as $y_p = y_0 + p \Delta y_0 + \dfrac{p(p-1)}{2} \Delta^2 y_0 + ...$

Observations. The converse of this theorem is also true i.e. *if the nth differences of a function tabulated at equally spaced intervals are constant, the function is a polynomial of degree n.* This fact is important in numerical analysis as it enables us to approximate a function by a polynomial of nth degree, if its nth order difference becomes nearly constant.

Example 1. *Find interpolating polynomial for y from the following data using Newton's forward formula:*

x	4	6	8	10
y	1	3	8	16

(*Bihar SBTE, 2010*)

Solution.

x	y	Δy	$\Delta^2 y$	$\Delta^3 y$
4	1			
		2		
6	3		3	
		5		0
8	8		3	
		8		
10	16			

$a + ph = x \qquad$ but $4 = x \qquad \Rightarrow \qquad p = \dfrac{x-4}{2}$

Using Newton's forward interpolation formula

$y = y_0 + p \Delta y_0 + \dfrac{p(p-1)}{2!} \Delta^2 y_0$

$y(x) = 1 + \dfrac{x-4}{2}(2) + \dfrac{\left(\dfrac{x-4}{2}\right)\left(\dfrac{x-4}{2}-1\right)}{2}(3)$

$9(x) = 1 + (x-4) + \dfrac{3}{8}(x^2 - 10x + 24); \quad y = 1 + (x-4) + \dfrac{3}{8}x^2 - \dfrac{15}{4}x + 3; \quad y = \dfrac{3}{8}x^2 - \dfrac{11}{4}x + 6$ **Ans.**

Example 2. *If $u_0 = 1$, $u_1 = 0$, $u_2 = 5$, $u_3 = 22$, $u_4 = 57$, find $u_{0.5}$.*

Solution. The Difference Table is as under :

x	u_x	Δu_x	$\Delta^2 u_x$	$\Delta^3 u_x$	$\Delta^4 u_x$
0	1				
		-1			
1	0		6		
		5		6	
2	5		12		0
		17		6	
3	22		18		
		35			
4	57				

Here, $a + ph = 0.5,$ $a = 0,$ $h = 1$

$0 + p(1) = 0.5$ $\Rightarrow$ $p = 0.5$

By Newton's Forward interpolation formula, we have

$$f(a + ph) = f(a) + p\,\Delta f(a) + \frac{p(p-1)}{2!}\Delta^2 f(a) + \frac{p(p-1)(p-2)}{3!}\Delta^3 f(a) + \ldots$$

$$u_{0.5} = u_0 + 0.5\,\Delta u_0 + \frac{0.5(0.5-1)}{2!}\Delta^2 u_0 + \frac{0.5(0.5-1)(0.5-2)}{3!}\Delta^3 u_0 + \ldots$$

$$= 0 + (0.5)(-1) + \frac{0.5(-0.5)}{2}(6) + \frac{0.5(-0.5)(-1.5)}{6}(6)$$

$$= 1 - 0.5 - 0.75 + 0.375 = 0.125$$ **Ans.**

Example 3. *State the appropriate interpolation formula which is to be used to calculate the value of exp (1.75) from the following data and hence evaluate it from the given data :*

x	1.7	1.8	1.9	2.0
$y = e^x$	5.474	6.050	6.686	7.389

Solution. The difference table is as under :

x	$y = f(x)$	Δy	$\Delta^2 y$	$\Delta^3 y$
1.7	5.474			
		0.576		
1.8	6.050		0.060	
		0.636		0.007
1.9	6.686		0.067	
		0.703		
2.0	7.389			

$a + ph = 1.75,$ $a = 1.7,$ $h = 0.1$

$1.7 + p(0.1) = 1.75,$ $p = 0.5$

By Newton's forward interpolation formula, we have

$$f(a + ph) = f(a) + p\,\Delta f(a) + \frac{p(p-1)}{2!}\Delta^2 f(a) + \frac{p(p-1)(p-2)}{3!}\Delta^3 f(a) + \ldots$$

$$f(1.75) = f(1.7) + 0.5\,\Delta f(1.7) + \frac{0.5(0.5-1)}{2!}\Delta^2 f(1.7) + \frac{0.5(0.5-1)(0.5-2)}{3!}\Delta^3 f(1.7) + \ldots$$

$$f(1.75) = f(1.7) + 0.5(0.576) + \frac{0.5(0.5-1)}{2!}(0.060) + \frac{0.5(0.5-1)(0.5-2)}{6}(0.007)$$

$$= 5.474 + 0.288 - 0.0075 + 0.0004375 = 5.7624375 - 0.0075 = 5.7549375$$

$$= 5.7549 \quad \text{(Rounded upto four decimal places)}$$ **Ans.**

Example 4. *Applying Newton's forward interpolation formula, compute the value of $\sqrt{5.5}$, given that*

$$\sqrt{5} = 2.236, \ \sqrt{6} = 2.449, \ \sqrt{7} = 2.646 \ and \ \sqrt{8} = 2.828, \ correct \ upto \ three \ places \ of$$
decimal.

Solution.   The difference table is as under :

x	y	Δy	$\Delta^2 y$	$\Delta^3 y$
5	2.236			
		0.213		
6	2.449		− 0.016	
		0.197		0.001
7	2.646		− 0.015	
		0.182		
8	2.828			

$$a + ph = 5.5, \qquad a = 5, \qquad h = 1$$
$$5 + p\,(1) = 5.5, \qquad p = 0.5$$

By Newton's forward interpolation formula, we have

$$f(a + ph) \;=\; f(a) + p\,\Delta f(a) + \frac{p(p-1)}{2!}\,\Delta^2 f(a) + \frac{p(p-1)\,(p-2)}{3!}\,\Delta^3 f(a) + \dots$$

$$f(5.5) \;=\; f(5) + 0.5\,\Delta f(5) + \frac{0.5\,(0.5-1)}{2!}\,\Delta^2 f(5) + \frac{0.5(0.5-1)\,(0.5-2)}{3!}\,\Delta^3 f(5) + \dots$$

$$f(5.5) \;=\; 2.236 + 0.5 \times 0.213 + \frac{0.5\,(0.5-1)}{2!}\,(-0.016) + \frac{0.5\,(0.5-1)\,(0.5-2)}{3!}\,(0.001)$$

$$= \; 2.236 + 0.1065 + 0.00200 + 0.0000625$$

$$= \; 2.3445625 = 2.345 \;\text{(upto three decimal places).}\qquad\qquad\textbf{Ans.}$$

Example 5. *For a table*

x	10	12	14	16	18
y	18	16	20	25	31

Using Newton's forward Interpolation formula find y for given x = 11.   *(Bihar SBTE, 2009)*

Solution.

x	y	Δy	$\Delta^2 y$	$\Delta^3 y$	$\Delta^4 y$
10	18				
		− 2			
12	16		6		
		4		−5	
14	20		1		5
		5		0	
16	25		1		
		6			
18	31				

Here $x_0 = 10;\; X = 11$

$$P = \frac{X - x_0}{h} = \frac{11-10}{2} = 0.5$$

By Newtons' forward Interpolation formula

$$y_{11} = y_{10} + P.\Delta.y_{10} + \frac{P(P-1)}{2!}\Delta^2 y_{10} + \frac{P(P-1)(P-2)}{3!}\Delta^3 y_{10} + \frac{P(P-1)(P-2)(P-3)}{4!}\Delta^4 y_{10} +$$

$$\frac{P(P-1)(P-2)(P-3)(P-4)}{5!}\Delta^5 y_{10}$$

$$= 18 + 0.5 \times (-2) + \frac{0.5(0.5-1)}{2} \times 6 + \frac{0.5(0.5-1)(0.5-2)}{6} \times (-5) + \frac{0.5(0.5-1)(0.5-2)(0.5-3)}{2 \times 3 \times 4} \times 5$$

$$= 18 - 1.0 + \frac{0.5 \times (-0.5)}{2} \times 6 + \frac{0.5(-0.5)(-1.5)}{6}(-5) + \frac{0.5(-0.5)(-1.5)-2.5}{24} \times 5$$

$$= 18 - 1.0 - 0.75 - 0.3125 - 0.1953125 = 15.7421875$$ **Ans.**

Example 6. *Given y (10) = 35.3, y (15) = 32.4, y (20) = 29.2, y (25) = 26, y (30) = 23.2 and*
y (35) = 20.5, find y (12).   *(Bihar SBTE, 2009)*

Solution.   We have the following forward difference table:

x	y	Δy	$\Delta^2 y$	$\Delta^3 y$	$\Delta^4 y$	$\Delta^5 y$
10	35.3					
		− 2.90				
15	32.4		− 0.3			
		− 3.20		0.3		
20	29.2		0		0.1	
		− 3.20		0.4		− 0.8
25	26		0.4		− 0.7	
		− 2.80		− 0.3		
30	23.2		0.1			
		− 2.70				
35	20.5					

Here $x_0 = 10; x = 12, h = 5$

$$P = \frac{x - x_0}{h} = \frac{12-10}{5} = \frac{2}{5} = 0.4$$

By Newtons' forward Interpolation formula, we have

$$y(12) = \Delta y_{12} + P\Delta y_{12} + \frac{P(P-1)\Delta^2 y_{12}}{2!} + \frac{P(P-1)(P-2)\Delta^3 y_{12}}{3!} + \frac{P(P-1)(P-2)(P-3)\Delta^4 y_{12}}{4!} +$$

$$\frac{P(P-1)(P-2)(P-3)(P-4)\Delta^5 y_{12}}{5!}$$

$$= 35.3 + (0.4) \times (-2.90) + \frac{0.4(0.4-1) \times (-0.3)}{2} + \frac{0.4(0.4-1)(0.4-2)}{3 \times 2} \times (0.3) +$$

$$\frac{0.4(0.4-1)(0.4-2)(0.4-3)}{4 \times 3 \times 2} \times (0.1) + \frac{0.4(0.4-1)(0.4-2)(0.4-3)(0.4-4)}{5 \times 4 \times 3 \times 2} \times (-0.8)$$

$$= 35.3 - 1.16 + \frac{0.072}{2} + \frac{0.1152}{6} - \frac{0.09984}{24} - \frac{2.875392}{120}$$

$$= 35.3 - 1.16 + 0.036 + 0.0192 - 0.00416 - 0.0240 = 35.3552 - 1.18816$$

$$= 34.16704 = 34.2$$ **Ans.**

Example 7. *Find the lowest degree polynomial y (x) that will fit the data*

x	0	2	4	6	8
y	5	9	61	209	501

Also find y (5).

Solution. The difference table is as shown :

$$a + ph = x, \quad a = 0, \quad h = 2 \quad \Rightarrow \quad 0 + p\,(2) = x \quad \Rightarrow p = \frac{x}{2}$$

By Newton's forward interpolation formula

$$y_p = f(a + ph) = f(a) + p\,\Delta f(a) + \frac{p(p-1)}{2!}\,\Delta^2 f(a) + \frac{p(p-1)(p-2)}{3!}\,\Delta^3 f(a)$$

$$+ \frac{p(p-1)(p-2)(p-3)}{4!}\,\Delta^4 f(a)$$

$$\Rightarrow \quad y(x) = 5 + \frac{x}{2}\,(4) + \frac{\dfrac{x}{2}\left(\dfrac{x}{2}-1\right)}{2!}\,.\,(48) + \frac{\dfrac{x}{2}\left(\dfrac{x}{2}-1\right)\left(\dfrac{x}{2}-2\right)}{3!}\,(48)$$

$$+ \frac{\dfrac{x}{2}\left(\dfrac{x}{2}-1\right)\left(\dfrac{x}{2}-2\right)\left(\dfrac{x}{2}-3\right)}{4!} \times 0$$

$$\Rightarrow \quad y(x) = 5 + 2x + 6x^2 - 12x + x^3 - 6x^2 + 8x = x^3 - 2x + 5.$$

This is the required polynomial

Now, $y(5) = (5)^3 - 2 \times 5 + 5 = 125 - 10 + 5 = 125 - 5 = 120$

Hence, $y(5) = 120$ **Ans.**

Example 8. *The number of members of a Civil Engineering Society are given below :*

x	1987	1988	1989	1990	1991
f (x)	150	192	241	—	374

Make the best estimate you can do of the number of members in 1990.

Solution. Since four enteries are given so y can be represented by the third degree polynomial.

Hence $\Delta^3 y = $ constant

$$\Delta^4 f(x) = 0 \qquad \text{for all } x \qquad\qquad [y = f(x)]$$

$$\Rightarrow \qquad (E-1)^4 f(x) = 0$$

$$\Rightarrow \qquad (E^4 - 4E^3 + 6E^2 - 4E + 1) f(x) = 0$$

$$\Rightarrow \qquad [E^4 - 4E^3 + 6E^2 - 4E + 1] f(1987) = 0$$

$\Rightarrow$ Now $E^4 f(1987) - 4E^3 f(1987) + 6 E^2 f(1987) - 4 E f(1987) + f(1987) = 0$

$$\Rightarrow \qquad \text{Again } f(1991) - 4f(1990) + 6f(1989) - 4f(1988) + f(1987) = 0$$

$$\Rightarrow \qquad 374 - 4f(1990) + 6(241) - 4(192) + 150 = 0$$

$$\Rightarrow \qquad 374 - 4f(1990) + 1446 - 768 + 150 = 0$$

$$\Rightarrow \qquad \text{Again } 4f(1990) = 374 + 1446 - 768 + 150$$

$$4f(1990) = 1202$$

$$f(1990) = 300.5$$

Number of members in 1990 is 301.

Ans.

Example 9. *Find the missing values in the following table :*

x	45	50	55	60	65
y	3	–	2	–	– 2.4

(*R.G.P.V., Bhopal, III Semester, June 2007*)

Solution. Here, three enteries are given so $f(x)$ can be represented by two degree polynomial.

$$\Delta^2 f(x) = \text{constant}$$

$$\Rightarrow \qquad \Delta^3 f(x) = 0$$

$$\Rightarrow \qquad (E-1)^3 f(x) = 0$$

$$\Rightarrow \qquad \{E^3 - 3E^2 + 3E - 1\} f(x) = 0$$

$$\Rightarrow \qquad E^3 f(x) - 3E^2 f(x) + 3E f(x) - f(x) = 0 \qquad\qquad \text{... (1)}$$

$$\Rightarrow \qquad f(60) - 3f(55) + 3f(50) - f(45) = 0$$

$$\Rightarrow \qquad f(60) - 3(2) + 3f(50) - 3 = 0$$

$$\Rightarrow \qquad f(60) + 3f(50) = 9 \qquad\qquad \text{... (2)}$$

Again from (1), we have

$$f(65) - 3f(60) + 3f(55) - f(50) = 0$$

$$-2.4 - 3f(60) + 3(2) - f(50) = 0$$

$$\Rightarrow \qquad 3f(60) + f(50) = 3.6 \qquad\qquad \text{... (3)}$$

Solving (2) and (3), we get $\qquad f(60) = 0.225 \text{ and } f(50) = 2.925$

Missing enteries are 2.925 and 0.225.

Ans.

Example 10. *Finding the missing values in the following table :*

x	0	5	10	15	20	25
y	6	10	–	17	–	31

(*R.G.P.V., Bhopal, III Semester, June 2006*)

Solution. Here, there are four given enteries so $f(x)$ can be represented by third degree polynomial.

Hence,
$$\Delta^3 f(x) = \text{constant}$$
$$\Delta^4 f(x) = 0$$
$$\Rightarrow \qquad (E-1)^4 f(x) = 0$$
$$\Rightarrow \qquad (E^4 - 4E^3 + 6E^2 - 4E + 1) f(0) = 0 \qquad \text{... (1)}$$
$$\Rightarrow E^4 f(0) - 4E^3 f(0) + 6E^2 f(0) - 4 E f(0) + f(0) = 0$$
$$\Rightarrow \qquad f(20) - 4 f(15) + 6 f(10) - 4 f(5) + f(0) = 0$$
$$\Rightarrow \qquad f(20) - 4(17) + 6 f(10) - 4(10) + 6 = 0$$
$$\Rightarrow \qquad f(20) - 68 + 6 f(10) - 40 + 6 = 0$$
$$\Rightarrow \qquad f(20) + 6 f(10) = 102 \qquad \text{... (2)}$$

Again from (1), we have
$$f(25) - 4 f(20) + 6 f(15) - 4 f(10) + f(5) = 0$$
$$\Rightarrow \qquad 31 - 4 f(20) + 6(17) - 4 f(10) + 10 = 0$$
$$\Rightarrow \qquad 31 - 4 f(20) + 102 - 4 f(10) + 10 = 0$$
$$\Rightarrow \qquad 4 f(20) + 4 f(10) = 143 \qquad \text{... (3)}$$

Solving (2) and (3), we get
$$f(10) = 13.25 \text{ and } f(20) = 22.5$$

Missing enteries are 143 and 22.5.   **Ans.**

Example 11. *Find out the missing values from the following :*

x	5	10	15	20	25	30
y	7	?	13	15	?	25

Solution. Since we are given four enteries so y can be represented by the third degree polynomial.

Hence
$$\Delta^3 f(x) = \text{constant}$$
$$\Rightarrow \qquad \Delta^4 f(x) = 0 \qquad \text{for all } x$$
$$\Rightarrow \qquad (E-1)^4 f(x) = 0$$
$$\Rightarrow \qquad (E^4 - 4E^3 + 6E^2 - 4E + 1) f(5) = 0$$
$$\Rightarrow f(25) - 4 f(20) + 6 f(15) - 4 f(10) + f(5) = 0$$
$$\Rightarrow \qquad f(25) - 4(15) + 6(13) - 4 f(10) + 7 = 0$$
$$\Rightarrow \qquad f(25) - 4 f(10) - 60 + 78 + 7 = 0$$
$$\Rightarrow \qquad f(25) - 4 f(10) = -25 \qquad \text{... (1)}$$

Again
$$(E-1)^4 f(10) = 0$$
$$\Rightarrow \qquad (E^4 - 4E^3 + 6E^2 - 4E + 1) f(10) = 0$$
$$\Rightarrow f(30) - 4 f(25) + 6 f(20) - 4 f(15) + f(10) = 0$$
$$\Rightarrow \qquad 25 - 4 f(25) + 6(15) - 4(13) + f(10) = 0$$
$$\Rightarrow \qquad -4 f(25) + f(10) + 25 + 90 - 52 = 0$$
$$\Rightarrow \qquad -4 f(25) + f(10) = -63 \qquad \text{... (2)}$$

Solving (1) and (2), we get $\qquad f(10) = 10.87$
and $\qquad\qquad\qquad\qquad\qquad\qquad f(25) = 18.47$

Missing enteries are 10.87 and 18.47.   **Ans.**

Example 12. *Write down the difference between Interpolation and extrapolation.*   *(Bihar SBTE, 2009)*

Solution.   **Interpolation:** The technique of determining an approximate value of $f(x)$ for a non-tabulesr valaue of x which lies in the interval (a, b) is called interpolation.

Extrapolation: The technique of determining the value of $f(x)$ for a value of x lying out side the interval $[a, b]$ is caalled extrapolation.

EXERCISE 5.1

1. Construct Newton's forward interpolation polynomial for the following data :

x	4	6	8	10
y	1	3	8	16

    Hence evaluate y for $x = 5$.   **Ans.** 1.625

2. Construct Newton's forward interpolation polynomial for the following data :

x	6	8	10	12	14	16
$y = f(x)$	15	10	20	25	22	35

    then find y for $x = 15$.   *(Bihar SBTE, 2014)* **Ans.** 1.625

3. Given the set of value:

x	5	6	7	8	9	10
y	10	13	14	16	20	25

4. Find the value of ∇y_{20} for the given set of values:

x	10	15	20	25	30	35
y	19.97	25.5	22.47	23.52	24.65	25.89

    (Bihar SBTE, 2009)

5. The following data gives the melting point of an alloy of lead and zinc :

Percentage of lead in the alloy (P):	50	60	70	80
Temperature (0°C)	205	225	248	274

    Find the melting point of the alloy containing 54% of lead, using appropriate interpolation formula.   **Ans.** 212.64

6. Express y as a polynomial in x by Newton's forward difference interpolation for the following table:

x	0	1	2	3	4
y	3	6	11	18	27

    Ans. $y = x^2 + 2x + 3$.

7. The following are data from the steam table. Find the pressure upto four places of decimal if the temperature of the steam is 145°C.

Temperature °C	140	150	160	170	180
Pressure kgf/cm^2	3.685	4.854	6.302	8.076	10.225

    Ans. 4.2375

8. A function $f(x)$ is given by the following table. Find $f(0.2)$ by a suitable formula :

x	0	1	2	3	4	5	6
$f(x)$	176	185	194	203	212	220	229

    Ans. 177.718036736

9. Using a polynomial of the third degree, complete the record given below of the export of a certain commodity during five years :

Year	1989	1990	1991	1992	1993
Export (in tons)	443	384	—	397	467

    Ans. 369 tons

10. Find the missing term in the following table :

x	0	1	2	3	4
y	1	3	9	—	81

Ans. 31

11. Obtain the estimate of the missing figures in the following table :

x	1	2	3	4	5	6	7	8
$f(x)$	1	8	?	64	—	216	343	512

Ans. $f(5) = 125$, $f(3) = 27$

12. Obtain the missing terms in the following table :

x	2.0	2.1	2.2	2.3	2.4	2.5	2.6
$f(x)$	0.135	—	0.111	0.100	—	0.080	0.074

Ans. $f(2.1) = 0.123, f(2.4) = 0.090$

13. In the table below the values of y are consecutive terms of a series of which the number 21.6 is the sixth term. Find the first and the tenth term of the series.

x	3	4	5	6	7	8	9
y	2.7	6.4	12.5	21.6	34.3	51.2	72.9

Ans. $y_1 = -19.2, y_{10} = 100$

14. Write down the difference between interpolation and extrapolation.   (*Bihar SBTE, 2009*)

Choose the correct answer:

15. In forward interpolation formula the difference operator is applied on .... entry.

(*a*) First   (*b*) Last   (*c*) Middle   (*d*) None of these

(*Bihar SBTE, 2003, 2011*)   **Ans.** (*a*)

16. If only two values of y i.e. y_0 and y_1 corresponding to x_0 and x_1 are given, then Newton's forward interpolation formula is called .... interpolation formula.

(*a*) linear   (*b*) parabolic   (*c*) elliptic   (*d*) hyperbolic

(*Bihar SBTE, 2010*) **Ans.** (*a*)

17. When a set of tabulated data of x and y for $y = f(x)$ are given, the rate of chnge in y can be obtained by ....
the interpolation formula.

(*a*) Differentiating   (*b*) Integrating   (*c*) Both of these   (*d*) None of these

(*Bihar SBTE, 2005*) **Ans.** (*a*)

18. The results obtained through the interpolation formula are always ....

(*a*) exact   (*b*) approximate   (*c*) Real   (*iv*) Imaginary

(*Bihar SBTE, 2004*) **Ans.** (*b*)

19. Given

x:	20	30	40
$y = f(x)$:	0.3420	0.5020	0.6428

Finding the value of y at x = 10 is called ....

(*a*) Initial polation   (*b*) Interpolation   (*c*) Central polation   (*d*) Extra polation

(*Bihar SBTE, 2010*) **Ans.** (*d*)

20. In Newton's Forward Interpolation formula we use the operator:

(*a*) E   (*b*) ∇   (*c*) D   (*d*) none of these   (*Bihar SBTE, 2014*)

21. The value of Dy_9 is ....

(a) 1   (b) 2   (c) 4   (d) 5   (*Bihar SBTE, 2015*)

5.5 NEWTON'S BACKWARD INTERPOLATION FORMULA

(Bihar, SBTE, 2010, R.G.P.V., Bhopal, III Semester, Dec. 2002)

Let the function be $y = f(x)$

x	x_0	x_1	x_2	....	....	x_n
y	y_0	y_1	y_2	....	....	y_n

Suppose it is required to evaluate $f(x)$ for $x = x_n + ph$, where p is any real number then, we have

$$y_p = f(x_n + ph) = E^p f(x_n) = (1-\nabla)^{-p} y_n$$

$$= \left[1 + p\nabla + \frac{p(p+1)}{2!} \nabla^2 + \frac{p(p+1)(p+2)}{3!} \nabla^3 + ... \right] y_n$$

$$y_p = y_n + p\nabla y_n + \frac{p(p+1)}{2!} \nabla^2 y_n + \frac{p(p+1)(p+2)}{3!} \nabla^3 y_n + ...$$

It is called Newton's Backward interpolation formula.

Note. This formula is used for finding the value of y for x, when x is near x_n (end).

It is also used for extrapolating values of y for x when x is slightly greater than x_n.

Example 13. *Find y (9) from the following data:*

x	4	6	8	10
y	1	3	8	16

(Bihar SBTE, 2008)

Solution.   We have the following backward difference table.

x	y	∇y_3	$\nabla^2 y_3$	$\nabla^3 y_3$
4	1			
		2		
6	3		3	
		5		0
8	8		3	
		8		
10	16			

Here, $x = 9$, $x_n = 10$, $h = 2$

$$\therefore \quad P = \frac{x - x_n}{h} = \frac{9 - 10}{2} = \frac{-1}{2} = -0.5$$

Using backward interpolation formula, we have

$$y(9) = y_3 + P\nabla y_3 + \frac{P(P+1)\nabla^2 y_3}{2!} + \frac{P(P+1)(P+2)\nabla^3 y_3}{3!}$$

$$= 16 + (-0.5) \times 8 + \frac{(-0.5)(-0.5+1) \times 3}{2} - \frac{0.5(-0.5)(-0.5+1) \times 5}{3 \times 2}$$

$$= 16 - 4 - 0.375 = 11.625 \qquad\qquad\qquad\qquad\qquad \textbf{Ans.}$$

Example 14. *Given*

x	1	2	3	4	5	6	7	8
f(x)	1	8	27	64	125	216	343	512

Estimate f (7.5).

Solution. Difference table is as under :

x	f(x)	∇f	$\nabla^2 f$	$\nabla^3 f$	$\nabla^4 f$	$\nabla^5 f$	$\nabla^6 f$	$\nabla^7 f$
1	1							
		7						
2	8		12					
		19		6				
3	27		18		0			
		37		6		0		
4	64		24		0		0	
		61		6		0		0
5	125		30		0		0	
		91		6		0		
6	216		36		0			
		127		6				
7	343		42					
		169						
8	512							

Here, $x = 7.5,$ $x_n = 8$

$x = x_n + Ph$

$7.5 = 8 + P (1)$ $\Rightarrow \quad P = -0.5$

By Newton's Backward difference formula, we have

$$y_p = y_n + P\nabla y_n + \frac{P(P+1)}{2!}\nabla^2 y_n + \frac{P(P+1)(P+2)}{3!}\nabla^3 y_n + \frac{P(P+1)(P+2)(P+3)}{4!}\nabla^4 y_n$$

$$= 512 + (-0.5)169 + \frac{(-0.5)(-0.5+1)}{2!}(42) + \frac{(-0.5)(-0.5+1)(-0.5+2)}{3!}(6) + 0$$

$$= 512 - (0.5)\,169 - (0.5)\,(0.5)\,(21) - (0.5)\,(0.5)\,(1.5) = 512 - 84.5 - 5.25 - 0.375$$

$$= 421.875 \qquad\qquad\qquad \textbf{Ans.}$$

Example 15. *Find the cubic polynomial which takes the following values :*

x	0	1	2	3
y	1	2	1	10

Hence or otherwise evaluate y (4).   (R.G.P.V., Bhopal, III Semester, Dec. 2001)

Solution.   Difference table is as under :

x	y	∇y	$\nabla^2 y$	$\nabla^3 y$
0	1			
		1		
1	2		-2	
		-1		12
2	1		10	
		9		
3	10			

Here $a = 3$, $a + ph = 4$, $h = 1, y_n = 10$

$$3 + p(1) = 4$$
$$\Rightarrow\ p = 1$$

By backward difference formula :

$$y_p = y_n + p\nabla y_n + \frac{p(p+1)}{2!}\nabla^2 y_n + \frac{p(p+1)(p+2)}{3!}\nabla^3 y_n + \dots$$

$$\Rightarrow\quad y_x = 10 + x(9) + \frac{x(x+1)}{2!}(10) + \frac{x(x+1)(x+2)}{3!}(12) = 10 + 9x + 5x^2 + 5x + 2(x^3 + 3x^2 + 2x)$$

$$= 10 + 9x + 5x^2 + 5x + 2x^3 + 6x^2 + 4x = 2x^3 + 11x^2 + 18x + 10 \qquad \dots (1)$$

Which is required polynomial.

On putting $x = 1$ in (1), we get

$$y_{(1)} = 2(1)^3 + 11(1)^2 + 18(1) + 10$$
$$= 2 + 11 + 18 + 10 = 41 \qquad\qquad \textbf{Ans.}$$

Example 16. *Find the cubic polynomial interpolation which takes on the values*
$$f_0 = 5,\quad f_1 = 1,\quad f_2 = 9,\quad f_3 = 25,\quad f_4 = 55.\quad \text{Hence find } f_5$$

Solution.   Difference table is as under :

x	$y = f(x)$	∇f	$\nabla^2 f$	$\nabla^3 f$	$\nabla^4 f$
0	5				
		-4			
1	1		12		
		8		-4	
2	9		8		10
		16		6	
3	25		14		
		30			
4	55				

Here, $x = 5$, $x_n = 4, y_n = 55$

$$x = x_n + ph$$
$$5 = 4 + p(1) \qquad\qquad \Rightarrow\qquad p = 1$$

Newton's Backward difference interpolation formula is

$$f_p = y_n + p \nabla y_n + \frac{p(p+1)}{2!} \nabla^2 y_n + \frac{p(p+1)(p+2)}{3!} \nabla^3 y_n + \frac{p(p+1)(p+2)(p+3)}{4!} \nabla^4 y_n$$

$$\Rightarrow \quad f_x = 55 + x\,(30) + \frac{x(x+1)}{2!}\,(14) + \frac{x\,(x+1)\,(x+2)}{3!}\,(6) + \frac{x\,(x+1)\,(x+2)\,(x+3)}{4!}\,(10)$$

$$= 55 + 30\,x + (7x^2 + 7x) + (x^3 + 3x^2 + 2x) + \left(\frac{5x^4 + 30x^3 + 55x^2 + 30x}{12} \right)$$

$$= \frac{1}{12}\,(660 + 360x + 84x^2 + 84x + 12x^3 + 36x^2 + 24x + 5x^4 + 30x^3 + 55x^2 + 30x)$$

$$= \frac{1}{12}\,(5x^4 + 42x^3 + 175x^2 + 498x + 660) = \frac{5}{12} x^4 + \frac{7}{2} x^3 + \frac{175}{12} x^2 + \frac{83}{2} x + 55 \quad ...(1)$$

Which is required cubic polynomial.

On putting $x = 5$ in (1), we get

$$f_5 = \frac{5}{12}\,(5)^4 + \frac{7}{2}\,(5)^3 + \frac{175}{12}\,(5)^2 + \frac{83}{2}\,(5) + 55$$

$$= \frac{3125}{12} + \frac{875}{2} + \frac{4375}{12} + \frac{415}{2} + 55 = 1325 \qquad \textbf{Ans.}$$

EXERCISE 5.2

1. Write Newton's backward interpolation formula and evaluate $f(21)$ from the following data:

x	0	5	10	15	20
y	1.0	1.6	3.8	8.2	15.4

Ans. 17.2288

2. The following data gives the melting point of an alloy of lead and zinc, where t is the temperature in °C and P is the percentage of lead in the alloy.

P	60	70	80	90
t	226	250	276	304

Find the melting point of the alloy containing 84 per cent of lead using Newton's interpolation method.

Ans. 286.96°C

3. From the following table, find the values of $\frac{dy}{dx}$ and $\frac{d^2y}{dx^2}$ at x = 2.03.

x	1.96	1.98	2.00	2.02	2.04
y	0.7825	0.7739	0.7651	0.7563	0.7473

Ans. – 0.06; 0.5

4. Calculate the first and second derivatives of the function tabulated below at the point x = 2.2 and also $\frac{dy}{dx}$ at x = 2.0.

x	1.0	1.2	1.4	1.6	1.8	2.0	2.2
y	2.7183	3.3201	4.0552	4.9530	6.0496	7.3891	9.0250

Ans. 9.0228; 8.992; 7.3896

5. Given the following table of values of x and y

x	1.00	1.05	1.10	1.15	1.20	1.25	1.30
y	1.0000	1.0247	1.0488	1.0723	1.0954	1.1180	1.1401]

Find $\dfrac{dy}{dx}$ and $\dfrac{d^2y}{dx^2}$ at $x = 1.25$

Ans. $0.4473;\ -0.1583$

6. From the following compute the value of sin 38° :

$x°$	0	10	20	30	40
$\sin x°$	0	0.17365	0.34202	0.50000	0.64279

Ans. 0.615227216

7. From the following evaluate e^x when $x = 0.38$

x	0	0.1	0.2	0.3	0.4
e^x	1	1.1052	1.2214	1.3499	1.4918

Ans. 1.488824529

8. The table below gives the value of tan x for $0.10 \le x \ge 0.30$.

x	0.10	0.15	0.20	0.25	0.30
$y = \tan x$	0.1003	0.1511	0.2027	0.2553	0.3093

Find the value of tan 0.26.

Ans. 0.26597168

9. Apply Newton's backward difference formula to the data below, to obtain a polynomial of degree 4 in x:

x	1	2	3	4	5
y	1	-1	1	-1	1

Ans. $y = \dfrac{2}{3}x^4 - 8x^3 + \dfrac{100}{3}x^2 - 56x + 31$

10. From the following data, estimate the number of persons having incomes between 2000 and 2500.

Income	Below 500	500-1000	1000-2000	2000-3000	3000-4000
No. of persons	6000	4250	3600	1500	650

Ans. 14706 (approx)

Choose the correct answer:

11. In backward interpolation formula, the backward difference operator is applied on .... entry.

(a) First   (b) Last   (c) Middle   (d) None of these

(*Bihar SBTE, 2004, 2008, 2009*) **Ans.** (b)

12. Newton's Backward Interplation formula is useful to interpolation near the .... of tabular form

(a) Beginning   (b) End   (c) Middle   (d) None of these

(*Bihar SBTE, 2014*) **Ans.** (b)

6

Numerical Differentiation

6.1 INTRODUCTION

So far we were finding the polynomial curve $y = f(x)$ passing through the ordered pairs $(x_0, y_0), (x_1, y_1) .... (x_n, y_n)$.

Now we are trying to find the derivative value of such polynomial. To get the derivative we first find out $y = f(x)$ through the points and then differentiate.

Numerical differentiation. Q. Obtain formula for $\dfrac{d^3 y}{dx^3}$ in terms of h and Δ of the function $y = f(x)$ at x_0.4

(Bihar SBTE, 2008)

Ans. Numerical differentiation: Numerical differentiation is the process of calculating the derivatives of a function at some particular value of the independent variable by means of a set of given values of the function.

6.2 NEWTON'S FORWARD DIFFERENCE FORMULA TO GET THE DERIVATIVE

(Bihar SBTE, 2009 2008, 2004)

By Newton's forward difference interpolation formula on page 65 is   *(Bihar, SBTE, 2009)*

$$f(x) = f(a + ph) = f(a) + p\Delta f(a) + \frac{p(p-1)}{2!}\Delta^2 f(a)$$

$$+ \frac{p(p-1)(p-2)}{3!}\Delta^3 f(a) + \frac{p(p-1)(p-2)(p-3)}{4!}\Delta^4 f(a) +$$

$$\frac{p(p-1)(p-2)(p-3)(p-4)}{5!}\Delta^5 f(a)+.... \qquad ... (1)$$

where $p = \dfrac{x-a}{h}$

Differentiating (1) w.r.t to p, we get

$$f'(x) = f'(a+ph)(h) = \Delta f(a) + \frac{2p-1}{2!}\Delta^2 f(a) + \frac{3p^2 - 6p + 2}{3!}\Delta^3 f(a) +$$

$$\frac{4p^3 - 18p^2 + 22p - 6}{4!}\Delta^4 f(a) + \frac{5p^4 - 40p^3 + 105p^2 - 100p + 24}{5!}\Delta^5 f(a)+...$$

$$\Rightarrow \boxed{\begin{aligned} f'(x) = f'(a+ph) = \frac{1}{h}\Bigg[&\Delta f(a) + \frac{2p-1}{2}\Delta^2 f(a) + \frac{3p^2 - 6p + 2}{6}\Delta^3 f(a) \\ &+ \frac{4p^3 - 18p^2 + 22p - 6}{4!}\Delta^4 f(a) + \frac{5p^4 - 40p^3 + 105p^2 - 100p + 24}{5!}\Delta^5 f(a)+... \Bigg] \end{aligned}} \quad ...(2)$$

If $x = a$, $x = a + p\,(h) \Rightarrow p = 0$

$$f'(x) = \frac{1}{h}\left[\Delta f(a) - \frac{1}{2}\Delta^2 f(a) + \frac{1}{3}\Delta^3 f(a) - \frac{1}{4}\Delta^4 f(a) + \ldots\right]$$

Again differentiating (2) w.r.t. 'p', we get

$$f''(a + ph)\,h = \frac{1}{h}\left[\Delta^2 f(a) + (p-1)\,\Delta^3\,f(a) + \frac{12p^2 - 36p + 22}{4!}\Delta^4 f(a) + \right.$$

$$\left. \frac{2p^3 - 12p^2 + 21p - 10}{12}\Delta^5\,f(a) + \ldots\right]$$

$$\boxed{f''(x) = f''(a + ph) = \frac{1}{h^2}\left[\Delta^2 f(a) + (p-1)\,\Delta^3\,f(a) + \frac{6p^2 - 18p + 11}{12}\Delta^4 f(a) \right.}$$
$$\boxed{\left. + \frac{2p^3 - 12p^2 + 21p - 10}{12}\Delta^5\,f(a) + \ldots\right]} \quad \ldots (3)$$

Equation (2) and (3) are the formulae to find out the derivatives.

Example 1. *Find the first and second derivatives of the function given below at the point* $x = 1.2$:

x	1	2	3	4	5
y	0	1	5	6	8

(R.G.P.V., Bhopal III Semester, June 2003)

Solution. Newton's Forward Difference Table

x	$f(x)$	$\Delta f(x)$	$\Delta^2 f(x)$	$\Delta^3 f(x)$	$\Delta^4 f(x)$
1	0				
		1			
2	1		3		
		4		−6	
3	5		−3		10
		1		4	
4	6		1		
		2			
5	8				

$x_p = 1.2$, $\quad a = 1$, $\quad h = 1$ $\qquad x_p = a + ph \Rightarrow 1.2 = 1 + p \Rightarrow p = 0.2$

By Newton's forward difference interpolation formula.

$$f'(x) = f'(a + ph) = \frac{1}{h}\left[\Delta f(a) + \frac{2p-1}{2}\Delta^2\,f(a) + \frac{3p^2 - 6p + 2}{6}\Delta^3\,f(a)\right.$$

$$\left. + \frac{4p^3 - 18p^2 + 22p - 6}{24}\Delta^4 f(a)\right]$$

$$f'(1.2) = \frac{1}{1}\left[1 + \frac{2 \times 0.2 - 1}{2}(3) + \frac{3(0.2)^2 - 6(0.2) + 2}{6}(-6)\right.$$

$$\left. + \frac{4(0.2)^3 - 18(0.2)^2 + 22(0.2) - 6}{24}(10)\right]$$

$$= \left[1 - \frac{0.6(3)}{2} - (0.12 - 1.2 + 2) + (0.032 - 0.72 + 4.4 - 6)\frac{5}{12}\right]$$

$$= [1 - 0.9 - 0.92 - 0.9533] = -1.7733 \qquad \text{Ans.}$$

Hence, first derivative of the function $f(x)$ is -1.7733 at $x = 1.2$.

Second derivative at x = 1.2

$$f''(x) = \frac{1}{h^2}\left[\Delta^2 f(a) + (p-1)\Delta^3 f(a) + \frac{6p^2 - 18p + 11}{12}\Delta^4 f(a)\right]$$

$$f''(1.2) = \frac{1}{(1)^2}\left[3 + (0.2 - 1)(-6) + \frac{6(0.2)^2 - 18(0.2) + 11}{12}(10)\right]$$

$$= 3 + 4.8 + \frac{0.24 - 3.6 + 11}{12}(10)$$

$$= 3 + 4.8 + 6.3667 = 14.1667$$

Hence, the second derivative of the function $f(x)$ is 14.1667 at $x = 1.2$ \qquad **Ans.**

Example 2. *Find f′ (5) and f″ (5) for the following table:*

x	2	7	12	17	22
y = f (x)	8	15	10	25	30

(Bihar SBTE, 2014)

Solution. $x_p = 5,$ \qquad $h = 5,$ \qquad $a = 2$

$x_p = a + ph$ \quad $\Rightarrow$ \quad $5 = 2 + p + 5$ \quad $\Rightarrow$ \quad $p = 0.6$

Newton Difference Table

x	f (x)	Δ f (x)	Δ² f (x)	Δ³ f (x)	Δ⁴ f (x)
2	8				
		7			
7	15		– 12		
		–5		32	
12	10		+ 20		– 62
		15		– 30	
17	25		– 10		
		5			
22	30				

Newton forward difference derivative formula

$$f'(x + ph) = \frac{1}{h}\left[\Delta f(a) + \frac{2p-1}{2!}\Delta^2 f(a) + \frac{3p^2 - 6p + 2}{3!}\Delta^3 f(a) + \frac{4p^3 - 18p^2 + 22p - 64}{4!}\Delta^4 f(a) + \dots\right]$$

$$= \frac{1}{0.6}\left[7 + \frac{2(.6) - 1}{2}(-12) + \frac{3(.6)^2 - 6(0.6) + 2}{6}(32) + \frac{4(0.6)^3 - 18(0.6)^2 + 22(0.6) - 6}{24}(-62)\right]$$

$$= \frac{5}{3}7 - 1.2 + \frac{10.8 - 3.6 + 2}{6}(32) + \frac{0.864 - 6.48 + 13.2 - 6}{24}(-62)$$

$$= \frac{5}{3}[1 - 1.2 + 49.0667 - 4.092]$$

$$= \frac{5}{3}(50.7747) = 84.6245$$

Ans.

Example 3. *Given that :*

x	1.0	1.1	1.2	1.3
y	0.841	0.891	0.932	0.963

Find $\dfrac{dy}{dx}$ at x = 1.0

(R.G.P.V., Bhopal, III Semester June 2007)

Solution. Newton's Forward difference table

x	f(x)	Δ f(x)	Δ² f(x)	Δ³ f(x)
1.0	0.841			
		0.050		
1.1	0.891		− 0.009	
		0.041		− 0.001
1.2	0.932		− 0.010	
		0.031		
1.3	0.963			

$x_p = 1, \quad a = 1, \quad h = 0.1 \qquad x = a + ph, \quad \Rightarrow \quad 1 = 1 + p(0.1) \quad \Rightarrow \quad p = 0$

By Newton's forward difference interpolation formula

$$f(x) = f(a + ph) = f(a) + p\Delta f(a) + \frac{p(p-1)}{2!}\Delta^2 f(a) + \frac{p(p-1)(p-2)}{3!}\Delta^3 f(a) + \dots$$

$$f'(a + ph) = \frac{1}{h}\left[\Delta f(a) + \frac{2p-1}{2}\Delta^2 f(a) + \frac{3p^2 - 6p + 2}{6}\Delta^3 f(a)\right]$$

$$f'(x) = \frac{1}{h}\left[\Delta f(a) - \frac{1}{2}\Delta^2 f(a) + \frac{1}{3}\Delta^3 f(a)\right] \qquad\qquad [p = 0]$$

$$f'(1) = \frac{1}{0.1}\left[\Delta f(1) - \frac{1}{2}\Delta^2 f(1) + \frac{1}{3}\Delta^3 f(1)\right]$$

$$= 10\left[0.05 - 0.5(-0.009) + \frac{1}{3}(-0.001)\right] \qquad\qquad [h = 0.1]$$

$$= 10\,[0.05 + 0.0045 - 0.0003] = 10\,[0.0545 - 0.0003] = 10\,(0.0542)$$

$$\frac{dy}{dx} = 0.542$$

Ans.

Hence, $\dfrac{dy}{dx} = 0.542$ at x = 1.0.

Example 4. *From the following table, obtain the value of* $\dfrac{d^2 y}{dx^2}$ *at the point x = 0.96 :*

x	0.96	0.98	1.00	1.02	1.04
y	0.7825	0.7739	0.7651	0.7563	0.7473

Solution.   Newton's Forward Difference Table

x	y	Δy	$\Delta^2 y$	$\Delta^3 y$	$\Delta^4 y$
0.96	0.7825				
		-0.0086			
0.98	0.7739		-0.0002		
		-0.0088		0.0002	
1.00	0.7651		0.0000		-0.0004
		-0.0088		-0.0002	
1.02	0.7563		-0.0002		
		-0.0090			
1.04	0.7473				

$$a = 0.96, \qquad\qquad h = 0.02$$

$$a + ph = 0.96 \quad\Rightarrow\quad 0.96 + p\,(0.02) = 0.96 \quad\Rightarrow\quad p = 0$$

$$f''(a + ph) = \frac{1}{h^2}[\Delta^2 f(a) + (p-1)\,\Delta^3 f(a) + \frac{1}{24}(12p^2 - 36p + 22)\,\Delta^4 f(a)]$$

$$= \frac{1}{(0.02)^2}\left[-0.0002 + (0-1)\,(0.0002) + \frac{1}{24}\,(0-0+22)\,(-0.0004)\right]$$

$$= 2500\,[-0.0002 - 0.0002 - 0.00037] = -2500 \times 0.00077 = -1.925 \qquad\qquad \textbf{Ans.}$$

Hence, $\dfrac{d^2 v}{dx^2} = -1.925$ at the point x = 0.96.

Example 5. *Find first, second derivatives of the following tabulated function at the point x = 1.5.*

x	1.5	2	2.5	3	3.5	4
f (x)	3.375	7.0	13.625	24	38.875	59.0

(Bihar SBTE, 2009, R.G.P.V., Bhopal, III Semester, June 2008)

Solution.

x	f (x)	$\Delta f(x)$	$\Delta^2 f(x)$	$\Delta^3 f(x)$	$\Delta^4 f(x)$
1.5	3.375				
		3.625			
2	7.0		3.00		
		6.625		0.75	
2.5	13.625		3.75		0
		10.375		0.75	
3	24		4.50		0
		14.875		0.75	
3.5	38.875		5.25		
		20.125			
4	59.0				

Here $a = 1.5$, $h = 0.5$,

$a + ph = 1.5 \Rightarrow 1.5 + p (0.5) = 1.5 \Rightarrow p = 0$

$$f'(a + ph) = \frac{1}{h}\left[\Delta f(a) + \frac{1}{2!}(2p-1)\,\Delta^2 f(a) + \frac{1}{3!}(3p^2 - 6p + 2)\,\Delta^3 f(a) + ...\right]$$

On putting $p = 0$, we get

$$f'(a) = \frac{1}{h}\left[\Delta f(a) - \frac{1}{2}\Delta^2 f(a) + \frac{1}{3}\Delta^3 f(a) + ...\right]$$

$$f'(1.5) = \frac{1}{0.5}\left[3.625 - \frac{1}{2}(3.00) + \frac{1}{3}(0.75) + ...\right] = 4.75$$

Hence, first derivative of the function $f(x)$ is 4.75 at $x = 1.5$.   **Ans.**

$$f''(a + ph) = \frac{1}{h^2}\left[\Delta^2 f(a) + (p-1)\,\Delta^3 f(a) + ...\right]$$

On substituting $p = 0$, we have

$$f''(a) = \frac{1}{h^2}\left[\Delta^2 f(a) - \Delta^3 f(a) + ...\right]$$

$$f''(1.5) = \frac{1}{(0.5)^2}[3.00 - 0.75] = 9.0$$

Hence, second derivative of the function $f(x)$ is 9.0 at $x = 1.5$.   **Ans.**

Example 6. *Find the first and second derivatives of the function f (x) at the point x = 1.1:*

x	1	1.2	1.4	1.6	1.8	2.00
f (x)	0.00	0.1280	0.5440	1.2960	1.4320	4.000

Solution.

x	f (x)	Δ	Δ²	Δ³	Δ⁴	Δ⁵
1.0	0.00					
		0.1280				
1.2	0.1280		0.2880			
		0.4160		0.0480		
1.4	0.5440		0.3360		− 1	
		0.7520		− 0.952		5
1.6	1.2960		−0.616		4	
		0.1360		3.048		
1.8	1.4320		2.4320			
		2.5680				
2.00	4.000					

Here $a = 1$, $h = 0.2$,

$a + ph = $ 1.1, $1 + p\,(0.2) = 1.1$, $p = 0.5$

$$f'(a+ph) = \frac{1}{h}\left[\Delta f(a) + \frac{2p-1}{2}\Delta^2 f(a) + \frac{3p^2-6p+2}{6}\Delta^3 f(a) + \frac{4p^3-18p^2+22p-6}{24}\Delta^4 f(a)\right.$$

$$\left. + \frac{5p^4-40p^3+105p^2-100p+24}{120}\Delta^5 f(a)\right]$$

$$f'(1.1) = \frac{1}{0.2}\left[0.1280 + \frac{1-1}{2}(0.2880) + \frac{3(0.5)^2-6\times 0.5+2}{6}(0.0480) + \frac{4(0.5)^3-18(0.5)^2+22(0.5)-6}{24}(-1)\right.$$

$$\left. + \frac{5(0.5)^4-40(0.5)^3+105(0.5)^2-100(0.5)+24}{120}(5)\right]$$

$$= \frac{1}{0.2}\left[0.1280 + 0 + \frac{0.75-3+2}{6}(0.048) - \frac{0.5-4.5+11-6}{24} + \frac{0.3125-5+26.25-50+24}{120}(5)\right]$$

$$= \frac{1}{0.2}\ [0.1280 - 0.002 - 0.0417 - 0.1849] = 5\ [-\ 0.1006] = -\ 0.5030 \qquad \textbf{Ans.}$$

Hence, first derivative of the function $f(x)$ is $-\ 0.5030$ when $x = 1.1$.

$$f''(a+ph) = \frac{1}{h^2}\left[\Delta^2 f(a) + \frac{6p-6}{6}\Delta^3 f(a) + \frac{12p^2-36p+22}{24}\Delta^4 f(a) + \frac{2p^3-12p^2+21p-10}{12}\Delta^5 f(a)\right]$$

$$f''(1.1) = \frac{1}{(0.2)^2}\left[0.2880 + \frac{0.5-1}{1}(0.0480) + \frac{6(0.5)^2-18(0.5)+11}{12}(-1) + \frac{2(0.5)^3-12(0.5)^2+21(0.5)-10}{12}(5)\right]$$

$$= 25\left[0.2880 - 0.5\,(0.0480) - \frac{3.5}{12} - \frac{2.25}{12}(5)\right] = 25\ [0.288 - 0.024 - 0.2917 - 0.9375]$$

$$= 25\ [-\ 0.9652] = -\ 24.13 \qquad\qquad \textbf{Ans.}$$

EXERCISE 6.1

1. Find $y'(0)$ and $y''(0)$ from the following table :

x	0	1	2	3	4	5
y	4	8	15	7	6	2

      Ans. –27.9, 117.67

2. The following data gives corresponding values of pressure and specific volume of a super heated steam.

v	2	4	6	8	10
p	105	42.7	25.3	16.7	13

Find the rate of change of pressure with respect to volume when $v = 2$.   **Ans.** – 52.4

3. Find the first and second derivatives of the function tabulated below at the point $x = 3.0$, using Newton's forward difference formula.

x	3.0	3.2	3.4	3.6	3.8	4.0
$f(x)$	–14.000	–10.032	–5.296	0.256	6.672	14.000

      Ans. $f'(3) = 18$, $f''(3) = 18$

4. The table given below reveals the velocity v of a body during the time 't' specified. Find its acceleration at $x = 1.1$ using Newton's forward formula.

t	1.0	1.1	1.2	1.3	1.4
v	43.1	47.7	52.1	56.4	60.8

Ans. 44.917

5. The population of a certain town is given below. Find the rate of growth of the pouplation in 1931.

Year (x)	1931	1941	1951	1961	1971
Population in Thousand (y)	40.62	60.80	79.95	103.56	132.65

Ans. 2.36425

6.3 NUMERICAL DIFFERENTIATION (BACKWARD DIFFERENCES)

Newton's formula for backward difference on page 69 is

$$y_p = y_n + p\nabla y_n + \frac{p(p+1)}{2!}\nabla^2 y_n + \frac{p(p+1)(p+2)}{3!}\nabla^3 y_n + ...$$

$$f(x) = f(x_n + ph) = f(x_n) + p\nabla f(x_n) + \frac{p(p+1)}{2!}\nabla^2 f(x_n) + \frac{p(p+1)(p+2)}{3!}\nabla^3 f(x_n) + ...$$

Differentiating with respect to p, we get

$$hf'(x_n + ph) = \nabla f(x_n) + \frac{2p+1}{2!}\nabla^2 f(x_n) + \frac{3p^2+6p+2}{3!}\nabla^3 f(x_n) + \frac{4p^3+18p^2+22p+6}{4!}\nabla^4 f(x_n) + ..$$

$$f'(x_n + ph) = \frac{1}{h}\left[\nabla f(x_n) + \frac{2p+1}{2}\nabla^2 f(x_n) + \frac{3p^2+6p+2}{6}\nabla^3 f(x_n)\right.$$

$$\left. + \frac{2p^3+9p^2+11p+3}{12}\nabla^4 f(x_n) + ...\right]$$

$$\boxed{f'(x) = f'(x_n + ph) = \frac{1}{h}\left[\nabla f(x_n) + \frac{2p+1}{2}\nabla^2 f(x_n) + \frac{3p^2+6p+2}{6}\nabla^3 f(x_n) + \frac{2p^3+9p^2+11p+3}{12}\nabla^4 f(x_n) + ...\right]}$$

$$f''(x_n + ph) = \frac{1}{h^2}\left[\nabla^2 f(x_n) + (p+1)\nabla^3 f(x_n) + \frac{6p^2+18p+11}{12}\nabla^4 f(x_n) + ...\right]$$

$$\boxed{f''(x_p) = f''(x_n + ph) = \frac{1}{h^2}\left[\nabla^2 f(x_n) + (p+1)\nabla^3 f(x_n) + \frac{6p^2+18p+11}{12}\nabla^4 f(x_n) + ...\right]}$$

Similarly, $f'''(x_n + ph) = \frac{1}{h^3}\left[\nabla^3 f(x_n) + \frac{2p+3}{2}\nabla^4 f(x_n) + ...\right]$ $[p = 0]$

Example 7. *Find the first derivative of the function y = f (x) at the point x = 3.5:*

x	1.5	2.0	2.5	3.0	3.5	4.0
y	3.375	7.0	13.625	24.0	38.875	59.0

(Bihar SBTE, 2010)

Solution.

x	y	∇y	$\nabla^2 y$	$\nabla^3 y$	$\nabla^4 y$
1.5	3.375				
		3.625			
2.0	7.0		3		
		6.625		0.75	
2.5	13.625		3.75		0
		10.375		0.75	
3.0	24.0		4.5		0
		14.875		0.75	
3.5	38.875		5.25		
		20.125			
4.0	59.0				

$$h = 0.5; \quad x = x_n + ph \quad \Rightarrow \quad 3.5 = 3.5 + p\,(0.5) \quad \Rightarrow p = 0$$

$$f'(x_n + ph) = \frac{1}{h}\left[\nabla f(x_n) + \frac{2p+1}{2!}\nabla^2 f(x_n) + \frac{3p^2+6p+2}{3!}\nabla^3 f(x_n) + \ldots\right]$$

$$\left(\frac{dy}{dx}\right)_{3.5} = \frac{1}{0.5}\left[\nabla y_n + \frac{1}{2}\nabla^2 y_n + \frac{1}{3}\nabla^3 y_n\right] = \frac{1}{0.5}\left[14.875 + \frac{1}{2}\times 4.5 + \frac{1}{3}\times 0.75\right] = 34.75$$ **Ans.**

Example 8. *The following table gives the values of a function at equal intervals :*

x	0.00	0.50	1.00	1.5	2.00
f (x)	0.3989	0.3521	0.2420	0.1295	0.0540

Evaluate f ′ (1.5), stating the formula used.

Solution. Newton's backward difference table

x	f (x)	$\nabla f(x)$	$\nabla^2 f(x)$	$\nabla^3 f(x)$	$\nabla^4 f(x)$
0.00	0.3989				
		− 0.0468			
0.50	0.3521		− 0.0633		
		− 0.1101		0.0609	
1.00	0.2420		− 0.0024		− 0.0215
		− 0.1125		0.0394	
1.50	0.1295		0.0370		
		− 0.0755			
2.00	0.0540				

$$x = x_n = 1.5, \quad h = 0.5$$

$$x = x_n + ph \implies 1.5 = 1.5 + p\,(0.5) \implies p = 0$$

By Newton's backward difference derivative formula,

$$f'\,(x_n + ph) = \frac{1}{h}\left[\nabla f\,(x_n) + \frac{2p+1}{2!}\nabla^2 f\,(x_n) + \frac{3p^2+6p+2}{3!}\nabla^3 f\,(x_n) + \frac{2p^3+9p^2+11p+3}{12}\nabla^4 f\,(x_n) + ...\right]$$

Putting the value of $p = 0$ in (1), we get

$$f'\,(x) = f'\,(x_n + ph) = \frac{1}{h}\left[\nabla f\,(x_n) + \frac{1}{2}\nabla^2 f\,(x_n) + \frac{1}{3}\nabla^3 f\,(x_n) + ...\right] \qquad \text{... (2)}$$

On putting the value of of h, $\nabla f\,(x_n)$, $\nabla^2 f\,(x_n)$, $\nabla^3 f\,(x_n)$ in (2), we get

$$f'\,(1.5) = \frac{1}{0.5}\left[-0.1125 + \frac{1}{2}(-0.0024) + \frac{1}{3}(0.0609) + ...\right] = \frac{1}{0.5}[(-0.0934) = -0.1868] \qquad \textbf{Ans.}$$

Example 9. *The population of a certain town (as obtained from census data) is shown in the following table:*

Year	1951	1961	1971	1981	1991
Population (in thousands)	19.96	36.65	58.81	77.21	94.61

Find the rate of growth of the population in the year 1981.

Solution. We have the following backward difference table.

x	y	∇y	$\nabla^2 y$	$\nabla^3 y$	$\nabla^4 y$
1951	19.96				
		16.69			
1961	36.65		5.47		
		22.16		−9.23	
1971	58.81		−3.76		11.99
		18.4		2.76	
1981	77.21		−1		
		17.4			
1991	94.61				

Here, $\quad x = 1981, \ h = 10$

$$x = x_n = 1981, \quad x = x_n + ph \implies 1981 = 1981 + p\,(10) \implies p = 0$$

$$f'\,(x_n + ph) = \frac{1}{h}\left[\nabla f\,(x_n) + \frac{2p+1}{2!}\nabla^2 f\,(x_n) + \frac{3p^2+6p+2}{3!}\nabla^3 f\,(x_n) + ...\right] \qquad \text{... (1)}$$

Putting $p = 0$ in (1), we get

$$f(x_3) = \frac{1}{h}\left[\nabla f\,(x_3) + \frac{\nabla^2 f\,(x_3)}{2} + \frac{\nabla^3 f\,(x_3)}{3}\right] = \frac{1}{10}\left[18.4 + \frac{-3.76}{2} + \frac{-9.23}{3}\right]$$

$$= \frac{1}{10}\,[18.4 - 1.88 - 3.0766] = \frac{1}{10} \times 13.4434 = 1.34$$

Therefore, the rate of growth of the population in year 1981 is 1.34 thousand per year. **Ans.**

Example 10.

x	6	8	10	12	14
f (x)	4	20	16	30	40

find $\dfrac{dy}{dx}$ for x = 13   (*Bihar SBTE, 2012*)

Solution. $x_p = 13,$ $x_n = 40$

$x_p = x_n + ph$

$\Rightarrow$ $13 = 14 + p\,(2)$ $\Rightarrow$ $p = -0.5$

x	f (x)	$\nabla\, f(x)$	$\nabla^2 f(x)$	$\nabla^3 f(x)$	$\nabla^4 f(x)$
6	4				
		16			
8	20		−20		
		−4		38	
10	16		18		−60
		14		−22	
12	30		−4		
		10			
14	40				

By Newton's Backward difference, derivative formula

$$f'\,(x_n + ph) = \frac{1}{h}\left[\nabla f(x_n) + \frac{2p+1}{2!}\nabla^2 f(x_n) + \frac{3p^2+6p+2}{3!}\nabla^3 f(x_n) + \frac{2p^3+9p^2+11p+3}{12}\nabla^4 f(x_n)\right]$$

$$f'\,(13) = \frac{1}{2}\left[10 + \frac{2\,(-0.5)+1}{2}\cdot(-4) + \frac{3(-0.5)^2+6(-0.5)+2}{6}(-22) + \frac{2\,(-0.5)^3+9\,(-0.5)^2+11\,(-0.5)+3}{12}\cdot(-60)\right]$$

$$= \frac{1}{2}\left[10 + 0 + \frac{3(0.25)-3+2}{6}(-22) + \frac{2\,(-0.125)+2.25-5.5+3}{12}(-60)\right]$$

$$= \frac{1}{2}\,[10 + 0.916667 + 2.5] = = \frac{1}{2}\,(13.416667) = 6.708334 \qquad \textbf{Ans.}$$

Example 11. *If*

x	4	8	12	16	20
f	9	12	17	25	30

find f ′ (18), f ″ (18)

Solution.   Now, to find out $f'\,(18)$ and $f''\,18)$ use Newton's backward interpolation formula.

x	y	∇y	$\nabla^2 y$	$\nabla^3 y$	$\nabla^4 y$
4	9				
		3			
8	12		2		
		5		1	
12	17		3		-7
		8		-6	
16	25		-3		
		5			
20	30				

$$x_p = 18 \qquad\qquad x_n = 20 \qquad\qquad x = 4, \quad x_p = x_n + ph \quad\Rightarrow\quad 18 = 20 + 4p$$

$$p = \frac{-20+18}{4} = \frac{-2}{4} = -0.5$$

$$y = 30; \qquad\qquad \nabla y = 5; \qquad\qquad \nabla^2 y = -3; \quad \nabla^3 y = -6 \qquad \nabla^4 y = -7$$

By Newton backward difference derivative formula

$$f'(x) = f'(x_n + ph) = \frac{1}{h}\left[\nabla y + \frac{2p+1}{2}\nabla^2 y + \frac{3p^2+6p+2}{6}\nabla^3 y + \frac{2p^3+9p^2+11p+3}{12}\nabla^4 y\right]$$

$$f'(18) = \frac{1}{4}\left[5 + \frac{2(-0.5)+1}{2}(-3) + \frac{3(-0.5)^2+6(-0.5)+2}{6}(-6) + \frac{2(-0.5)^3+9(-0.5)^2+11(-0.5)+3}{12}\times(-7)\right]$$

$$= \frac{1}{4}\left[4+0+\frac{1}{4}+\frac{7}{24}\right] = \frac{133}{96} = 1\frac{37}{96} \qquad\qquad\qquad \textbf{Ans.}$$

and $f''(x) = f''(x_n + ph) = \dfrac{1}{h^2}\left[\nabla^2 y + (p+1)\nabla^3 y + \dfrac{6p^2+18p+11}{12}\nabla^4 y\right]$

$$= \frac{1}{4^2}\left[-3 + (-0.5+1)(-6) + \frac{6(-0.5)^2+18(-0.5)+11}{12}(-7)\right]$$

$$= \left[\frac{1}{16}(-3-3-\frac{49}{24}\right] = \frac{-193}{384} = -0.50 \qquad\qquad \textbf{Ans.}$$

Example 12. *Find the numerical value of the first and second derivatives at x = 0.4 of the function f (x) defined as under :*

x	0.1	0.2	0.3	0.4
$f(x)$	1.10517	1.22140	1.34986	1.49182

(Bihar SBTE, 2003, 2011)

Solution.   Here we shall used Newton's backward interpolation formula as x = 0.4 lies at the end of the table.

x	$f(x)$	$\nabla f(x)$	$\nabla^2 f(x)$	$\nabla^3 f(x)$
0.1	1.10517			
		0.11623		
0.2	1.22140		0.01223	
		0.12846		0.00127
0.3	1.34986		0.01350	
		0.14196		
0.4	1.49182			

Here, $x_p = 0.4$, $h = 0.1$, $x = x_n + ph$, $0.4 = 0.4 + ph$ $\Rightarrow$ $p = 0$

Newton's backward difference formula

$$f'(x_n + ph) = \frac{1}{h}\left[\nabla f(x_n) + \frac{2p+1}{2!}\nabla^2 f(x_n) + \frac{3p^2 + 6p + 2}{3!}\nabla^3 f(x_n) + ...\right]$$

$$\Rightarrow f'(x_n) = \frac{1}{0.1}\left[\nabla f(x_n) + \frac{1}{2}\nabla^2 f(x_n) + \frac{1}{3}\nabla^3 f(x_n)\right] \qquad [p = 0]$$

$$f(0.4) = 10\left[0.14196 + \frac{1}{2}(0.01350) + \frac{1}{3}(0.00127)\right]$$

$$= 10\,[0.14196 + 0.00675 + 0.000423] = 10 \times 0.149133 = 1.49133.$$

Hence, the first derivative of the function $f(x)$ is 1.49133 at $x = 0.4$ **Ans.**

Second derivative

$$f''(x_n + ph) = \frac{1}{h^2}\left[\nabla^2 f(x_n) + (p+1)\nabla^3 f(x_n) + \frac{6p^2 + 18p + 11}{12}\nabla^4 f(x_n) + ...\right]$$

$$f''(0.4) = \frac{1}{(0.1)^2}[(0.01350) + (0.00127)] = \frac{1}{0.01}[0.01477]$$

$$= 1.4770$$

Hence, the second derivative of the function $f(x)$ is 1.4770 at $x = 0.4$.   **Ans.**

Example 13. *Given that*

x	1.0	1.1	1.2	1.3	1.4	1.5	1.6
y	7.989	8.403	8.781	9.129	9.451	9.750	10.031

Find $\dfrac{dy}{dx}$ *and* $\dfrac{d^2y}{dx^2}$ *at* $x = 1.6$.   (*R.G.P.V., Bhopal, III Semester, Dec. 2007*)

Solution. The difference table is as under :

x	y	Δy	$\Delta^2 y$	$\Delta^3 y$	$\Delta^4 y$	$\Delta^5 y$	$\Delta^6 y$
1.0	7.989						
		0.414					
1.1	8.403		− 0.036				
		0.378		0.006			
1.2	8.781		− 0.030		− 0.002		
		0.348		0.004		0.001	
1.3	9.129		− 0.026		− 0.001		0.002
		0.322		0.003		0.003	
1.4	9.451		− 0.023		+ 0.002		
		0.299		0.005			
1.5	9.750		− 0.018				
		0.281					
1.6	10.031						

Since 1.6 is the end of the given data, we use Newton's Backward Derivative formula

$$\frac{dy}{dx} = \frac{1}{h}\left[\nabla f(x_n) + \frac{2p+1}{2!}\nabla^2 f(x_n) + \frac{3p^2+6p+2}{3!}\nabla^3 f(x_n) + \frac{2p^3+9p^2+11p+3}{12}\nabla^4 f(x_n) + ...\right]$$

When $p = 0$,

$$\frac{dy}{dx} = \frac{1}{h}\left[\nabla f(x_n) + \frac{1}{2}\nabla^2 f(x_n) + \frac{1}{3}\nabla^3 f(x_n) + \frac{1}{4}\nabla^4 f(x_n) + \frac{1}{5}\nabla^5 f(x_n) + \frac{1}{6}\nabla^6 f(x_n)\right] \qquad ... (4)$$

On putting the values of h, $\nabla f(x_n)$, $\nabla^2 f(x_n)$, $\nabla^3 f(x_n)$, $\nabla^4 f(x_n)$, $\nabla^5 f(x_n)$ in (4), we get:

$$\left(\frac{dy}{dx}\right)_{1.16} = \frac{1}{0.1}\left[0.281 + \frac{1}{2}(-0.018) + \frac{1}{3}(0.005) + \frac{1}{4}(0.002) + \frac{1}{5}(0.003) + \frac{1}{6}(0.002)\right]$$

$$= \frac{1}{0.1}[0.281 - 0.009 + 0.0017 + 0.0005 + 0.0006 + 0.0003] = \frac{1}{0.1}[0.2751] = 2.751 \text{ Ans.}$$

$$\frac{d^2y}{dx^2} = \frac{1}{h^2}\left[\nabla^2 y + (p+1)\nabla^3 y + \frac{6p^2+18p+11}{12}\nabla^4 y + ...\right]$$

When $p = 0$, $\left(\dfrac{d^2y}{dx^2}\right) = \dfrac{1}{h^2}\left[\nabla^2 y + \nabla^3 y + \dfrac{11}{12}\nabla^4 y + \dfrac{5}{6}\nabla^5 y + \dfrac{137}{180}\nabla^6 y\right]$

$$= \frac{1}{(0.1)^2}\left[-0.018 + 0.005 + \frac{11}{12}(0.002) + \frac{5}{6}(0.003) + \frac{137}{180}(0.002)\right]$$

$$= \frac{1}{(0.01)}[-0.018 + 0.005 + 0.0018 + 0.0025 + 0.0015]$$

$$= \frac{1}{(0.01)}[-0.0072] = -0.72 \qquad\qquad\qquad\qquad\qquad \text{Ans.}$$

EXERCISE 6.2

1. Find the first two derivatives of $(x)^{\frac{1}{3}}$ at $x = 56$ from the table given below. (Using Newton's backward difference formula) :

x	50	51	52	53	54	55	56
$y = x^{\frac{1}{3}}$	3.6840	3.7084	3.7325	3.7563	3.7798	3.8030	3.8259

Ans. $y'(56) = 0.02275$

$y''(56) = -0.0003$

2. A rocket is launched from the ground. It's velocity is registered during the first 80 seconds and is given in the table below:

Time (t)	0	10	20	30	40	50	60	70	80
Velocity (m/sec)	30.00	31.63	33.44	35.47	37.75	40.33	43.25	46.69	50.67

Find the acceleration of the rocket at time $t = 80$

Ans. 0.38829 m/sec^2

7

Numerical Integration

7.1 INTRODUCTION

Numerical integration is the process of evaluating a definite integral from the values of the functions given in a tabular form. When the above process applied to the integration of a function of single variable, the process is called quadrature. The numerical integration of the tabulated values is solved by representing $f(x)$ by an interpolation formula and then integrating it between the given limits.

7.2 A GENERAL QUADRATURE FORMULA   (Bihar SBTE, 2003, 2010)

Let $I = \int_a^b f(x)\,dx$

where $f(x)$ is obtained by an interpolation formula in the interval (a, b) is divided into n sub-intervals of equal width h.

Let $x_0 = a$, $x_1 = a + h$, $x_2 = a + 2h$, ... $x_n = a + nh = b$ and corresponding values of $f(x)$ be

y_0, y_1, y_2, ... y_n respectively.

The number of ordinates is $(n + 1)$.

If $y = f(x) = f(x_0 + ph)$, then $\dfrac{x - x_0}{h} = p$

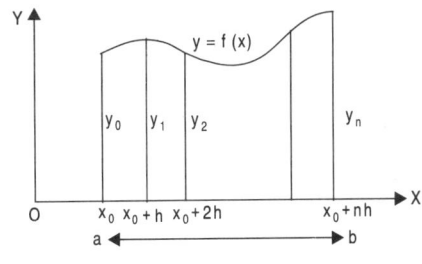

$\Rightarrow \quad \dfrac{dx}{h} = dp \quad \Rightarrow dx = h\,dp$

By Newton's Forward interpolation formula,

$$y = y_0 + p\Delta y_0 + \frac{p(p-1)}{2!}\Delta^2 y_0 + \frac{p(p-1)(p-2)}{3!}\Delta^3 y_0 + ...$$

Integrating both sides

$$\int_{x_0}^{x_0+nh} y\,dx = \int_0^n \left[y_0 + p\Delta y_0 + \frac{p(p-1)}{2!}\Delta^2 y_0 + \frac{p(p-1)(p-2)}{3!}\Delta^3 y_0 \right.$$

$$+ \frac{p(p-1)(p-2)(p-3)}{4!}\Delta^4 y_0 + \frac{p(p-1)(p-2)(p-3)(p-4)}{5!}\Delta^5 y_0$$

$$\left. + \frac{p(p-1)(p-2)(p-3)(p-4)(p-5)}{6!}\Delta^6 y_0 + ... \right] h\,dp \quad ... (1)$$

7.3 TRAPEZOIDAL RULE   (Bihar SBTE, 2012, 2011, 2003)

We have to find the area of the region ABCD by trapezoidal rule. Base AB is divided into n equal parts. First of all we find out the area of the first division and then area of the remaining divisions.

On taking $n = 1$ in the equation (1), the figure will be a trapezium and the curve will be a straight line passing through (x_0, y_0) and (x_1, y_1) and other integrals will be zero.

On putting $n = 1$ in (1), we get

$$\int_{x_0}^{x_0+h} f(x)\,dx = h\left[y_0 + \frac{1}{2}\Delta y_0\right] = h\left[y_0 + \frac{1}{2}(y_1 - y_0)\right]$$

$$= h\left[\frac{y_0}{2} + \frac{y_1}{2}\right]$$

$$\Rightarrow \qquad \int_{x_0}^{x_0+h} f(x)\,dx = \frac{h}{2}(y_0 + y_1) \qquad \qquad ...\,(i)$$

Similarly $\qquad \int_{x_0+h}^{x_0+2h} f(x)\,dx = \frac{h}{2}(y_1 + y_2) \;...\,(ii)$

$$\int_{x_0+2h}^{x_0+3h} f(x)\,dx = \frac{h}{2}(y_2 + y_3) \qquad \qquad (iii)$$

....

....

$$\int_{x_0+(n-1)h}^{x_0+nh} f(x)\,dx = \frac{h}{2}(y_{n-1} + y_n) \qquad \qquad (n)$$

On adding these above n integrals, we get

$$\int_{x_0}^{x_0+h} f(x)\,dx + \int_{x_0+h}^{x_0+2h} f(x)\,dx + \int_{x_0+2h}^{x_0+3h} f(x)\,dx + ... \int_{x_0+(n-1)h}^{x_0+nh} f(x)\,dx$$

$$= \frac{h}{2}\left[(y_0 + y_1) + (y_1 + y_2) + (y_2 + y_3) + ... +(y_{n-1} + y_n)\right] = \frac{h}{2}\left[(y_0 + y_n) + 2(y_1 + y_2 + ... + y_{n-1})\right]$$

$$\boxed{\int_{x_0}^{x_0+nh} f(x)\,dx = \frac{h}{2}\left[(y_0 + y_n) + 2(y_1 + y_2 + y_3 + ... + y_{n-1})\right]}$$

This is known as trapezoidal rule.

Geometrical Proof :

The Simple Trapezoidal Rule is based on approximating $f(x)$ by the straight line joining$(x_1, \quad y_1)$ and (x_2, y_2)..........

Area of trapezium $= \dfrac{1}{2}$ (Sum of the parallel sides) $\times$ (Perpendicular distance between them)

Area of I trapezium $= \dfrac{1}{2}(y_1 + y_2)\,.\,h$

$$.....(1)$$

Area of II trapezium $= \dfrac{1}{2}(y_2 + y_3)\,.\,h$

$$.....(2)$$

Area of III trapezium $= \dfrac{1}{2}(y_3 + y_4)\,.\,h$

$$.....(3)$$

....

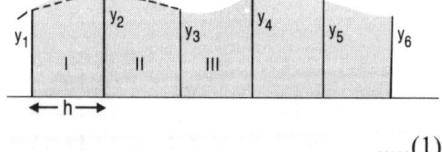

Area of $(n-1)$th trapezium $= \dfrac{1}{2} \{y_{n-1} + y_n\} \cdot h$

Adding (1), (2), (3) ....and so on, we get

Area $= \dfrac{h}{2} [(y_1 + y_n) + 2 (y_2 + y_3 + y_4 + \ldots\ldots\ldots + y_{n-1})]$ **Proved.**

Note. This formula gives the area of the trapezium ABCD.

Example 1. *Find from the following table, the area bounded by the curve and the x-axis from x = 0 to x = 4.*

x	0	1	2	3	4
y	1	2.72	7.39	20.09	54.60
	y_0	y_1	y_2	y_3	y_4

Solution. h = difference between the values of $x = 1$

Applying Trapezoidal Rule

$$\text{Area} = \int_0^4 y\, dx = \dfrac{h}{2} [(y_0 + y_4) + 2 (y_1 + y_2 + y_3)] = \dfrac{1}{2} [(1 + 54.60) + 2 (2.72 + 7.39 + 20.09)]$$

$$= \dfrac{1}{2} [55.60 + 2 (31.01)] = \dfrac{1}{2} (55.60 + 62.02) = 58.81 \text{ sq. units} \qquad \textbf{Ans.}$$

Example 2. *Evaluate* $\int_0^6 \dfrac{dx}{1+x^2}$ *by using Trapezoidal rule.*   *(Bihar SBTE, 2004, 2005, 2009)*

Solution.   Let $y = \dfrac{1}{1+x^2}$

x	0	1	2	3	4	5	6
$y = \dfrac{1}{1+x^2}$	1	0.5	0.2	0.1	0.06	0.04	0.03
	y_0	y_1	y_2	y_3	y_4	y_5	y_6

Here, $y_0 = 1$, $y_1 = 0.5$, $y_2 = 0.2$, $y_3 = 0.1$, $y_4 = 0.06$, $y_5 = 0.04$, $y_6 = 0.03$, $h = 1$

By using Trapezoidal Rule, we have

$$\int_0^6 \dfrac{dx}{1+x^2} = \dfrac{h}{2} [y_0 + y_6 + 2 (y_1 + y_2 + y_3 + y_4 + y_5)]$$

$$= \dfrac{1}{2} [i + 0.03 + 2 (0.5 + 0.2 + 0.1 + 0.06 + 0.04)]$$

$$= \dfrac{1}{2} [1.03 + 1 + 0.4 + 0.2 + 0.12 + 0.08] = \dfrac{1}{2} \times 2.83 = 1.415 \qquad \textbf{Ans.}$$

Example 3. *Calculate an approximate value of* $\int_0^{\frac{\pi}{2}} \sin x\, dx$ *by Trapezoidal rule.*

Solution.   We divide the range $\left(0, \dfrac{\pi}{2}\right)$ into ten equal parts, so that $h = \dfrac{\pi}{20}$, and form trigonometric tables :

x	0	$\dfrac{\pi}{20}$	$\dfrac{\pi}{10}$	$\dfrac{3\pi}{20}$	$\dfrac{\pi}{5}$	$\dfrac{\pi}{4}$	$\dfrac{3\pi}{10}$	$\dfrac{7\pi}{20}$	$\dfrac{2\pi}{5}$	$\dfrac{9\pi}{20}$	$\dfrac{\pi}{2}$
$\sin x$	0.0000	0.1564	0.3090	0.4540	0.5878	0.7071	0.8090	0.8910	0.9511	0.9877	1.0000
y	y_0	y_1	y_2	y_3	y_4	y_5	y_6	y_7	y_8	y_9	y_{10}

Using the Trapezoidal rule, approximate value of the integral

$$= h\left\{\frac{1}{2}(y_0 + y_{10}) + (y_1 + y_2 + \ldots + y_9)\right\}$$

$$= \frac{\pi}{20}\left[\frac{1}{2}(0+1) + 0.1564 + 0.3090 + 0.4540 + 0.5878 + 0.7071 + 0.8090 + 0.8910 + 0.9511 + 0.9877\right]$$

$$= \frac{\pi}{20}(0.5 + 5.8531) = 0.9979 \qquad\qquad (\pi = 3.141592654) \qquad\qquad \textbf{Ans.}$$

EXERCISE 7.1

1. Use Trapezoidal rule to evaluate $\int_0^1 x^3 dx$ considering five sub-intervals. **Ans.** 0.26

2. Compute the value of $\int_1^2 \dfrac{dx}{x}$ using Trapezoidal Rule, take $h = 0.25$. **Ans.** 0.6971

3. Evaluate $\int_1^2 \dfrac{dx}{1+x^2}$ taking h = 0.2, using Trapezoidal Rule. **Ans.** 0.32284

4. Evaluate $\int_0^1 e^x dx$ taking $h = 0.05$, using Trapezoidal rule. **Ans.** 1.713870

Choose the correct alternative :

5. The Trapezoidal rule integrates exactly polynomials of order

   (a) 1   (b) 2   (c) 3   (d) 4   **Ans.** (a)

6. Which of the following is a Compact Trapezoidal Rule ?

   (a) $\dfrac{h}{2}[f_0 + f_1 + f_2 + \ldots + f_N]$ (b) $\dfrac{h}{2}\left[f_0 + 2(f_1 + f_2 + \ldots + f_{N-1}) + f_N\right]$

   (c) $\dfrac{h}{2}\left[f_0 + 4(f_1 + f_3 + \ldots + f_{2N-1}) + 2(f_2 + f_4 + \ldots + f_{2N-2}) + f_{2N}\right]$

   (d) $\dfrac{h}{2}\left[f_0 + 4(f_1 + f_2 + \ldots + f_{2N-1}) + f_{2N}\right]$ **Ans.** (b)

7. Using Trapezoidal rule with only two sub-intervals, an approximate value for the integral

   $\int_0^{10} \sin(1 - 0.1x)dx$ is

   (a) 4.0   (b) 4.5   (c) 5.0   (d) 5.5   **Ans.** (b)

8. To apply trapezoidal rule the interval of interpolation is divided into ,............ number of sub-intervals.

   (a) any   (b) even   (c) odd   (d) multiple of 3   *(Bihar SBTE, 2010)* **Ans.** (a)

Fill up the blanks:

9. The concept of establishing the trapezoidal rule of integration is based on the fact that the whole bounded region is divided into a number of ....   (Trapezium / Rectangle)   **Ans.** Trapezium

10. Evaluate $\int_{-3}^3 x^4 dx$ by using trapazoidal rule.   *(Bihar SBTE, 2011)* **Ans.** 9.354

7.4 SIMPSON'S ONE THIRD RULE

(Bihar SBTE, 2003, 2009, 2010, 2011)

On putting $P = 2$ in Art. 7.2, the curve through (x_0, y_0), (x_1, y_1), and (x_2, y_2) is a parabola (polynomial of second degree).

Other integrals vanish.

$$\int_{x_0}^{x_0+2h} f(x)\,dx = h\left[2y_0 + \frac{(2)^2}{2}\Delta y_0 + \left(\frac{(2)^3}{6} - \frac{(2)^2}{4}\right)\Delta^2 y_0\right]$$

$$= h\left[2y_0 + 2(y_1 - y_0) + \left(\frac{4}{3} - 1\right)(\Delta y_1 - \Delta y_0)\right]$$

$$= h\left[2y_0 + 2y_1 - 2y_0 + \frac{1}{3}\{y_2 - y_1 - (y_1 - y_0)\}\right]$$

$$= h\left[2y_0 + 2y_1 - 2y_0 + \frac{1}{3}y_2 - \frac{1}{3}y_1 - \frac{1}{3}y_1 + \frac{1}{3}y_0\right] = h\left[\frac{1}{3}y_0 + \frac{4}{3}y_1 + \frac{1}{3}y_2\right]$$

$$\int_{x_0}^{x_0+2h} f(x)\,dx = \frac{h}{3}[y_0 + 4y_1 + y_2] \qquad\qquad \dots (i)$$

Similarly, $\int_{x_0+2h}^{x_0+4h} f(x)\,dx = \frac{h}{3}[y_2 + 4y_3 + y_4] \qquad\qquad \dots (ii)$

$$\int_{x_0+4h}^{x_0+6h} f(x)\,dx = \frac{h}{3}[y_4 + 4y_5 + y_6] \qquad\qquad \dots (iii)$$

....

....

$$\int_{x_0+(n-2)h}^{x_0+nh} f(x)\,dx = \frac{h}{3}\left[y_{n-2} + 4y_{n-1} + y_n\right] \qquad\qquad \dots\left(\frac{n}{2}\right)$$

Adding (i), (ii), (iii), ... $\left(\dfrac{n}{2}\right)$, we get

$$\int_{x_0}^{x_0+2h} f(x)\,dx + \int_{x_0+2h}^{x_0+4h} f(x)\,dx + \int_{x_0+4h}^{x_0+6h} f(x)\,dx + \dots + \int_{x_0+(n-2)h}^{x_0+nh} f(x)\,dx$$

$$= \frac{h}{3}\left[(y_0 + 4y_1 + y_2) + (y_2 + 4y_3 + y_4) + (y_4 + 4y_5 + y_6) + \dots + (y_{n-2} + 4y_{n-1} + y_n)\right]$$

$$\int_{x_0}^{x_0+nh} f(x)\,dx = \frac{h}{3}\left[(y_0 + y_n) + 2(y_2 + y_4 + y_6 + \dots y_{n-2}) + 4(y_1 + y_3 + y_5 + \dots y_{n-1})\right]$$

$$\int f(x)\,dx = \frac{h}{3}\left[(y_0 + y_n) + 2\Sigma y_e + 4\Sigma y_o\right]$$

This is known as the Simpson's one-third rule. It is mostly called simply Simpson's rule.

Geometrical Proof.

Assumption. Any curve can be built up from arcs of parabolas of vertical axis.

Proof. Let the given curve AG be

$$y = f(x) \qquad \qquad \qquad ...(1)$$

Draw $\perp$s AP and BO to x-axis.

Divide PU into even numbers of equal parts say h and draw ordinates $y_1, y_2, y_3, \ldots$

Let O be the origin. The co-ordinates of P and Q are $(-h, 0)$, $(h, 0)$.

$y = ax^2 + bx + c$ represents a parabola with its axis vertical.

$$\text{Area } APQC = \int_{-h}^{+h} y\, dx = \int_{-h}^{h} (ax^2 + bx + c)\, dx$$

$$= \left[\frac{ax^3}{3} + \frac{bx^2}{2} + cx \right]_{-h}^{h} = \frac{2h}{3}(ah^2 + 3c) \qquad \qquad ...(2)$$

$A(-h, y_0)$, $B(0, y_1)$, $C(h, y_2)$ lie on the parabola.

$\therefore$

$$y_0 = ah^2 - bh + c \qquad \qquad ...(3)$$
$$y_1 = c \qquad \qquad ...(4)$$
$$y_2 = ah^2 + bh + c \qquad \qquad ...(5)$$

On adding (3) and (5) $y_0 + y_2 = 2ah^2 + 2c$

$$y_0 + y_2 = 2ah^2 + 2y_1 \quad \Rightarrow \quad 2ah^2 = y_0 + y_2 - 2y_1$$

$$ah^2 = \frac{1}{2}(y_0 + y_2 - 2y_1)$$

Put the value of ah^2 and c in (2), we get

$$\text{Area } APQC = \frac{2h}{3}\left[\frac{1}{2}(y_0 + y_2 - 2y_1) + 3y_1 \right] = \frac{h}{3}[y_0 + y_2 - 2y_1 + 6y_1]$$

$$= \frac{h}{3}[y_0 + 4y_1 + y_2] \qquad \qquad ...(6)$$

Similarly the area of $CQSE = \frac{h}{3}[y_2 + 4y_3 + y_4] \qquad \qquad ...(7)$

Similarly the area of $ESUG = \frac{h}{3}[y_4 + 4y_5 + y_6]$ and so on $\qquad ...(8)$

Adding (6), (7) and (8), we get

$$\text{Whole Area } = \frac{h}{3}[(y_0 + 4y_1 + y_2') + (y_2 + 4y_3 + y_4) + (y_4 + 4y_5 + y_6) + \ldots]$$

$$= \frac{h}{3}[(y_0 + y_6) + 2(y_2 + y_4 + \ldots) + 4(y_1 + y_3 + y_5 + \ldots)]$$

$$\boxed{\text{Area } = \frac{h}{3}[X + 2E + 4O]}$$

Ans.

where $\qquad$ X = Sum of the first and the last ordinates.
$\qquad \qquad$ O = Sum of odd ordinates.

E = Sum of the even ordinates.

Note. The given interval AB must be divided into even number of equal sub-divisions while applying the Simpson's rule, since we find the integral, taking two strips at a time.

Example 4. *A curve is drawn to pass through the points given by the following table:*

x	1	1.5	2	2.5	3	3.5	4
y	2	2.4	2.7	2.8	3	2.6	2.1

Estimate the area bounded by the curve, x-axis and the lines x = 1, x = 4.

(R.G.P.V., Bhopal, III Semester, June 2007)

Solution. We have, $h = 1.5 - 1 = 0.5$

x	1	1.5	2	2.5	3	3.5	4
y	2	2.4	2.7	2.8	3	2.6	2.1
y	y_0	y_1	y_2	y_3	y_4	y_5	y_6

By Simpson's $\dfrac{1}{3}$ rule, we have

$$\text{Area} = \frac{h}{3} \left[(y_0 + y_6) + 2(y_2 + y_4) + 4(y_1 + y_3 + y_5) \right] \qquad \text{... (1)}$$

On putting the values of $h, y_0, y_1, y_2 \ldots\ldots y_6$ in (1), we get

$$\text{Area} = \frac{0.5}{3} \left[(2 + 2.1) + 2(2.7 + 3) + 4(2.4 + 2.8 + 2.6) \right]$$

$$= \frac{0.5}{3} \left[4.1 + 2(5.7) + 4(7.8) \right]$$

$$= \frac{0.5}{3} \left[4.1 + 11.4 + 31.2 \right]$$

$$= \frac{0.5 \times 46.7}{3} = 7.7833$$

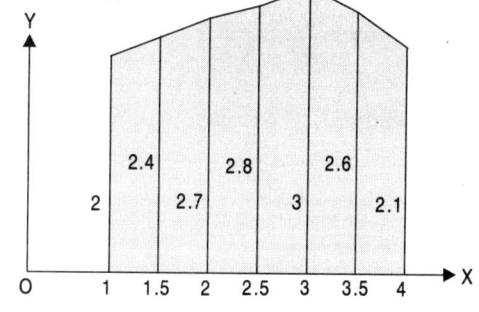

Hence, the required area is 7.7833 sq. units **Ans.**

Example 5. *Evaluate* $\int_0^4 e^x \, dx$, *by Simpson's rule, using data* $e = 2.72$, $e^2 = 7.39$, $e^3 = 20.09$, $e^4 = 54.60$, *and compare it with the actual value.*

Solution. We divide the whole range (0,4) into four equal parts by taking $h = 1$. By Simpson's one rule, we have

$$\int_0^4 e^x \, dx = \int_{x_0}^{x_0 + 4h} y \, dx = \frac{h}{3} \left[y_0 + y_4 + 4(y_1 + y_3) + 2y_2 \right]$$

$$= \frac{1}{3} \left[1 + 54.60 + 4(2.72 + 20.09) + 2(7.39) \right] = \frac{1}{3} \left[55.60 + 91.24 + 14.78 \right]$$

$$= \frac{1}{3} \left[161.62 \right] = 53.873 \qquad \qquad \textbf{Ans.}$$

The actual value of the integral is

$$\int_0^4 e^x \, dx = e^4 - e^0 = 54.60 - 1.00 = 53.60$$

Example 6. *Calculate by Simpson's Rule an approximate value of $\int_{-3}^{+3} x^4 \, dx$ by taking seven equidistant intervals.*

Solution. The seven equidistant parts are $-3, -2, -1, 0, 1, 2, 3$ and the length of each interval is 1. The values of the ordinates are obtained by putting these values in the integrand x^4.

x	-3	-2	-1	0	1	2	3
x^4	81	16	1	0	1	16	81
y	y_0	y_1	y_2	y_3	y_4	y_5	y_6

$$\int_{x_0}^{x_0+4h} y \, dx = \frac{h}{3}\left[(y_0+y_6)+2(y_2+y_4)+4(y_1+y_3+y_5)\right]$$

$$\int_{-3}^{3} x^4 \, dx = \frac{1}{3}\left[(81+81)+2(1+1)+4(16+0+16)\right]$$

$$= \frac{1}{3}[162+4+128] = \frac{1}{3} \times 294 = 98$$

Hence, the approximate value of $\int_{-3}^{3} x^4 \, dx$ is 98.

Exact value $= \int_{-3}^{3} x^4 \, dx = \left[\frac{x^5}{5}\right]_{-3}^{3} = \frac{1}{5}[243+243] = \frac{1}{5} \times 486 = 97.2$ **Ans.**

Example 7. *Show that $\int_{0}^{1} \frac{dx}{1+x} = \log_e 2 = 0.69315$.* *(Bihar SBTE, 2014)*

Solution. We divide the whole range (0, 1) into 10 equal parts making use of Simpson's $\left(\frac{1}{3}\right)'$ rule, we have

x	0	$\frac{1}{10}$	$\frac{2}{10}$	$\frac{3}{10}$	$\frac{4}{10}$	$\frac{5}{10}$	$\frac{6}{10}$	$\frac{7}{10}$	$\frac{8}{10}$	$\frac{9}{10}$	1
$\frac{1}{1+x}$	1	$\frac{10}{11}$	$\frac{10}{12}$	$\frac{10}{13}$	$\frac{10}{14}$	$\frac{10}{15}$	$\frac{10}{16}$	$\frac{10}{17}$	$\frac{10}{18}$	$\frac{10}{19}$	$\frac{1}{2}$
y	y_0	y_1	y_2	y_3	y_4	y_5	y_6	y_7	y_8	y_9	y_{10}

$$\int_{0}^{1} \frac{dx}{1+x} = \int_{0}^{x_0+10h} y \, dx = \frac{h}{3}\left[(y_0+y_{10})+4(y_1+......y_9)+2(y_2+....+y_8)\right]$$

$$= \frac{1}{30}\left[1+\frac{1}{2}+4\left(\frac{10}{11}+\frac{10}{13}+\frac{10}{15}+\frac{10}{17}+\frac{10}{19}\right)+2\left(\frac{10}{12}+\frac{10}{14}+\frac{10}{16}+\frac{10}{18}\right)\right] \quad \left[\text{Here } h = \frac{1}{10}\right]$$

$$= \frac{1}{30}[1.5+4 \times 3.45954+2 \times 2.72817] = \frac{1}{30}[20.7945] = 0.69315 \quad .... (1)$$

$$= 0.69315 \quad \text{upto five decimal places}$$

Again $\int_{0}^{1} \frac{1}{1+x} dx = \left[\log(1+x)\right]_{0}^{1} = \log 2$ ... (2)

From (1) and (2), we have

$$\log_e 2 = 0.69315$$ **Proved.**

Example 8. *Evaluate the value of $\log_e 2$ by finding $\int_0^1 \frac{2x}{1+x^2} dx$,*

using Simpson's rule, by dividing the interval into four equal parts.

Solution. $\int_0^1 \frac{2x}{1+x^2} dx = \left[\log_e (1+x^2) \right]_0^1 = \log_e 2.$

$$f(x) = \frac{2x}{1+x^2}.$$

The following is the table of values of x and $f(x)$.

x	0	0.25	0.50	0.75	1.00
$f(x) = \dfrac{2x}{1+x^2}$	0	0.47	0.80	0.96	1.00
y	y_0	y_1	y_2	y_3	y_4

Here $h = 0.25$. Applying Simpson's one third rule,

$$\int_0^1 \frac{2x}{1+x^2} dx = \frac{h}{3} [(y_0 + y_4) + 2y_2 + 4(y_1 + y_3)]$$

$$\left[\log (1+x^2) \right]_0^1 = \frac{0.25}{3} [(0 + 1) + 2 \times 0.8 + 4 (0.47 + 0.96)]$$

Hence, $\log_e 2 = 0.69$ **Ans.**

Example 9. *Evaluate* $\int_0^6 \frac{dx}{1+x^2}$ *by using Simpson's $\left(\frac{1}{3}\right)$ rule. Hence obtain the*

approximate value of π dividing the range into 6 equal parts.

(Bihar, SBTE, 2009, 2004, 2005)

Solution.   We divide the range of integration into 6 equal parts by taking $h = \dfrac{6-0}{6} = 1.$ Now, the values

of the given function $y = \left\{ \dfrac{1}{1+x^2} \right\}$ is given as below for each point of sub-division.

x	0	1	2	3	4	5	6
$\dfrac{1}{1+x^2}$	1	0.500	0.200	0.100	0.058	0.038	0.027

By Simpson's one third Rule

$$\int_0^6 \frac{1}{1+x^2} dx = \frac{h}{3} [(y_0 + y_6) + 4 (y_1 + y_3 + y_5) + 2 (y_2 + y_4)]$$

$$= \frac{1}{3} \ [(1 + 0.027) + 4 \ (0.5 + 0.1 + 0.038) + 2 \ (0.2 + 0.058)]$$

$$= \frac{1}{3} \ [1.027 + 4 \ (0.638) + 2 \ (0.258)]$$

$$= \frac{1}{3} \ [1.027 + 2.552 + 0.516]$$

$$= \frac{1}{3} \ (4.095) = 1.365 \qquad\qquad\qquad \textbf{Ans.}$$

Example 10. *Calculate an approximate value of $\int_0^{\frac{\pi}{2}} \sin x \, dx$ by Simpson's Rule using 11 ordinates.*

Solution. We divide the range $\left(0, \frac{\pi}{2}\right)$ into ten equal parts, so that $h = \frac{\pi}{20}$, and form trigonometric tables:

x	0	$\frac{\pi}{20}$	$\frac{\pi}{10}$	$\frac{3\pi}{20}$	$\frac{\pi}{5}$	$\frac{\pi}{4}$	$\frac{3\pi}{10}$	$\frac{7\pi}{20}$	$\frac{2\pi}{5}$	$\frac{9\pi}{20}$	$\frac{\pi}{2}$
$\sin x$	0.0000	0.1564	0.3090	0.4540	0.5878	0.7071	0.8090	0.8910	0.9511	0.9877	1.0000
y	y_0	y_1	y_2	y_3	y_4	y_5	y_6	y_7	y_8	y_9	y_{10}

Using Simpson's rule, the approximate value of the integral

$$= \frac{h}{3} \ [(y_0 + y_{10}) + 2(y_2 + y_4 + y_6 + y_8) + 4(y_1 + y_3 + y_5 + y_7 + y_9)]$$

$$= \frac{\pi}{60} \ [0 + 1 + 2 \ (0.3090 + 0.5878 + 0.8090 + 0.9511) + 4 \ (0.1564 + 0.4540 + 0.7071 + 0.8910 + 0.9877)]$$

$$= \frac{\pi}{60} \ [1 + 2 \ (2.6569) + 4 \ (3.1962)] \ = \ \frac{\pi}{60} \ [1 + 5.3138 + 12.7848]$$

$$= \frac{\pi}{60} \ [19.0986] = 1.000000358$$

Actual value of $\int_0^{\frac{\pi}{2}} \sin x \, dx = \left[-\cos x\right]_0^{\frac{\pi}{2}} = 1$

Here, error in calculation by Simpson's rule = 1.000000358 − 1 = 0.000000358 in excess.

Hence, error in calculation by trapezoidal rule = 1 − 0.9979 = 0.0021 by defection.

Therefore, there is a greater accuracy in the calculation by Simpson's rule. **Ans.**

EXERCISE 7.2

1. Using Simpson's $\left(\frac{1}{3}\right)$ rule to find the value of $\int_1^5 f(x) \, dx$ given

x	1	2	3	4	5
$f(x)$	10	50	70	80	100

2. Use Simpson's rule to prove that $\log_e 7$ is approximately 1.9587 using $\int_1^7 \frac{dx}{x}$. **Ans.** 1.9587

3. When a train is moving at 30 m/sec steam is shut of and brakes are applied. The speed of the train per second after t second is given by

Time (t)	0	5	10	15	20	25	30	35	40
Speed (v)	30	24	19.5	16	13.6	11.7	10.0	8.5	7.5

Using Simpson's rule, determine the distance moved by the train in 40 seconds. **Ans.** 606.66 m

4. Evaluate $\int_0^6 \frac{1}{(1+x)^2}\, dx$ by Simpson's $\left(\frac{1}{3}\right)$ rule. **Ans.** 0.9082

5. Apply Simpson's rule to find the value of $\int_0^2 \frac{dx}{1+x^3}$ dividing into 4 equal parts. **Ans.** 1.096

6. Find the value of log 3 from $\int_0^1 \frac{x^2}{1+x^3}\, dx$, using Simpson's $\frac{1}{3}$ rule by dividing the range into four equal parts.

 Ans. 0.23108

7. Calculate $\int_0^1 e^{-x}\, dx$ with 10 intervals by Trapezoidal and Simpson's methods.

 Ans. 0.6686, 0.6321

8. Evaluate $\int_1^2 \frac{\sin x}{x}\, dx$ taking 6 intervals **Ans.** 0.65901, 0.65933

9. The velocity v of a particle at distance s from a point on its path is given by the table below :

S in metre	0	10	20	30	40	50	60
V m/sec.	47	58	64	65	61	52	38

Estimate the time taken to travel 60 metres by using Simpson's one-third rule.

$$\left[\textbf{Hint.}\ v = \frac{ds}{dt}\right]$$ **Ans.** 1.0635166 sec;

10. Deduce Trapezoidal Rule & Simpson's $\frac{1}{3}$ rule for numerical integration. (*Bihar SBTE, 2011*)

Choose the correct answer:

11. Which of the following is a Composit Simpson's Rule?

(a) $\frac{h}{3}\left[f_0 + f_1 + f_2 + \ldots\ldots\ldots + f_N\right]$ (b) $\frac{h}{3}\left[f_0 + 4\,(f_1 + f_3 + \ldots\ldots + f_{2N-1}) + 2\,(f_2 + f_4 + \ldots\ldots + f_{2N-2}) + f_{2N}\right]$

(c) $\frac{h}{2}\left[f_0 + 2(f_1 + f_2 + \ldots\ldots + f_{N-1}) + f_N\right]$ (d) $\frac{h}{3}\left[f_0 + 4(f_1 + f_2 + \ldots\ldots + f_{2N-1}) + f_{2N}\right]$ **Ans.** (b)

12. In applying Simpson's 1/3 rule the integration range must be divided into .... sub-intervals.

(a) odd (b) even (c) zero (d) none (*Bihar SBTE, 2003*) **Ans.** (b)

13. In Simpson's $\frac{3}{8}$ rule, $f(x)$ is a polynomial of third degree *i.e.* $f(x) = ax^3 + bx^2 + cx + d$. To apply this rule, the number of intervals n must be a multiple of ....

(a) 2 (b) 3 (c) 5 (d) None of these

 (*Bihar SBTE, 2005*) **Ans.** (b)

14. To compute the integral $\int_{0.5}^{1.1} x^2 . dx$ by Trapezoidal rule we divide the interval (0.5, 1. 1) into n number of sub interval which must be multiple of :

   (a) 2   (b) 3   (c) 4   (d) 5   (*Bihar SBTE, 2009*) **Ans.** (*a*)

15. Numerical Integration of the polynomial of degree two by Simpson's $\frac{1}{3}$ rd rule will differ from its exact integration.

   (a) True   (b) False   (c) Can't say   (d) May or may not   **Ans.** (*d*)

16. The value of the integral $\int_{2}^{4} (x-2) (4 - x) (x^2-1)^{\frac{1}{3}} dx$ as evaluated by Simpson's rule with $h = 1$ is

   (a) 0   (b) 8/3   (c) 4/3   (d) None of these   **Ans.** (*b*)

17. The value of $\int_{0}^{6} f(x) dx$ by Simpson's rule using $f(0), f(3)$ and $f(6)$, where

   $f(x) = -4 + \left[25 - (x-3)^2 \right]^{\frac{1}{2}}$ is

   (a) $\frac{\pi}{3}$ (b) 4   (c) 6   (d) 10   **Ans.** (*b*)

18. The value of $\int_{2}^{6} x^3 dx$ by Simpson's Rule is

   (a) 56   (b) 80   (c) 60   (d) none of these   **Ans.** (*d*)

19. The value of $\int_{0}^{6} \frac{dx}{1+x^2}$ using Simpson's $\frac{1}{3}$ rd rule with six intervals is ....

   (a) 1.9292   (b) – 1.929   (c) 2.929   (d) 1.366   **Ans.** (*d*)

20. Simpson's rule is used to evaluate the integral $\int_{0}^{1} \frac{2x \, dx}{1+x^2}$

   If $h = \frac{1}{2}$ is used, then the value of the integral is

   (a) log 2   (b) $\frac{1}{2}$ (c) $\frac{7}{10}$ (d) $\frac{3}{10}$ **Ans.** (*c*)

21. By Simpson's Rule the value of the integral

   $\int_{1}^{3} \left\{ 5x^4 + \lambda (x-1) (x-2) (x-3) \right\} dx$ with $h = 1$ is

   (a) 242   (b) $\lambda + 242$ (c) $\frac{730}{3}$ (d) $\lambda + \frac{730}{3}$ **Ans.** (*c*)

22. Which method is used to evaluate $\int_{x_0}^{x_0+nh} f(x) dx$

   (a) Gauss-Jordan Method   (b) Euler's method

   (c) Simpson $\frac{1}{3}$ Rule   (d) Hungarian method

   (*Bihar SBTE, 2011*) **Ans.** (c)

Fill up the blanks:

23. In applying Simpson's $\frac{1}{3}$ rule the integration range must be divided into .... sub intervals.   (Odd/Even)

   Ans. Even

CHAPTER
8
Difference Equations

8.1 DIFFERENCE EQUATION

Difference equation is the equation between the differences of an unknown function.
For example;

$$\Delta y_n + 2 y_n = 0 \qquad \qquad ...(1)$$

$$\Delta^2 y_n + 5 \Delta y_n + 6y_n = 0 \qquad \qquad ...(2)$$

Second way to express the difference equation.

Putting the value of $\qquad \Delta = E - 1$

(1) becomes $\qquad (E-1)y_n + 2y_n = 0$

or $\qquad \qquad \qquad E y_n + y_n = 0 \qquad \qquad ...(3)$

(2) becomes $\qquad (E-1)^2 y_n + 5 (E-1) y_n + 6y_n = 0$

or $\qquad (E^2 - 2E + 1) y_n + 5 (E-1) y_n + 6y_n = 0$

or $\qquad \qquad E^2 y_n + 3Ey_n + 2y_n = 0 \qquad \qquad ...(4)$

Third way

(3) can be written $y_{n+1} + y_n = 0 \qquad \qquad ...(5)$

(4) can be written $y_{n+2} + 3y_{n+1} + 2y_n = 0 \qquad \qquad ...(6)$

8.2 ORDER OF A DIFFERENCE EQUATION = HIGHEST POWER OF E

Order of a difference equation is the difference between the largest and the smallest arguments involved in the difference equation, divided by the unit of interval.

$$\text{Thus the order of (5)} = \frac{\text{Largest argument} - \text{smallest argument}}{\text{Unit of interval}} = \frac{(n+1) - n}{1} = 1$$

$$\text{Similarly,} \quad \text{order of (6)} = \frac{(n+2) - n}{1} = 2$$

Example 1. *From the equation* $y_n = A . 2^n + B . 3^n$, *derive a difference equation by eliminating the arbitrary constants A and B. What is the order of the difference equation ?*

Solution. $\quad y_n = A . 2^n + B . 3^n \qquad \qquad ...(1)$

$$y_{n+1} = A . 2^{n+1} + B . 3^{n+1} = 2A . 2^n + 3B . 3^n \qquad \qquad ...(2)$$

$$y_{n+2} = A . 2^{n+2} + B . 3^{n+2} = 4A . 2^n + 9B . 3^n \qquad \qquad ...(3)$$

Eliminating A and B from (1), (2) and (3), we get

$$\begin{vmatrix} y_n & 1 & 1 \\ y_{n+1} & 2 & 3 \\ y_{n+2} & 4 & 9 \end{vmatrix} = 0$$

$\Rightarrow$ $6y_n - 5y_{n+1} + y_{n+2} = 0$

$\Rightarrow$ $y_{n+2} - 5y_{n+1} + 6y_n = 0$

which is the desired equation.   **Ans.**

This equation can also be written as

$$(E^2 - 5E + 6)\, y_n = 0$$

The highest power of E is the order of the equation.

Here the equation is of second order.   **Ans.**

8.3 SOLUTION OF A DIFFERENCE EQUATION

The complete solution of a difference equation

= complementary function + particular integral

$$y_n = C.F. + P.I.$$

8.4 COMPLEMENTARY FUNCTION

(i) Write down the difference equation in terms of E.

i.e. $a_0 E^2 y_n + a_1\, E y_n + a_2 y_n = 0$

(ii) Write down the *auxiliary equation* as $f(E) = 0$, solve $f(E) = 0$

i.e. $(a_0 E^2 - a_1\, E + a_2) = 0$

The rules of obtaining the complementary function depend upon the nature of the roots of the auxiliary equation.

Case 1. *Roots are real and different* say $m_1 . m_2$

Then $C.F. = C_1 (m_1)^n + C_2 (m_2)^n$

Case 2. *Roots are equal* m_1 , m_1 .

Then $C.F. = (C_1 + C_2\, n)\, (m_1)^n$

Case 3. *Roots are imaginary* say $\alpha \pm i\beta$

Then $C.F. = r^n [C_1 \cos n\theta + C_2 \sin n\theta]$

where $r = \sqrt{\alpha^2 + \beta^2}, \theta = \tan^{-1} \dfrac{\beta}{\alpha}$

Example 2. *Solve* $(\Delta^2 - 3\Delta + 2)\, y_n = 0$.

Solution. $(\Delta^2 - 3\Delta + 2)\, y_n = 0$

$\Delta = E - 1$ (By definition)

The given equation becomes

$[(E-1)^2 - 3(E-1) + 2]\, y_n = 0$

$\Rightarrow$ $[E^2 - 5E + 6]\, y_n = 0$

A.E. is $m^2 - 5m + 6 = 0 \Rightarrow (m-3)(m-2) = 0 \Rightarrow m = 3, 2$

$$C.F. = C_1 \, 2^n + C_2 \, 3^n$$

$$y_n = C_1 \, 2^n + C_2 \, 3^n \qquad \qquad \textbf{Ans.}$$

Example 3. *Solve* $\quad (E^2 + 6E + 9) \, y_n = 0.$

Solution. $(E^2 + 6E + 9) \, y_n = 0$

A.E. is $\qquad m^2 + 6m + 9 = 0 \implies (m+3)^2 = 0 \qquad \implies \qquad m = -3, -3$

$$C.F. = (C_1 + C_2 \, n)(-3)^n$$

$$y_n = (C_1 + C_2 \, n)(-3)^n \qquad \qquad \textbf{Ans.}$$

Example 4. *Show that the solution of the difference equation* $y_{n+2} + 2y_{n+1} + 4y_n = 0$ *is*

$$y_n = 2^n \, A \cos \frac{2n\,\pi}{3} + 2^n \, B \sin \frac{2n\,\pi}{3}$$

where A and B are arbitrary constants.

Solution. $y_{n+2} + 2y_{n+1} + 4y_n = 0$

$\implies \qquad E^2 y_n + 2Ey_n + 4y_n = 0 \qquad \implies \qquad (E^2 + 2E + 4) \, y_n = 0$

A.E. is $\qquad m^2 + 2m + 4 = 0$

$$m = \frac{-2 \pm \sqrt{4-16}}{2} = -1 \pm \sqrt{3} \, i$$

$\therefore \qquad y_n = r^n \, [A \cos n\theta + B \sin n\theta] \quad \text{where} \quad r = \sqrt{(-1)^2 + (\sqrt{3})^2} = 2$

$$\theta = \tan^{-1}\left(\frac{\sqrt{3}}{-1}\right) = \frac{2\pi}{3}$$

$\therefore \qquad y_n = 2^n \left[A \cos \frac{2n\pi}{3} + B \sin \frac{2n\pi}{3} \right] \qquad \qquad \textbf{Proved.}$

Example 5. *Solve* $y_{n+2} - y_{n+1} + y_n = 0,$ *given that* $y_0 = 1$ *and* $y_1 = \dfrac{1 + \sqrt{3}}{2}.$

(*A.M.I.E., Summer 2004*)

Solution. $y_{n+2} - y_{n+1} + y_n = 0.$

The above equation in the operator form is

$$E^2 y_n - E y_n + y_n = 0 \qquad \implies \qquad (E^2 - E + 1) \, y_n = 0$$

A.E. is $\qquad m^2 - m + 1 = 0 \qquad \implies \qquad m = \dfrac{1 \pm \sqrt{1-4}}{2} \implies m = \dfrac{1}{2} \pm \dfrac{\sqrt{3}}{2} \, i$

$\therefore \qquad C.F. = r^n \, (A \cos n\theta + B \sin n\theta), \quad \text{where} \quad r = \sqrt{\left(\dfrac{1}{2}\right)^2 + \left(\dfrac{\sqrt{3}}{2}\right)^2} = 1$

$$\theta = \tan^{-1} \frac{\sqrt{3}}{1} = \frac{\pi}{3}$$

$$y_n = (1)^n \left(A \cos \frac{n\,\pi}{3} + B \sin \frac{n\,\pi}{3} \right) \qquad \qquad \dots(1)$$

Initial conditions $\qquad y_0 = 1 \quad \text{and} \qquad y_1 = \dfrac{1 + \sqrt{3}}{2} \qquad \qquad \text{Put } n = 0 \text{ in (1)}$

$\therefore \qquad y_0 = A \qquad \qquad \therefore \qquad 1 = A \qquad \qquad \text{Put } n = 0 \text{ in (1)}$

$\therefore$ $y_1 = A \cos \dfrac{\pi}{3} + B \sin \dfrac{\pi}{3}$

$$\frac{1+\sqrt{3}}{2} = 1 \times \frac{1}{2} + B \frac{\sqrt{3}}{2} \Rightarrow B = 1$$

Putting the values of A and B in (1), the required solution is

$$y_n = \cos \frac{n\pi}{3} + \sin \frac{n\pi}{3}$$

                                                          Ans.

Example 6. *Solve the difference equation* $2y_{n+2} - 5y_{n+1} + 2y_n = 0$ *subject to the conditions*

$$y_0 = 0 \quad and \quad y_1 = 1$$

Solution. $2y_{n+2} - 5y_{n+1} + 2y_n = 0 \quad \Rightarrow \quad 2E^2 y_n - 5E y_n + 2 y_n = 0 \quad \Rightarrow \quad (2E^2 - 5E + 2) y_n = 0$

A.E. is $2m^2 - 5m + 2 = 0 \quad \Rightarrow \quad (2m - 1)(m - 2) = 0 \quad \Rightarrow \quad m = 2, \dfrac{1}{2}$

C. F is $y_n = C_1 (2)^n + C_2 \left(\dfrac{1}{2}\right)^n$ ... (1)

Put $n = 0$ and $y_n = 0$ in (1), we have $0 = C_1 + C_2$ ... (2)

Put $n = 1$, $y_n = 1$ in (1), we get $1 = 2C_1 + \dfrac{1}{2} C_2$ ... (3)

On solving (2) and (3), we get $C_1 = \dfrac{2}{3}$, $C_2 = -\dfrac{2}{3}$

On putting the values of C_1 and C_2, we get $y_n = \dfrac{2}{3} \cdot 2^n - \dfrac{2}{3} \cdot 2^{-n}$

$$y_n = \frac{2}{3}(2^n - 2^{-n})$$

                                                          Ans.

Example 7. *Solve the following difference equations (any two)*

$$\left(E - 2\cos\alpha + \frac{1}{E}\right) y_n = 0$$

                                               (Bihar SBTE, 2012)

Solution.   Here we have,

$$\left(E - 2\cos\alpha + \frac{1}{E}\right) y_n = 0 \qquad \Rightarrow \qquad (E^2 - 2E \cos \alpha + 1) y_n = 0$$

A.E is $E^2 - 2E \cos \alpha + 1 = 0$

$$E = \frac{2\cos\alpha \pm \sqrt{4\cos^2\alpha - 4}}{2} = \cos\alpha \pm \sqrt{\cos^2\alpha - 1} = \cos\alpha + i \sin\alpha$$

C.F. $= y_n = C_1 \cos n\alpha + C_2 \sin n\alpha.$ **Ans.**

EXERCISE 8.1

Solve the following difference equations :

     1.   Form the difference equation, given $y_n = (A_n + B) 2^n$.   *(A.M.I.E., Summer 2001)*

                                                            Ans. $y_{n-2} - 4y_{n+1} + 4y_n = 0$

     2. $(E^2 - 8E + 12) y_n = 0$ *(A.M.I.E., Summer 2002)*   **Ans.** $y_n = C_1 (2)^n + C_2 (6)^n$

     3. $\Delta^2 u_n - 2\Delta u_n + u_n = 0$ **Ans.** $u_n = (c_1 + c_2 n) 2^n$

4. $u_{n+3} - 3u_{n+2} + 4u_n = 0$ **Ans.** $u_n = c_1(-1)^n + (c_2 + c_3 n)(2)^n$

5. $y_{n+3} - 2y_{n+2} - 5y_{n+1} + 6y_n = 0$ (*A.M.I.E., winter 2001*)   **Ans.** $y_n = A + B(-2)^n + C(3)^n$

6. $u_{n+3} - 2u_{n+2} - 5u_{n+1} + 6u_n = 0$ (*A.M.I.E., Summer 2001*)   **Ans.** $u_n = c_1(1)^n + c_2(-2)^n + c_3(3)^n$

7. $y_{n+2} + a y_{n+1} + b y_n = 0$, Given that $y_0 = 0, y_1 = 1, y_2 = y_3 = 2$ **Ans.** $y_n = 2^{n/2} \sin\dfrac{n\pi}{4}$.

8. $y_{n+2} - 2y_{n+1} - 8y_n = 0$ subject to the conditions $y_0 = 0, y_1 = 1$.   **Ans.** $y_n = \dfrac{1}{6}[(4)^n - (-2)^n]$

Choose the correct alternative:

9. The order of difference equation

$\Delta y_{n+1} + 4\,\Delta y_n = 0$ is

(a) 2   (b) 3   (c) 4   (d) None of these   (*Bihar SBTE, 2011*) **Ans.** (a)

10. What is the order of difference equation $4y_n + 4y_n = 0$?

(a) 0   (b) 1   (c) 2   (d) None of these   (*Bihar SBTE, 2009*) **Ans.** (a)

11. Order of difference equation $y_{n+2} + 3y_{n+1} + 2y_n = 0$ is ...

(a) 0   (b) 1   (c) 2   (d) None of these   (*Bihar SBTE, 2014*) **Ans.** (c)

8.5 PARTICULAR INTEGRAL

Consider the difference equation $a_0 E^2 y_n + a_1 E y_n + a_2 y_n = \phi(n)$

$\Rightarrow$ $(a_0 E^2 + a_1 E + a_2) y_n = \phi(n)$

$\Rightarrow$ $f(E) y_n = \phi(n) \Rightarrow \text{P.I} = \dfrac{1}{f(E)} . \phi(n)$

Case I. When $\phi(n) = a^n$

(a) P.I. $= \dfrac{1}{f(E)} a^n$ Put $E = a$

$= \dfrac{a^n}{f(a)}$ if $f(a) \neq 0$

(b) If $f(a) = 0$

(i) $(E - a) y_n = a^n$

P.I. $= \dfrac{1}{E - a} . a^n = n a^{n-1}$

(ii) $(E - a)^2 y_n = a^n$

P.I. $= \dfrac{1}{(E-a)^2} . a^n = \dfrac{n(n-1)}{2!} a^{n-2}$

Example 8. *Solve the following difference equation:*

$y_{n+2} + 7 . y_{n+1} + 12y_n = 5^n.$ (*Bihar SBTE, 2009*)

Solution.   The given equation in symbolic form is

$(E^2 + 7E + 12) y_n = 5^n$

$\therefore$ The auxiliary equation is $m^2 + 7m + 12 = 0$

$\Rightarrow$ $m^2 + 3m + 4m + 12 = 0$

$$\Rightarrow \qquad m(m+3)+4(m+3)=0$$
$$\Rightarrow \qquad (m+3)(m+4)=0$$
$$\therefore \qquad m=-3,-4$$
$$\therefore \qquad C.F.= C_1(-3)^n + C_2(-4)^n$$

and

$$P.I. = \frac{1}{f(E)}\, 5^n = \frac{1}{E^2+7E+12}.5^n$$

$$= \frac{1}{(5)^2+7(5)+12}.5^n$$

$$= \frac{1}{25+35+12}.5^n = \frac{1}{72}5^n$$

Thus the complete solution is $= C.F.+P.I. \qquad y_n = [C_1(-3)^n + C_2.(-4)^n] + \dfrac{5^n}{72}$ **Ans.**

Example 9.   *Solve the difference equation*
$$u_{n+2} \quad -7u_{n+1}+10u_n = 12e^{3n}+4^n.$$

Solution. $u_{n+2} \quad -7u_{n+1}+10u_n = 12e^{3n}+4^n.$
$$(E^2-7E+10)u_n = 12e^{3n}+4^n$$

A.E. is $\qquad m^2-7m+10=0 \;\Rightarrow\; (m-5)(m-2)=0 \Rightarrow m=2,5$
$$C.F. = C_1(2)^n + C_2(5)^n$$

$$P.I. = \frac{1}{E^2-7E+10}(12e^{3n}+4^n) = 12\frac{1}{E^2-7E+10}e^{3n} + \frac{1}{E^2-7E+10}.4^n$$

$$=12\frac{e^{3n}}{e^6-7e^3+10} + \frac{4^n}{(4)^2-7(4)+10} = 12\frac{e^{3n}}{e^6-7e^3+10} - \frac{4^n}{2}$$

Complete solution is : $\qquad u_n = C_1(2)^n + C_2(5)^n + \dfrac{12e^{3n}}{e^6-7e^3+10} - 2^{2n-1}$ **Ans.**

Example 10. *Solve the following difference equations:* $(E^2-4E+3)\,y_n = 3^n$ \hfill *(Bihar SBTE, 2014)*

Solution.   A.E. is $m^2-4m+3=0$
$$\Rightarrow (m-1)(m-3)=0$$
$$\Rightarrow m=1,3$$
$$C.F.=C_1(1)^n+C_2(3)^n$$

$$P.I. = \frac{1}{E^2-4E+3}3^n = \frac{1}{(E-1)(E-3)}3^n = \frac{1}{(3-1)(E-3)}3^n = \frac{1}{2(E-3)}3^n = \frac{1}{2}n\,e^{n-1}$$

$$y_n = C_1(1)^n + C_2(3)^n + \frac{n}{2}.3^{n-1}$$ **Ans.**

Example 11.   *Solve* $u_{n+2}-4u_{n+1}+4u_n = 2^n.$ \hfill *(Bihar SBTE, 2012, 2010)*

Solution. $u_{n+2}-4u_{n+1}+4u_n = 2^n$
$$(E^2-4E+4)u_n = 2^n$$

A.E. is $\qquad m^2-4m+4=0$

$\Rightarrow$ $$(m-2)^2 = 0$$

$\Rightarrow$ $$m = 2, 2$$

$$\text{C.F.} = (C_1 + C_2 n) \cdot 2^n$$

$$\text{P.I.} = \frac{1}{E^2 - 4E + 4} \cdot 2^n = \frac{1}{(E-2)^2} \cdot 2^n = \frac{n(n-1) \cdot 2^{n-2}}{2!} = n(n-1) \cdot 2^{n-3}$$

Complete solution is : $$u_n = (C_1 + C_2 n) 2^n + n(n-1) \cdot 2^{n-3}$$ **Ans.**

Example 12. *Solve the following difference equation*

$$(E-5)(E-7)(E+6)(E-1) = 2^n$$ *(Bihar SBTE, 2012)*

Solution. $(E-5)(E-7)(E+6)(E-1) = 2^n$

A.E. is $(m-5)(m-7)(m+6)(m-1) = 0$

$\Rightarrow$ $m = 5, 7, -6, 1$

$$\text{C.F.} = C_1(5)^n + C_2(7)^n + C_3(-6)^n + C_4(1)^n$$

$$\text{P.I} = \frac{1}{(E-5)(E-7)(E+6)(E-1)} 2^n = \frac{1}{(2-5)(2-7)(2+6)(2-1)} 2^n = \frac{1}{-3 \times -5 \times 8 \times 1} 2^n = \frac{1}{120} 2^n$$

$$y_n = C_1(5)^n + C_2(7)^n + C_3(-6)^n + C_4(1)^n + \frac{1}{120} 2^n$$ **Ans.**

Example 13. *Solve the difference equation* $y_{n+2} - 4y_n = 2^n$

Solution. $y_{n+2} - 4y_n = 2^n$

Given equation in symbolic form is $(E^2 - 4) y_n = 2^n$

$\therefore$ The auxiliary equation is $m^2 - 4 = 0$, $m = +2, -2$

$\therefore$ $$\text{C. F.} = C_1(2)^n + C_2(-2)^n$$

$$\text{P. I.} = \frac{1}{E^2 - 4} 2^n = \frac{1}{4}\left[\frac{1}{E-2} - \frac{1}{E+2}\right] 2^n = \frac{1}{4}\left[\frac{1}{E-2} 2^n - \frac{1}{E+2} 2^n\right]$$

$$= \frac{1}{4}\left[n \cdot 2^{n-1} - \frac{2^n}{2+2}\right] = \frac{1}{4}\left[n 2^{n-1} - \frac{1}{2} 2^{n-1}\right] = \frac{1}{8}(2n-1) 2^{n-1}$$

The complete solution is

$$y_n = C_1(2)^n + C_2(-2)^n + (2n-1) 2^{n-4}$$ **Ans.**

Example 14. *Solve* $(y_{n+2} - 4y_{n+1} + y_n) = 5 e^x$ *(Bihar SBTE, 2014)*

Solution. $E^2 y_n - 4E y_n + y_n = 5e^x$

A.E. is $m^2 - 4m + 1 = 0$

$$m = \frac{4 \pm \sqrt{16-4}}{2}$$

C.F. $= C_1(2 + \sqrt{3})^n + C_2(2 - \sqrt{3})$

$$= \frac{1}{E^2 - 4E + 1} 5 = \frac{1}{E^2 - 4E + 1} 5(1)^n = 5 \frac{1}{1-4+1} (1)^n = -\frac{5}{2}$$

$$y_n = C_1(2 + \sqrt{3})^n + C_2(2 - \sqrt{3})^n - \frac{5}{2}$$ **Ans.**

EXERCISE 8.2

Solve the following difference equations :

1. $y_{n+2} - 4y_{n+1} + 3y_n = 5^n$ **Ans.** $c_1 + c_2 . 3^n + \dfrac{5^n}{8}$

2. $(E^3 - 3E^2 + 4)y_n = 2^n$ **Ans.** $c_1(-1)^n + (c_2 + c_3 n)(2)^n + \dfrac{n(n-1)}{6}.2^{n-2} + \dfrac{2^n}{27} - \dfrac{n 2^{n-1}}{9}$

3. $y_{n+2} - 4y_{n+1} + 3y_n = 7^n$ **Ans.** $y_n = c_1(3)^n + c_2(1)^n + \dfrac{7^n}{24}$

4. $(E^2 - 4E + 4)y_n = 3^n$ **Ans.** $(c_1 + n c_2) 2^n + 3^n$

5. $y_{n+2} - 5y_{n+1} - 6y_n = (12)^n$ (*Bihar SBTE, 2012*) **Ans.** $C_1(-1)^n + C_2(6)^n + \dfrac{12^n}{78}$

6. $u_{n+2} - 3u_{n+1} - 4u_n = 3^n$ **Ans.** $c_1(-1)^n + c_2(4)^n - \dfrac{1}{4}.3^n$

7. $y_{n+2} - 6y_{n+1} + 8y_n = 2^n$, (*Nagpur, Summer 2001*) **Ans.** $c_1(2)^n + c_2(4)^n - \dfrac{1}{36}2^n - \dfrac{n}{6}.2^{n-1}$

8. Obtain the function whose difference is $x^2 - 3x + 2$ (*Nagpur, Summer 2001*) **Ans.** $\dfrac{x}{3}(x^2 - 6x + 11)$

9. Complementary function of difference equation $(E^2 - 1) y_n = 0$ is ....
   (a) (1, –1) (b) (2, 1) (c) (1, 1) (d) (1, 0)   (*Bihar SBTE, 2012*) **Ans.** (d)

Case II. When $\phi(n) = \sin kn$

$$\text{P. I.} = \frac{1}{f(E)} \sin kn = \frac{1}{f(E)}\left(\frac{e^{ikn} - e^{-ikn}}{2i}\right) = \frac{1}{2i}\left[\frac{1}{f(E)}.e^{ink} - \frac{1}{f(E)}.e^{-ikn}\right] = \frac{1}{2i}\left[\frac{e^{ink}}{f(e^{ik})} - \frac{e^{-ink}}{f(e^{-ik})}\right]$$

as we have in case **1**.

Example 15. Solve $u_{x+2} - 7u_{x+1} + 12u_x = \cos x$.

Solution. $u_{x+2} - 7u_{x+1} + 12u_x = \cos x$

$\Rightarrow$ $E^2 u_x - 7 E u_x + 12 u_x = \cos x$

$\Rightarrow$ $[E^2 - 7E + 12] u_x = \cos x$

A. E. is $m^2 - 7m + 12 = 0 \Rightarrow (m-3)(m-4) = 0 \Rightarrow m = 3, 4$

C. F. = $C_1 (3)^x + C_2 (4)^x$

$$\text{P. I.} = \frac{1}{E^2 - 7E + 12} \cos x = \frac{1}{E^2 - 7E + 12} \frac{e^{ix} + e^{-ix}}{2} = \frac{1}{2}\left[\frac{1}{E^2 - 7E + 12}e^{ix} + \frac{1}{E^2 - 7E + 12}.e^{-ix}\right]$$

$$= \frac{1}{2}\left[\frac{e^{ix}}{e^{2i} - 7e^i + 12} + \frac{e^{-ix}}{e^{-2i} - 7e^{-i} + 12}\right]$$

$u_x = $ C.F. + P.I.

$$u_x = C_1(3)^x + C_2(4)^x + \frac{1}{2}\left[\frac{e^{ix}}{e^{2i} - 7e^i + 12} + \frac{e^{-ix}}{e^{-2i} - 7e^{-i} + 12}\right]$$ **Ans.**

Example 16. Solve $y_{x+2} + 5y_{n+1} + 6y_n = n + 2^n$ (*Nagpur University, Summer 2002*)

Solution. We have, $(E^2 + 5E + 6) y_n = n + 2^n$

The auxiliary equation is $m^2 + 5m + 6 = 0$

$\Rightarrow$ $(m + 2)(m + 3) = 0$ $\Rightarrow$ $m = -2, -3$

C.F. $= c_1 (-2)^n + c_2 (-3)^n$

P.I. $= \dfrac{1}{E^2 + 5E + 6} n + \dfrac{1}{E^2 + 5E + 6} 2^n = \dfrac{1}{(1+\Delta)^2 + 5 (1+\Delta) + 6} n + \dfrac{1}{2^2 + 5 (2) + 6} 2^n$

$= \dfrac{1}{\Delta^2 + 7\Delta + 12} n + \dfrac{1}{20} 2^n = \dfrac{1}{12} \left[1 + \left(\dfrac{\Delta^2 + 7\Delta}{12}\right)\right]^{-1} n^{(1)} + \dfrac{1}{20} 2^n$

$= \dfrac{1}{12} \left[1 - \left(\dfrac{\Delta^2 + 7\Delta}{12}\right) + ...\right] n^{(1)} + \dfrac{1}{20} 2^n = \dfrac{1}{12} \left[1 - \dfrac{7}{12} \Delta\right] n^{(1)} + \dfrac{1}{20} 2^n$

$= \dfrac{1}{12} \left(n - \dfrac{7}{12} \Delta\, n^{(1)}\right) + \dfrac{1}{20} 2^n = \dfrac{1}{12} \left(n - \dfrac{7}{12}\right) + \dfrac{1}{12} 2^n = \dfrac{1}{144} (12n - 7) + \dfrac{1}{20} 2^n$

Hence the complete solution is

$y_n = \text{C.F.} + \text{P.I.}$

$= C_1 (-2)^n + C_2 (-3)^n + \dfrac{1}{20} 2^n + \dfrac{1}{44} (12n - 7)$ **Ans.**

Example 17. *Solve the difference equation* $(\Delta^2 - 2\Delta + 1) u_n = n^2 + 4$ *(Nagpur University, Summer 2002)*

Solution. Here, we have $(\Delta^2 - 2\Delta + 1) u_n = n^2 + 4$

$\Rightarrow$ $[(E - 1)^2 - 2 (E - 1) + 1] u_n = n^2 + 4$

$\Rightarrow$ $[E^2 - 2E + 1 - 2E + 2 + 1] u_n = n^2 + 4$

$\Rightarrow$ $(E^2 - 4E + 4) u_n = n^2 + 4$

A.E. is $m^2 - 4m + 4 = 0$ $\Rightarrow$ $(m - 2)^2 = 0$ $\Rightarrow$ $m = 2, 2$

These roots are real but same

C.F. $= (C_1 + C_2 n) 2^n$

P.I. $= \dfrac{1}{E^2 - 4E + 4} (n^2 + 4) = \dfrac{1}{E^2 - 4E + 4} [(n (n-1) + n + 4]$

$= \dfrac{1}{(1+\Delta)^2 - 4 (1+\Delta) + 4} (n^{(2)} + n^{(1)} + 4) = \dfrac{1}{\Delta^2 - 2\Delta + 1} (n^{(2)} + n^{(1)} + 4)$

$= \dfrac{1}{(1 - 2\Delta + \Delta^2)} (n^{(2)} + n^{(1)} + 4) = [1 - 2 \Delta + \Delta^2]^{-1} (n^{(2)} + n^{(1)} + 4)$

$= [1 - (2 \Delta - \Delta^2)]^{-1} (n^{(2)} + n^{(1)} + 4) = [1 + (2 \Delta - \Delta^2) + (2 \Delta - \Delta^2)^2 + ...] (n^{(2)} + n^{(1)} + 4)$

$= [1 + 2 \Delta - \Delta^2 + 4 \Delta^2 - 4\Delta^3 + \Delta^4 + ...] (n^{(2)} + n^{(1)} + 4)$

$= [1 + 2 \Delta + 3\Delta^2 - 4\Delta^3 + ...] (n^{(2)} + n^{(1)} + 4)$

$= [(n^{(2)} + n^{(1)} + 4) + 2 (2n + 1) + 3 (2)]$

$= n (n - 1) + n + 4 + 4n + 2 + 6$

$= n^2 - n + n + 4 + 4n + 2 + 6$

$= n^2 + 4n + 12$

Thus, the complete solution is

$$y_n = \text{C.F.} + \text{P.I.}$$ **Ans.**

$$\Rightarrow \quad y_n = (c_1 + c_2\, n)\, 2^n + n^2 + 4n + 12$$

Example 18. *Solve:* $\quad (\Delta^2 + \Delta + 1)\, y_x = x^2$

Solution. $\qquad\qquad (\Delta^2 + \Delta + 1)\, y_x = x^2$

$$[(E-1)^2 + (E-1) + 1]\, y_x = x^2$$
$$[E^2 - E + 1]\, y_x = x^2$$

A.E. is $\qquad m^2 - m + 1 = 0 \Rightarrow m = \dfrac{1 \pm \sqrt{1-4}}{2} = \dfrac{1}{2} \pm i\,\dfrac{\sqrt{3}}{2}$

$$\text{C.F.} = r^n\, (C_1 \cos n\,\theta + C_2 \sin n\,\theta)$$

where $\qquad r = \sqrt{\dfrac{1}{4} + \dfrac{3}{4}} = 1\quad \theta = \tan^{-1} \dfrac{\sqrt{3}}{1} = \dfrac{\pi}{3}$

$$\text{C.F.} = \left(C_1 \cos \dfrac{n\pi}{3} + C_2 \sin \dfrac{n\pi}{3} \right)$$

$$\text{P.I.} = \dfrac{1}{\Delta^2 + \Delta + 1}\cdot x^2 = (1 + \Delta + \Delta^2)^{-1}\, x^2$$

$$= \left[1 - (\Delta + \Delta^2) + \dfrac{(-1)\,(-2)}{2}\,(\Delta + \Delta^2)^2 + \ldots\ldots \right][x\,(x-1) + x]$$

$$= [1 - \Delta - \Delta^2 + \Delta^2 + \ldots\ldots]\,\{[x]^2 + [x]\} = [1 - \Delta + \ldots\ldots\ldots]\,\{[x]^2 + [x]\}$$

$$= [[x]^2 + [x] - 2\,[x] - 1] = [x\,(x-1) + x - 2x - 1] = [x^2 - 2x - 1]$$

Complete solution is $y_n = \text{C.F.} + \text{P.I.}$

$$y_n = C_1 \cos \dfrac{n\pi}{3} + C_2 \sin \dfrac{n\pi}{3} + x^2 - 2x - 1.$$ **Ans.**

Example 19. *Solve the following difference equation*

$$y_{n+2} - 4y_{n+1} + y_n = 3$$

Solution. $\qquad y_{n+2} - 4y_{n+1} + y_n = 3$

The above equation, in the operator form, is

$$E^2 y_n - 4Ey_n + y_n = 3$$

$$\Rightarrow \qquad (E^2 - 4E + 1)y_n = 3$$

A.E. is $\qquad m^2 - 4m + 1 = 0$

$$\Rightarrow \qquad\qquad m = \dfrac{4 \pm \sqrt{16 - 4}}{2} = 2 \pm \sqrt{3}$$

$$\text{C.F.} = c_1(2 + \sqrt{3})^n + c_2(2 - \sqrt{3})^n$$

$$\text{P.I.} = \dfrac{1}{E^2 - 4E + 1}\, 3 = 3.\dfrac{1}{E^2 - 4E + 1}(1)^n = 3.\dfrac{1}{1 - 4 + 1} = -\dfrac{3}{2}$$

Hence the complete solution is $\quad y_n = c_1(2 + \sqrt{3})^n + c_2(2 - \sqrt{3})^n - \dfrac{3}{2}$ **Ans.**

Example 20. *Solve* $u_{n+2} + u_{n+1} + u_n = n^2 + n + 1.$ *(Nagpur University, Summer 2002)*

Solution.   The given difference equation is

$$(E^2 + E + 1)\, u_n = n^2 + n + 1$$

A.E. is $m^2 + m + 1 = 0$

$$\therefore \qquad m = \frac{-1 \pm \sqrt{-3}}{2} = \frac{-1}{2} \pm \frac{i\sqrt{3}}{2}$$

Where $r = \sqrt{\alpha^2 + \beta^2} = \sqrt{\left(-\frac{1}{2}\right)^2 + \left(\frac{\sqrt{3}}{2}\right)^2} = \sqrt{\frac{1}{4} + \frac{3}{4}} = 1$

And $\theta = \tan^{-1} \dfrac{\beta}{\alpha} = \tan^{-1}\left(\dfrac{\sqrt{3}/2}{-\dfrac{1}{2}}\right) = \tan^{-1}(-\sqrt{3}) = \dfrac{2\pi}{3}$

C.F. $= r^n (c_1 \cos n\theta + c_2 \sin n\theta)$

C.F. $= c_1 \cos \dfrac{2n\pi}{3} + c_2 \sin \dfrac{2n\pi}{3}$

P.I. $= \dfrac{1}{E^2 + E + 1}(n^2 + n + 1) = \dfrac{1}{(1+\Delta)^2 + (1+\Delta) + 1}\,[\, n(n-1) + 2n + 1\,]$

$= \dfrac{1}{3 + 3\Delta + \Delta^2}\,[n^{(2)} + 2n^{(1)} + 1] = \dfrac{1}{3\left[1 + \left(\Delta + \dfrac{\Delta^2}{3}\right)\right]}\,[n^{(2)} + 2n^{(1)} + 1]$

$= \dfrac{1}{3}\left[1 + \left(\Delta + \dfrac{\Delta^2}{3}\right)\right]^{-1}[n^{(2)} + 2n^{(1)} + 1]$

$= \dfrac{1}{3}\left[1 - \left(\Delta + \dfrac{\Delta^2}{3}\right) + \left(\Delta + \dfrac{\Delta^2}{3}\right)^2 + \ldots\ldots\right][n^{(2)} + 2n^{(1)} + 1]$

$= \dfrac{1}{3}\left[1 - \Delta + \dfrac{2}{3}\Delta^2\right][n^{(2)} + 2n^{(1)} + 1] = \dfrac{1}{3}\left[(n^{(2)} + 2n^{(1)} + 1 - (2n^{(1)} + 2) + \dfrac{2}{3}(2)\right]$

$= \dfrac{1}{3}\left[n^{(2)} + \dfrac{1}{3}\right] = \dfrac{1}{3}\left[n(n-1) + \dfrac{1}{3}\right] = \dfrac{1}{3}\left(n^2 - n + \dfrac{1}{3}\right)$

$\therefore$ The complete solution is

$$u_n = \text{C.F.} + \text{P.I.} = c_1 \cos \frac{2n\pi}{3} + c_2 \sin \frac{2n\pi}{2} + \frac{1}{3}\left(n^2 - n + \frac{1}{3}\right) \qquad \textbf{Ans.}$$

Case IV. When $\phi(n) = a^n \psi(n)$ where $\psi(n)$ is a polynomial

$$\text{P.I.} = \frac{1}{f(E)}\, a^n\, \psi(n) = a^n\, \frac{1}{f(aE)}\, \psi(n)$$

Now proceed as in Case III.

Example 21. *Solve* $y_{n+2} - 2y_{n+1} + y_n = 2^n . n^2$ (*Nagpur University, Winter 2004, Summer 2004, 2003, 2001*)

Solution. $(E^2 - 2E + 1)y_n = 2^n . n^2$

A.E. is $m^2 - 2m + 1 = 0 \implies (m-1)^2 = 0 \implies m = 1, 1$

$$C.F. = (c_1 + c_2 n)(1)^n = C_1 + C_2 n$$

$$P.I. = \frac{1}{(E-1)^2} 2^n . n^2$$ [*E* is replaced by 2 *E*.]

$$= 2^n \frac{1}{(2E-1)^2} n^2 = 2^n \frac{1}{(2+2\Delta-1)^2} . n^2 = 2^n \frac{1}{(1+2\Delta)^2} . n^2 = 2^n .(1+2\Delta)^{-2}[n(n-1)+n]$$

$$= 2^n \left[1 - 4\Delta + \frac{-2(-3)}{2} 4\Delta^2 + ...\right]\{[n]^2 + [n]\} = 2^n \left[1 - 4\Delta + 12\Delta^2 ....\right]\{[n]^2 + [n]\}$$

$$= 2^n [[n]^2 + [n] - 4 \times 2[n] - 4 + 12 \times 2] = 2^n [n(n-1) + n - 8n - 4 + 24]$$

$$= 2^n [n^2 - 8n + 20]$$

$$y_n = C_1 + C_2 n + 2^n [n^2 - 8n + 20]$$ **Ans.**

Example 22. *Solve* $4y_{n+2} - 4y_{n+1} + y_n = \dfrac{n}{2^n}$

with $y_0 = 0, y_1 = 1.$ (*Nagpur University, Summer 2004, Winter 2003*)

Solution. $4y_{n+2} - 4y_{n+1} + y_n = \dfrac{n}{2^n}$

$$(4E^2 - 4E + 1) y_n = \frac{n}{2^n}$$

Auxiliary equation is $4m^2 - 4m + 1 = 0 \qquad \implies (2m-1)^2 = 0 \qquad \implies m = \frac{1}{2}, \frac{1}{2}$

∴ $$C.F. = (C_1 + C_2 n)\left(\frac{1}{2}\right)^n$$

P.I. $$= \frac{1}{4E^2 - 4E + 1} n\left(\frac{1}{2}\right)^n = \frac{1}{2^n} \frac{1}{4\left(\dfrac{E}{2}\right)^2 - 4\left(\dfrac{E}{2}\right) + 1} n = \frac{1}{2^n} \left\{\frac{1}{E^2 - 2E + 1} n^{(1)}\right\}$$

$$= \frac{1}{2^n} \left[\frac{1}{(E-1)^2} n^{(1)}\right]$$ [∵ $\Delta = E - 1$]

$$= \frac{1}{2^n} \left\{\frac{1}{\Delta^2} n^{(1)}\right\} = \frac{1}{2^n} \frac{1}{\Delta}\left(\frac{n^{(2)}}{2}\right) = \frac{1}{2^n} \frac{n^{(3)}}{6} = \frac{1}{2^n} \frac{n(n-1)(n-2)}{6}$$

The complete solution is $y_n = C.F. + P.I.$

$$\implies \qquad y_n = (C_1 + C_2 n)\left(\frac{1}{2}\right)^n + \frac{1}{2^n} . \frac{n(n-1)(n-2)}{6}$$ ... (1)

On putting $n = 0$ in (1), we get

$$y_0 = C_1$$
$$0 = C_1$$ (Given $y_0 = 0$)

On putting the value of C_1 in (1), we get

$$y_n = C_2 n \left(\frac{1}{2}\right)^n + \frac{1}{2^n} \frac{n(n-1)(n-2)}{6}$$ ... (2)

On putting $n = 1$ in (2), we get

$$y_1 = \frac{C_2}{2} \qquad \text{But } y_1 = 1 \qquad \text{(Given)}$$

$$1 = \frac{C_2}{2} \quad \Rightarrow \quad C_2 = 2$$

On putting the value of C_2 in (2), we get

$$y_n = 2n\left(\frac{1}{2}\right)^n + \frac{1}{2^n} n(n-1)(n-2) \quad \Rightarrow \quad y_n = \frac{n}{2^n}[2 + (n-1)(n-2)]$$

$$\Rightarrow \qquad y_n = \frac{n}{2^n}(n^2 - 3n + 4) \qquad\qquad\qquad\qquad\qquad \textbf{Ans.}$$

Example 23. Solve $u_{n+2} - 7u_{n+1} - 8u_n = 2^n n(n-1)$.   *(Nagpur University, Winter 2000)*

Solution. Here, we have

$$[E^2 - 7E - 8]\, u_n = 2^n n(n-1)$$

A.E. is $(m^2 - 7m - 8) = 0$

i.e. $(m-8)(m+1) = 0 \qquad\qquad \Rightarrow \quad m = 8, -1$

$$\text{C.F.} = C_1(-1)^n + C_2(8)^n$$

$$\text{P.I.} = \frac{1}{E^2 - 7E - 8} 2^n n(n-1) = 2^n \frac{1}{4E^2 - 14E - 8} n(n-1)$$

$$= 2^n \cdot \frac{1}{4(1+\Delta)^2 - 14(1+\Delta) - 8} n^{(2)} = 2^n \cdot \frac{1}{4\Delta^2 - 6\Delta - 18} n^{(2)}$$

$$= 2^n \cdot \frac{1}{-18\left(1 - \dfrac{(4\Delta^2 - 6\Delta)}{18}\right)} n^{(2)} = -\frac{2^n}{18}\left[1 - \left(\frac{4\Delta^2 - 6\Delta}{18}\right)\right]^{-1} n^{(2)}$$

$$= -\frac{2^n}{18}\left[1 + \frac{(4\Delta^2 - 6\Delta)}{18} + \left(\frac{(4\Delta^2 - 6\Delta)^2}{18}\right) + \dots\right] n^{(2)}$$

$$= -\frac{2^n}{18}\left[1 + \frac{4\Delta^2}{18} - \frac{6\Delta}{18} + \frac{36\Delta^2}{(18)^2} + \dots\right] n^{(2)} = -\frac{2^n}{18}\left[1 - \frac{1}{3}\Delta + \frac{1}{3}\Delta^2\right] n^{(2)}$$

$$= -\frac{2^n}{18}\left[n^{(2)} - \frac{1}{3}\Delta\, n^{(2)} + \frac{1}{3}\Delta^2\, n^{(2)}\right] = -\frac{2^n}{18}\left[n(n-1) - \frac{2}{3}n + \frac{2}{3}\right]$$

$$= -\frac{2^n}{54}[3n^2 - 3n - 2n + 2] = \frac{-2^n}{54}(3n^2 - 5n + 2)$$

Complete solution is $u_n = C_1(-1)^n + C_2(8)^n - \dfrac{2^n}{54}(3n^2 - 5n + 2)$ **Ans.**

Example 24. Solve the difference equation $y_{x+3} - 8y_x = (x^2 + 1)\, 2^x$

Solution. Here, we have

$$(E^3 - 8)\, y_x = (x^2 + 1)\, 2^x$$

A.E. is $m^3 - 8 = 0$, $m = 2$ is a root

$$\Rightarrow \qquad (m - 2)(m^2 + 2m + 4) = 0$$

$$\Rightarrow \qquad m = 2, \ m = \frac{-2 \pm \sqrt{4-16}}{2} = \frac{-2 \pm 2i\sqrt{3}}{2} = -1 \pm i\sqrt{3}$$

The complex root is $-1 \pm i\sqrt{3}$ $\qquad \therefore \quad \alpha = -1, \ \beta = \sqrt{3}$

$$r = \sqrt{\alpha^2 + \beta^2} = \sqrt{3+1} = 2$$

$$\theta = \tan^{-1}\left(\frac{\beta}{\alpha}\right) = \tan^{-1}(-\sqrt{3}) = \frac{2\pi}{3}$$

$$\text{C.F.} = C_2 \, 2^x + 2^x \left(C_2 \cos \frac{2\pi x}{3} + C_3 \sin \frac{2\pi x}{3} \right)$$

$$\text{P.I.} = \frac{1}{E^3 - 8}(x^2+1)\,2^x = 2^x \frac{1}{(2E)^3 - 8}(x^2+1) = 2^x \cdot \frac{1}{8(E^3-1)}(x^2+1) = \frac{2^{x-3}}{(1+\Delta)^3 - 1}(x^2+1)$$

$$= 2^{x-3} \frac{1}{\Delta^3 + 3\Delta^2 + 3\Delta + 1 - 1}(x^2+1) = 2^{x-3} \frac{1}{-3\Delta\left(1 - \Delta + \dfrac{\Delta^2}{3}\right)}[x(x-1)+x+1]$$

$$= 2^{x-3} \frac{1}{3\Delta\left[1 + \left(\Delta + \dfrac{\Delta^2}{3}\right)\right]}[x^{(2)} + x^{(1)} + 1] \qquad = 2^{x-3} \frac{1}{3\Delta}\left[1 + \left(\Delta + \dfrac{\Delta^2}{3}\right)\right]^{-1}[x^{(2)} + x^{(1)} + 1]$$

$$= 2^{x-3} \frac{1}{3\Delta}\left[1 - \left(\Delta + \frac{\Delta^2}{3}\right) + \left(\Delta + \frac{\Delta^2}{3}\right)^2 + \dots\right][x^{(2)} + x^{(1)} + 1]$$

$$= 2^{x-3} \frac{1}{3\Delta}\left(1 - \Delta + \frac{2}{3}\Delta^2\right)[x^{(2)} + x^{(1)} + 1]$$

$$= \frac{2^{x-3}}{3}\frac{1}{\Delta}\left[x^{(2)} + x^{(1)} + 1 - \Delta(x^{(2)} + x^{(1)} + 1) + \frac{2}{3}\Delta^2(x^{(2)} + x^{(1)} + 1)\right]$$

$$= \frac{2^{x-3}}{3}\frac{1}{\Delta}\left(x^{(2)} + x^{(1)} + 1 - 2x - 1 + \frac{2}{3}\cdot 2\right) = \frac{2^{x-3}}{3}\frac{1}{\Delta}\left(x^{(2)} - x + \frac{4}{3}\right) = \frac{2^{x-3}}{3}\left(\frac{x^{(3)}}{3} - \frac{x^{(2)}}{2} + \frac{4}{3}x\right)$$

$$= \frac{2^{x-3}}{3 \times 6}[2x(x-1)(x-2) - 3x(x-1) + 8x] = \frac{2^{x-3}}{3}\cdot\frac{1}{6}[2(x^2 - 3x^2 + 2x) - 3(x^2 - x) + 8x]$$

$$= \frac{2^{x-4}}{9}(2x^3 - 9x^2 + 15x)$$

Solution is $y_x = \text{C.F.} + \text{P.I.}$

$$= C_1 \, 2^x + 2^x\left(C_2 \cos \frac{2\pi x}{3} + C_3 \sin \frac{2\pi x}{3}\right) + \frac{2^{x-4}}{9}(2x^3 - 9x^2 + 15x) \qquad \textbf{Ans.}$$

EXERCISE 8.3

Solve the following difference equations:

1. $y_{n+3} - 3y_{n+2} - 4y_{n+1} + 12y_n = 4 - n$ **Ans.** $y_n = c_1(2)^n + c_2(-2)^n + c_3(3)^n + \dfrac{17 - 6n}{36}$

2. $y_{n+2} - 2y_{n+1} + y_n = 3n + 5$ **Ans.** $y_n = c_1 + c_2 n + \dfrac{n}{2}(n-1)(n+3)$

3. $u_{n+2} - 2u_{n+1} + 6u_n = 4$ **Ans.** $6^{\frac{n}{2}}[c_1 \cos n\theta + c_2 \sin n\theta] + \dfrac{4}{5}$

4. $y_{n+2} - 6y_{n+1} + 8y_n = 2^n + 6n$ **Ans.** $y_n = c_1 4^n + (c_2 - \dfrac{n}{4})2^n + 2n + \dfrac{8}{3} - \dfrac{2^n}{4}$

5. $y_{x+2} + y_{x+1} + y_x = x^2 + x$

6. $(E^3 - 5E^2 + 8E - 4)y_n = 2^n.n^2$ **Ans** $y_n = c_1(-1)^n + (c_2 + c_3 n)2^n + \dfrac{2^{n-4}}{3}(n^4 - 12n^3 + 65n^2 - 54n)$

7. $y_{n+2} + 3y_{n+1} + 2y_n = \sin \dfrac{n\pi}{2}$ **Ans.** $y_n = c_1(-1)^n + c_2(-2)^n + \dfrac{1}{10}\sin \dfrac{n\pi}{2} - \dfrac{3}{10}\cos \dfrac{n\pi}{2}$

8. $y_{n+2} - y_{n+1} = n + \sin n$ **Ans.** $y_n = c + \dfrac{1}{2}n(n-3) - \dfrac{\cos(n - \frac{3}{2})}{2\sin \frac{1}{2}}$

9. $y_{n+2} - 7y_{n+1} - 8y_n = 2^n.n^2$ *(A.M.I.E., Summer 2001)*   **Ans.** $y_n = c_1(8)^n + c_2(-1)^n - \dfrac{2^n}{8}(n^2 + n + 2)$

10. Fill in the blanks :

    (a) The solution of $y_{n+2} - 7y_{n+1} + 10y_n = 0$ is ....   **Ans.** $c_1(2)^n + c_2(5)^n$

    (b) The solution of $(E^2 - 8E + 16)y_n = 0$ is ....   **Ans.** $(c_1 + c_2 n)(4)^n$

    (c) Particular integral of $\dfrac{1}{E-3}3^n$ is ....   **Ans.** $n.3^{n-1}$

    (d) Particular integral of $(E^2 - 7E + 12)y_n = 2^n$ is ....   **Ans.** 2^{n-1}

    (e) The solution of $\Delta^2 y = [n]^2$ is ....   **Ans.** $\dfrac{[n]^4}{12}$

    (f) The order of the difference equation $\Delta y_{n+2} + \Delta^2 y_n = 1$ is ....   *(A.M.I.E., Winter 2000)* **Ans.** 3

11. Prove that the first order difference of $\dfrac{2^x}{(x+1)!}$ is $\left[\left\{\dfrac{-x}{(x+2)!}\right\}.2^x\right]$, if the interval of difference being 1:

(A.M.I.E., Winter 2003)

    [Hint : $y(x+1) - y(x) = \dfrac{2^{x+1}}{(x+2)!} - \dfrac{2^x}{(x+1)!} = \dfrac{-x}{(x+2)!}.2^x]$

12. Solve the difference equation

    $y_{n+1} - 3y_n = n, \quad y_0 = 1$ *(A.M.I.E., Winter 2003)* **Ans.** $\dfrac{5}{4}3^n - \dfrac{1}{2}\left(n + \dfrac{1}{2}\right)$

CHAPTER 9

Introduction to Statistics

(Measure of Central Tendencies, Measures of dispersion)

9.1 STATISTICS

Statistics is defined as a subject which deals with the collection, classification and interpretation of some facts. It is used to draw conclusions in situation of uncertainty. It is very useful in the field of our life and all types of economic computations, industry, business, administration and other fields. In industry where there is a mass production of items. For inspection of each item is not possible. So, just a sample of few of them are inspected and conclusions are drawn about the whole production. Conclusions may not be 100% correct. In engineering statistics is used in mass production, inspection of quality of products or raw material comparison of machines and tools and so on.

9.2 STATISTICAL METHOD (Bihar SBTE, 2012)

(1) Collection and organisation of data

The first step is the collection of data. Data must be edited to avoid wrong conclusions. Then this data is classified and tabulated.

(2) Presentation of data.

The mass data collected should be presented in a suitable, consice form i.e., in tabular or graphic form.

(3) Analysis of Data

Analysis includes condensation, summarisation, conclusion through the measure of central tendencies, dispersion, skewness, kurtosis correlation, regression etc.

(4) Interpretation of data

Valid conclusion must drawn on the basis of analysis. A high degree of skill and experience is necessary for the interpretation correct interpretation leads to valid conclusion otherwise fallacious conclusions could be drawn.

9.3 EXPLAIN THE TERMS PARAMETER AND STATISTIC (Bihar SBTE, 2009)

(a) Parameter and statistic:

The mean and the standard deviation are usually referred to as statistical constants. In order to avoid verbal confusion, the statistical constants for the population are termed as parameter and those for the samples are termed as statistic.

In principle, parameter values are not known and their estimate based on the sample values are generally used.

So, the statistic may be regarded as an estimate of parameter and is a function of sample values only. A statistic varies from samle to sample. Fluctuation of statistic is one of the fundamental problems of the sampling theory.

9.4 FREQUENCY DISTRIBUTION

Frequency distribution is the arranged data, summarised by distributing it into classes or categories with their frequencies.

Wages of 100 workers

Wages in Rs.	0-10	10-20	20-30	30-40	40-50
Number of workers	12	23	35	20	10

9.5 MEASURES (AVERAGE) OF CENTRAL TENDENCIES

An average is a value which is representative of a set of data. Average value may also be termed as measures of central tendency. There are five types of averages in common.

(*i*) Arithmetic average or mean (*ii*) Median (*iii*) Mode
(*iv*) Geometric Mean (*v*) Harmonic Mean

9.6 ARITHMETIC MEAN

(*a*) **Direct Method.** If $x_1, x_2, x_3, ..., x_n$ are n numbers, then their arithmetic mean (A.M.) is defined by

$$A.M. = \frac{x_1 + x_2 + x_3 + ... + x_n}{n} = \frac{\Sigma x}{n}$$

If the number x_1 occurs f_1 times x_2 occurs f_2 times and so on, then

$$A.M. = \frac{f_1 x_1 + f_2 x_2 + f_3 x_3 + ... + f_n x_n}{f_1 + f_2 + ... + f_n} = \frac{\Sigma f x}{\Sigma f}$$

This is known as direct method.

Example 1. *Find the mean of 20, 22, 25, 28, 30.*

Solution. $A.M. = \dfrac{20 + 22 + 25 + 28 + 30}{5} = \dfrac{125}{5} = 25$ **Ans.**

Example 2. *Find the mean of the following*

Numbers	8	10	15	20
Frequency	5	8	8	4

Solution. Direct method $\Sigma fx = 8 \times 5 + 10 \times 8 + 15 \times 8 + 20 \times 4$
$$= 40 + 80 + 120 + 80 = 320$$
$$\Sigma f = 5 + 8 + 8 + 4 = 25$$

$$A.M. = \frac{\Sigma fx}{\Sigma f} = \frac{320}{25} = 12.8$$ **Ans.**

(*b*) **Short cut method**

Let a be the assumed mean, d the deviation of the variate x from a. Then

$$\frac{\Sigma fd}{\Sigma f} = \frac{\Sigma f(x-a)}{\Sigma f} = \frac{\Sigma fx}{\Sigma f} - \frac{\Sigma fa}{\Sigma f} = A.M. - \frac{a\Sigma f}{\Sigma f} = A.M. - a$$

$$\therefore \qquad \boxed{A.M. = a + \frac{\Sigma fd}{\Sigma f}}$$

Example 3. *Prove that algebraic sum of the deviations of a set of numbers from their arithmetic mean is zero.* *(Bihar SBTE, 2008)*

Solution. We know that,

$$\text{Arithmetic mean} = \text{Assumed mean} + \frac{\Sigma fd}{\Sigma f} \qquad \qquad \text{... (1)}$$

If assumed mean = Arithmetic mean

So, (1) becomes, $AM = AM + \dfrac{\Sigma fd}{\Sigma f} \;\Rightarrow\; \dfrac{\Sigma fd}{\Sigma f} = 0 \;\Rightarrow\; \Sigma fd = 0$ **Proved.**

Example 4. *Find the arithmetic mean for the following distribution:*

Class	0-10	10-20	20-30	30-40	40-50
Frequency	7	8	20	10	5

Solution. Let assumed mean $(a) = 25$.

Class	Class-mark x	Frequency f	$x - 25 = d$	$f.d$
0 – 10	5	7	– 20	– 140
10 – 20	15	8	– 10	– 80
20 – 30	25	20	0	0
30 – 40	35	10	+ 10	+ 100
40 – 50	45	5	+ 20	+ 100
Total		$\Sigma f = 50$		$\Sigma fd = -20$

$$A.M. = a + \frac{\Sigma fd}{\Sigma f} = 25 + \frac{-20}{50} = 24.6 \qquad \textbf{Ans.}$$

(*c*) **Step deviation method**

Let a be the assumed mean, i the class length then

$$D = \frac{x-a}{i}, \quad A.M. = a + \frac{\Sigma fD}{\Sigma f} i$$

Example 5. *Find the arithmetic mean of the data given in example 3 by step deviation method.*

Solution. Let assumed mean $(a) = 25$, $i = 10$ (given)

Class	class-mark x	Frequency f	$D = \dfrac{x-a}{i}$	$f.D$
0 – 10	5	7	– 2	– 14
10 – 20	15	8	– 1	– 8
20 – 30	25	20	0	0
30 – 40	35	10	+ 1	+ 10
40 – 50	45	5	+ 2	+ 10
Total		$\Sigma f = 50$		$\Sigma fD = -2$

$$A.M. = a + \frac{\Sigma fD}{\Sigma f} . i = 25 + \frac{-2}{50} \times 10 = 24.6 \qquad \textbf{Ans.}$$

Example 6. *The marks (out of 100) obtained by different students in an examination are shown in the following frequency table. Compute the mean by step deviation method.*

Marks	17.5 to 22.5	22.5 to 27.5	27.5 to 32.5	32.5 to 37.5	37.5 to 42.5	42.5 to 47.5	47.5 to 52.5	52.5 to 57.5	57.5 to 62.5	62.5 to 67.5	67.5 to 72.5
No. of Students	5	8	33	80	170	243	213	145	67	35	4

(Bihar SBTE, 2008)

Solution. Let A = 45

Marks	No. of students	Class Marks (x)	$d = x - A$	$f \times d$	fd / i
17.5–22.5	2	20	–25	– 50	– 10
22.5–27.5	8	25	– 20	– 160	– 32
27.5–32.5	33	30	–15	–495	–99
32.5–37.5	80	35	–10	–800	–160
37.5–42.5	170	40	–5	–850	–170
42.5–47.5	243	45	0	0	0
47.5–52.5	213	50	5	1065	213
52.5–57.5	145	55	10	1450	290
57.5–62.5	67	60	15	1005	201
62.5–67.5	35	65	20	700	140
67.5–72.5	4	0	25	100	20
	$\Sigma f = 1000$			$\Sigma fd = 1965$	$\Sigma fD = 393$

By using step deviation method we have

$$\text{A.M.} = A + \frac{\Sigma f D}{\Sigma f} \cdot i = 45 + \frac{393}{1000} \times 5 = 45 + \frac{1965}{1000} = 45 + 1.965 = 46.965 \qquad \textbf{Ans.}$$

9.7 MEDIAN

Median is defined as the measure of the central item when they are arranged in ascending or descending order of magnitude.

(a) When the total number of the items is odd and equal to say n, then the value of $\frac{1}{2}(n+1)$th item gives the median.

(b) When the total number of the frequencies is even, say n, then there are two middle items, and so the mean of the values of $\frac{1}{2}n$th and $\left(\frac{1}{2}n+1\right)$th items is the median.

Example 7. *Find the median of 6, 8, 9, 10, 11, 12, 13.*

Solution. Total number of items = $n = 7$

Since, n is an odd number. Hence,

The middle item = $\frac{1}{2}(7+1)^{th} = 4th$

Median = Value of the 4th item = 10 .   **Ans.**

(c) For grouped data,

$$\text{Median} = l + \frac{\frac{1}{2}N - C}{f} \cdot i$$

where l is the lower limit of the median class, f is the frequency of the class, i is the class-length, C is the cumulative frequency of the class preceding the median-class and N is the cumulative frequency of the data.

Example 8. *Find the value of Median from the following data:*

No. of days for which absent (less than)	5	10	15	20	25	30	35	40	45
No. of students	29	224	465	582	634	644	650	653	655

(Bihar SBTE, 2009)

Solution. The given cumulative frequency distribution will first be converted into ordinary frequency as under

Class-Interval	Cumulative frequency	Ordinary frequency
0 – 5	29	29 = 29
5 – 10	224 = C	224 – 29 = 195
10 – 15	465	465 – 224 = 241 f
15 – 20	582	582 – 465 = 117
20 – 25	634	634 – 582 = 52
25 – 30	644	644 – 634 = 10
30 – 35	650	650 – 644 = 6
35 – 40	653	653 – 650 = 3
40 – 45	655 = N	655 – 653 = 2

Here,

$$\frac{N}{2} = \frac{655}{2} = 327.5$$

Hence, Median class = class having $c.f.$ just more than $\dfrac{N}{2}$ i.e. $327.5 = 10 - 15$

Now,

$$\text{Median} = l + \frac{\frac{N}{2} - C}{f} i$$

where l stands for lower limit of median class,

N stands for the total frequency,

C stands for the cumulative frequency of the class just preceding the median class,

i stands for width of class interval

f stands for frequency of the median class.

$$\text{Median} = 10 + \frac{\frac{655}{2} - 224}{241} \times 5 = 10 + \frac{103.5 \times 5}{241} = 10 + 2.15 = 12.15 \quad \textbf{Ans.}$$

9.8 QUARTILES

Quartiles are the values of the variate which divide the total frequency into four equal parts. When the lower half before the median is divided into two equal parts, the value of the dividing variate is called **lower Quartile** and is denoted by Q_1. The value of the variate dividing the upper half into two equal parts is called the **upper Quartile** and is denoted by Q_3. Q_2 is the median.

The formulae for computation of Q_1 and Q_3 are,

$$Q_1 = l + \frac{i}{f}\left(\frac{N}{4} - C\right)$$

$$Q_3 = l + \frac{i}{f}\left(\frac{3N}{4} - C\right).$$

9.9 DECILES

Deciles are those values of the variate which divide the total frequency into 10 equal parts.

$$D_1 = l + \frac{i}{f}\left(\frac{N}{10} - C\right)$$

$$D_2 = l + \frac{i}{f}\left(\frac{2N}{10} - C\right)$$

$$D_3 = l + \frac{i}{f}\left(\frac{3N}{10} - C\right) \quad \text{and so on}$$

D_5, the fifth decile is the median.

9.10 PERCENTILES

Percentiles are those values of the variate which divide the total frequency into 100 equal parts.

$$P_1 = l + \frac{i}{f}\left(\frac{N}{100} - C\right)$$

$$P_2 = l + \frac{i}{f}\left(\frac{2N}{100} - C\right) \quad \text{and so on.}$$

P_{50}, the 50th percentile is the median.

Note. In case of series where frequency is not given,

P_{10} = value of $\frac{10}{100}(n+1)^{th}$ observation, P_{50} = value of $\frac{50}{100}(n+1)^{th}$ observations etc.

9.11 MODE

Mode is defined to be the size of the variable which occurs most frequently.

In case of continuous frequency distribution.

$$\text{Mode} = l + \left(\frac{f - f_{-1}}{2f - f_{-1} - f_{+1}}\right)i$$

where l is lower limit, i is the class length, f is the frequeny of the modal class, f_{-1} and f_1 are the frequencies of the classes preceding and succeeding the modal class respectively.

The following points must be taken care of while calculating mode :

1.   . The values (or classes of values) of the variable must be in ascending order of magnitude.

2.   If the classes are in inclusive form, then the actual limits of the modal class are to be taken for finding l and i.

3.   The classes must be of equal width.

Example 9. *Find the mode of the following items : 0, 1, 6, 7, 2, 3, 7, 6, 6, 2, 6, 0, 5, 6, 0.*

Solution. Since 6 occurs 5 times and no other item occurs 5 or more than 5 times, hence the mode is 6.

Ans.

Emperical formula | **Mean – Mode = 3 [Mean – Median]**

Example 10. *Find the mode from the following data:*

Age	0 – 6	6 – 12	12 – 18	18 – 24	24 – 30	30 – 36	36 – 42
Frequency	6	11	25	35	18	12	6

(Bihar SBTE, 2010)

Solution.

Age	Frequency	Cumulative frequency
0 – 6	6	6
6 – 12	11	17
12 – 18	$25 = f_{-1}$	42
18 – 24	$35 = f$	77
24 – 30	$18 = f_1$	95
30 – 36	12	107
36 – 42	6	113

Here, max. frequency of any item is 35.

Hence modal class is 18–24

$$\text{Mode} = l + \frac{f - f_{-1}}{2f - f_{-1} - f_1} \times i = 18 + \frac{35 - 25}{70 - 25 - 18} \times 6 = 18 + \frac{60}{27} = 18 + 2.22 = 20.22$$

Ans.

Example 11. *Find the mean, median and mode of the following data:*

Class	11-15	16-20	21-25	26-30	31-35	36-40	41-45
Frequency	7	10	13	26	55	22	11

(Bihar SBTE, 2011)

Solution.

Class	Mid value (x)	Freq. (f)	$f(x)$	Cumulative freq.
10.5 – 15.5	13	7	91	7
15.5 – 20.5	18	10	180	17
20.5 – 25.5	23	13	299	30
25.5 – 30.5	28	26	728	56 = C
30.5 – 35.5	33	55	1815	111
35.5 – 40.5	38	22	836	133
40.5 – 45.5	43	11	473	144
		$\Sigma f = 144$	$\Sigma f(x) = 4422$	

$$\text{Mean} = \frac{\Sigma f(x)}{\Sigma f} = \frac{4422}{144} = 30.71$$

Here, $\frac{N}{2} = \frac{144}{2} = 72$

$$\text{Median} = l + \frac{\frac{N}{2} - C.f.}{f} = 30.5 + \frac{72-56}{55} = 30.5 + \frac{16}{55} = 30.5 + 0.29 = 30.79$$

$$f = 55, \quad f_{-1} = 26, \quad f_1 = 22$$

$$\text{Mode} = l + \frac{f - f_{-1}}{2f - f_{-1} - f_1} = 30.5 + \frac{55-26}{2 \times 55 - 26 - 22} = 30.5 + 0.47 = 30.97 \qquad \textbf{Ans.}$$

9.12 GEOMETRIC MEAN

If $x_1, x_2, x_3, \dots, x_n$ be n values of variates x, then the geometric mean $G = (x_1 \times x_2 \times x_3 \times x_4 \times \dots \times x_n)^{1/n}$

Example 12. *Find the geometric mean of 4, 8, 16.*

Solution. $G.M. = (4 \times 8 \times 16)^{1/3} = 8$ **Ans.**

9.13 HARMONIC MEAN

Harmonic mean of a series of values is defined as the reciprocal of the arithmetic mean of their reciprocals. Thus if H be the harmonic mean, then

$$\frac{1}{H} = \frac{1}{n}\left[\frac{1}{x_1} + \frac{1}{x_2} + \dots + \frac{1}{x_n} \right]$$

Example 13. *Calculate the harmonic mean of 4, 8, 16.*

Solution. Let, $\dfrac{1}{H} = \dfrac{1}{3}\left[\dfrac{1}{4} + \dfrac{1}{8} + \dfrac{1}{16}\right] = \dfrac{7}{48};\quad H = \dfrac{48}{7} = 6.857$ **Ans.**

MEASURES OF DISPERSIONS

9.14 AVERAGE DEVIATION OR MEAN DEVIATION

It is the mean of the absolute values of the deviations of a given set of numbers from their arithmetic mean.

If $x_1, x_2, x_3, \dots, x_n$ be a set of numbers with frequencies $f_1, f_2, \dots f_n$ respectively. Let $\bar{x}$ be the arithmetic mean of the numbers $x_1, x_2, \dots, x_n$, then

$$\text{Mean deviation} = \frac{\sum f_i |x_i - \bar{x}|}{\sum f_i}$$

Example 14. *Find the mean deviation of the following frequency distribution.*

Class	0 – 6	6 – 12	12 – 18	18 – 24	24 – 30
Frequency	8	10	12	9	5

Solution. $a = 15$

Class	Mid-value x	Frequency f	$d = x - a$	fd	$\|x - 14\|$	$f\|x - 14\|$
0 – 6	3	8	– 12	– 96	11	88
6 – 12	9	10	– 6	– 60	5	50
12 – 18	15	12	0	0	1	12
18 – 24	21	9	+ 6	54	7	63
24 – 30	27	5	+ 12	60	13	65
		$\sum f = 44$		$\sum fd = -42$		$\sum f\|x-14\| = 278$

$$\text{Mean} = \bar{x} = a + \frac{\sum fd}{\sum f} = 15 - \frac{42}{44} = 14 \text{ (approx.)}$$

$$\text{Mean or Average deviation} = \frac{\sum f|x - \bar{x}|}{\sum f} = \frac{278}{44} = 6.32 \qquad \textbf{Ans.}$$

9.15 STANDARD DEVIATION

Standard deviation is defined as the square root of the mean of the square of the deviation from the arithmetic mean.

$$S.D. = \sigma = \sqrt{\frac{\Sigma f(x-\bar{x})^2}{\Sigma f}}$$

Note. 1. The square of the standard deviation *i.e.*; σ^2 is called variance.

2. σ^2 is called the second moment about the mean and is denoted by μ_2.

9.16 SHORTEST METHOD FOR CALCULATING STANDARD DEVIATION

We know that $\sigma^2 = \dfrac{1}{N}\Sigma f(x-\bar{x})^2 = \dfrac{1}{N}\Sigma f(x-a-\overline{x-a})^2$

$$= \frac{1}{N}\Sigma f(d-\overline{x-a})^2 \qquad \text{where } x - a = d$$

$$= \frac{1}{N}\Sigma fd^2 - 2(\bar{x}-a)\frac{1}{N}\Sigma fd + (\bar{x}-a)^2 \frac{1}{N}\Sigma f$$

$$= \frac{1}{N}\Sigma fd^2 - 2(\bar{x}-a)\frac{1}{N}\Sigma fd + (\bar{x}-a)^2 \qquad [\because \Sigma f = N] \quad \left[\bar{x} = a + \frac{\Sigma fd}{N} \text{ or } \bar{x}-a = \frac{\Sigma fd}{N}\right]$$

$$\sigma^2 = \frac{1}{N}\Sigma f d^2 - 2\left(\frac{\Sigma f d}{N}\right)\left(\frac{\Sigma f d}{N}\right) + \left(\frac{\Sigma f d}{N}\right)^2 = \frac{1}{N}\Sigma f d^2 - 2\left(\frac{\Sigma f d}{N}\right)^2 + \left(\frac{\Sigma f d}{N}\right)^2$$

$$\sigma^2 = \frac{1}{N}\Sigma fd^2 - \left(\frac{\Sigma fd}{N}\right)^2$$

$$S.D. = \sigma = \sqrt{\frac{\Sigma fd^2}{N} - \left(\frac{\Sigma fd}{N}\right)^2}$$

Note. Coefficient of variation $= \dfrac{\sigma}{x} \times 100$

Example 15. *Find the standard deviation of the following distribution:*

x	0	1	2	3	4	5	6	7
f	14	21	25	43	51	40	29	12

(Bihar SBTE, 2009)

Solution. Let us arrange the calculations in the form of following table for getting the deviations:

Let us calculate $\bar{x}$. $(\bar{x} = 4)$

x	f	f.x	$d = x-\bar{x}$	$f\lvert d\rvert$	d^2	$f.d^2$
0	14	0	-4	56	16	224
1	21	21	-3	63	9	189
2	25	50	-2	50	4	100
3	43	129	-1	43	1	43
4	51	204	0	0	0	0
5	40	200	1	40	1	40
6	29	174	2	58	4	116
7	12	84	3	36	9	108
	$\Sigma f = 235$	$\Sigma fx = 862$		$\Sigma f\lvert d\rvert = 346$		$\Sigma f d^2 = 820$

Hence, standard deviation:

$$= \sqrt{\frac{\Sigma fd^2}{\Sigma f} - \left(\frac{\Sigma fd}{\Sigma f}\right)^2} = \sqrt{\frac{820}{235} - \left(\frac{346}{235}\right)^2}$$

$$\sqrt{3.489 - 2.166} = \sqrt{1.323} = 1.19 \qquad \textbf{Ans.}$$

Example 16. *Find the standard deviation of the following distribution:*

Class interval	4 – 6	6 – 8	8 – 10	10 – 12	12 – 14	14 – 16
Frequency	13	10	9	5	8	5

(Bihar SBTE, 2005, 2011)

Solution. Let a = assumed mean = 9

Let us arrange the calculations in the form of following table for getting the deviations:

Class	Mid value x	Frequency f	Deviation d = x – a	f. d	fd²
4 – 6	5	13	– 4	– 52	208
6 – 8	7	10	– 2	– 20	40
8 – 10	9	9	0	0	0
10 – 12	11	5	2	10	20
12 – 14	13	8	4	32	128
14 – 16	15	5	6	30	180
		$\Sigma f = 50$		$\Sigma fd = 0$	$\Sigma fd^2 = 576$

$$\text{Standard deviation} = \sqrt{\frac{\Sigma fd^2}{\Sigma f} - \left(\frac{\Sigma fd}{\Sigma f}\right)^2} = \sqrt{\frac{576}{50} - \left(\frac{0}{50}\right)^2} = \sqrt{11.52} = 3.394 \qquad \textbf{Ans.}$$

Example 17. *Calculate the mean and standard deviation for the following table, given the age distribution of 542 members.*

Age in years	20 – 30	30 – 40	40 – 50	50 – 60	60 – 70	70 – 80	80 – 90
No. of members	3	61	132	153	140	51	2

Solution. Assumed mean = 55

Here, we take $d = \dfrac{x - a}{i} = \dfrac{x - 55}{10}$

Age grouped	Mid value (x)	Frequency (f)	$d = \dfrac{x - 55}{10}$	fd	fd²
20 – 30	25	3	– 3	– 9	27
30 – 40	35	61	– 2	– 122	244
40 – 50	45	132	– 1	– 132	132
50 – 60	55	153	0	0	0
60 – 70	65	140	1	140	140
70 – 80	75	51	2	102	204
80 – 90	85	2	3	6	18
		$\Sigma f = 542$		$\Sigma fd = -15$	$\Sigma fd^2 = 765$

$$\text{Mean} = \bar{x} = a + \frac{\Sigma f d}{\Sigma f} \cdot i = 55 + \frac{(-15)\,10}{542} = 55 - 0.28 = 54.72$$

$$\text{Variance} = \sigma^2 = i^2 \left[\frac{1}{N} \Sigma f d^2 - \left(\frac{\Sigma f d}{N} \right)^2 \right] = 100 \left[\frac{765}{542} - (0.028)^2 \right] = 100 \times 1.4107 = 141.07$$

$$\text{S.D.} = \sigma = 11.9 \text{ years}$$

Ans.

9.17 SYMMETRY

A distribution is said to be symmetrical when its mean, median and mode are identical. *i.e.*;

Mean = Median = Mode.

In other words, a distribution is said to be symmetric when the frequencies are symmetrically distributed about the mean (or when the values of the variable are equidistant from the mean and have the same frequency).

Consider the following frequency distribution:

x	10	20	30	40	50	60	70
f	2	6	10	14	10	6	2

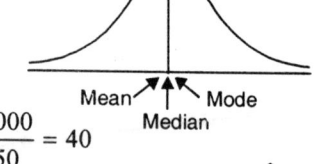

$$\text{Mean} = \bar{x} = \frac{10 \times 2 + 20 \times 6 + 30 \times 10 + 40 \times 14 + 50 \times 10 + 60 \times 6 + 70 \times 2}{2 + 6 + 10 + 14 + 10 + 6 + 2} = \frac{2000}{50} = 40$$

In this distribution, we observe that the values 20 and 60 are equidistant from the mean, viz. 40 with the same frequency 6.

A symmetrical distribution when plotted on a graph will give a perfectly bell-shaped curve, which is known as normal curve.

9.18 SKEWNESS

Skewness denotes the opposite of symmetry. It is lack of symmetry,

Skew symmetrical Distribution   *(U.P. III Semester, 2006)*

A distribution which is not symmetrical is said to be skew symmetrical distribution. In skew symmetrical distribution the left tail and the right tail are not of equal length. One tail will be longer than the other.

(*a*) **Negatively skew distribution.** In negatively skew distribution the left tail is longer than the right tail.

(*b*) **Positively skew distribution.** In positively skew distribution the right tail of the curve will be longer than the left.

In skew distribution mean, median and mode are not equal.

9.19 TEST OF SKEWNESS

1. There is no skewness in the distribution if AM = Mode = Median
2. There is no skewness in the distribution if, Third quartile - Median = Median – First quartile.
3. There is no skewness if
   The sum of the frequencies which are less than Mode = Sum of the frequencies which are greater than Mode
4. There is no skewness if quartiles are equidistant from the median.

5. The distribution is negatively skewed if A.M. is less than Mode.

6. The curve is not symmetrical about the median if AM ≠ Median ≠ Mode.

9.20 USES OF SKEWNESS

1. It gives the nature of the curve.

2. It gives nature and concentration of observations about the mean.

9.21 TYPES OF DISTRIBUTION

1. Fairly symmetrical   2. Positively skewed   3. Negatively skewed.

9.22 MEASURE OF SKEWNESS

Measure of skewness is known as the measure of symmetry.

There are two types of measure of skewness.

1. Absolute measure: Absolute measure = Mean – Mode

2. Relative measure : These are four types of relative measure of skewness.

   (i) Karl Pearson's Coefficient of Skewness

   (ii) Bowley's Coefficient of Skewness.

   (iii) Kelly's Coefficient of Skewness and

   (iv) Measure of skewness based on the moments.

   (Mode = 3 Median – 2 Mode)

9.23 KARL PEARSON'S COEFFICIENT OF SKEWNESS:

$$\text{Karl Pearson's Coefficient of Skewness} = \boxed{\frac{\text{Mean} - \text{Mode}}{\text{Standard deviation}}}$$

$$= \frac{\text{Mean} - (3 \text{ Median} - 2 \text{ Mean})}{\text{Standard deviation}} = \boxed{\frac{3 (\text{Mean} - \text{Median})}{\text{Standard deviation}}}$$

It generally lies between – 1 and 1.

If its value is zero then there is no skewness.

$$\boxed{\text{Standard deviation} = \sqrt{\frac{\Sigma f (x - \bar{x})^2}{\Sigma f}}}$$

$$\boxed{\text{Standard deviation} = \sqrt{\frac{\Sigma f d^2}{\Sigma f} - \left(\frac{\Sigma f d}{\Sigma f}\right)^2}}$$

9.24 TYPES OF SKEWNESS IN TERMS OF MEAN AND MODE

1. There is no skewness in the distribution

   $$S_k = 0$$

   $$\Rightarrow \quad \frac{\text{Mean} - \text{Mode}}{\text{S.D.}} = 0 \qquad \Rightarrow \quad \text{Mean} - \text{Mode} = 0 \qquad \Rightarrow \quad \text{Mean} = \text{Mode}.$$

2. The distribution is positively skewed if $S_k > 0$.

   $$\frac{\text{Mean} - \text{Mode}}{\text{S.D.}} > 0 \quad \Rightarrow \quad \text{Mean} - \text{Mode} > 0 \quad \Rightarrow \quad \text{Mean} > \text{Mode}.$$

3. The distribution is negatively skewed if
$$S_k' < 0$$

$$\Rightarrow \qquad \frac{\text{Mean} - \text{Mode}}{\text{S.D.}} < 0 \quad \Rightarrow \quad \text{Mean} - \text{Mode} < 0 \quad \Rightarrow \quad \text{Mean} < \text{Mode}$$

Example 18. *Compute the standard deviation and coeffient of Skewness from the following data:*

x	6	7	8	9	10	11	12
f	3	6	9	13	8	5	4

(*Bihar SBTE, 2008, U.P. III Semester, 2009-2010*)

Solution. Let $a = 9$

x	f	$d = x - 9$	$f d$	$f d^2$	c.f.
6	3	− 3	− 9	27	3
7	6	− 2	− 12	24	9
8	9	− 1	− 9	9	18
9	13	0	0	0	31
10	8	·1	8	8	39
11	5	2	10	20	44
12	4	3	12·	36	48
	$\Sigma f = 48$		$\Sigma fd = 0$	$\Sigma fd^2 = 124$	

$$\text{Mean} = a + \frac{\Sigma fd}{\Sigma f} = 9 + \frac{0}{48} = 9$$

Mode = Item of maximum frequency (13) = 9

$$\text{S.D.} = \sqrt{\frac{\Sigma fd^2}{\Sigma f} - \left(\frac{\Sigma fd}{\Sigma f}\right)^2} = \sqrt{\frac{124}{48} - \left(\frac{0}{48}\right)^2} = \sqrt{\frac{124}{48}} = 1.61$$

Karl Pearson's Coefficient of Skewness = $\dfrac{\text{Mean} - \text{Mode}}{\text{S.D.}} = \dfrac{9-9}{1.61} = \dfrac{0}{1.61} = 0$ **Ans.**

Example 19. *Calculate Karl Pearson's Coefficient of Skewness from the table given below:*

x	14.5	15.5	16.5	17.5	18.5	19.5	20.5	21.5
f	35	40	48	100	125	87	43	22

Solution. Let $a = 17.5$

x	f	$d = x - 17.5$	fd	fd^2	c.f.
14.5	35	− 3	− 105	315	35
15.5	40	− 2	− 80	160	75
16.5	48	− 1	− 48	48	123
17.5	100	0	0	0	223
18.5	125	1	125	125	348
19.5	87	2	174	348	435
20.5	43	3	129	387	478
21.5	22	4	88	352	500
	$\Sigma f = 500$		$\Sigma fd = 283$	$\Sigma fd^2 = 1735$	

Mean $= a + \dfrac{\Sigma fd}{\Sigma f} = 17.5 + \dfrac{283}{500} = 17.5 + 0.566 = 18.066$

Mode = Item of maximum frequency = 18.5

$$\text{S.D.} = \sqrt{\dfrac{\Sigma fd^2}{\Sigma f} - \left(\dfrac{\Sigma fd}{\Sigma f}\right)^2} = \sqrt{\dfrac{1735}{500} - \left(\dfrac{283}{500}\right)^2}$$

$$= \sqrt{3.470 - 0.32} = \sqrt{3.15} = 1.77$$

Karl Pearson's Coefficient of Skewness $= \dfrac{\text{Mean} - \text{Mode}}{\text{S.D.}} = \dfrac{18.066 - 18.5}{1.77}$

$$= \dfrac{-0.434}{1.77} = -0.245 \qquad\qquad \textbf{Ans.}$$

Example 20. *Calculate Karl Pearson's Coefficient of Skewness from the table given below*

Wages of day	55 – 58	58 – 61	61 – 64	64 – 67	67 – 70
No. of workers	12	17	23	18	11

Solution. Let $a = 62.5$

Wages of day	No. of workers f	Mid value x	$d = x - 62.5$	fd	fd^2	c.f
55 – 58	12	56.5	– 6	– 72	432	12
58 – 61	17	59.5	– 3	– 51	153	29
61–64	23	62.5	0	0	0	52
64 – 67	18	65.5	3	54	162	70
67 – 70	11	68.5	6	66	396	81
	81			– 3	1143	

Mean $= a + \dfrac{\Sigma fd}{\Sigma f} = 62.5 + \dfrac{-3}{81} = 62.5 - \dfrac{1}{27} = 62.46$

Median $= l + \dfrac{\dfrac{N}{2} - c.f.}{f} i = 61 + \dfrac{\dfrac{81}{2} - 29}{23}(3)$

$$= 61 + \dfrac{23}{2\,(23)}(3) = 62.5$$

$$\text{S.D.} = \sqrt{\dfrac{\Sigma fd^2}{\Sigma f} - \left(\dfrac{\Sigma fd}{\Sigma f}\right)^2} = \sqrt{\dfrac{1143}{81} - \left(\dfrac{-3}{81}\right)^2}$$

$$= \sqrt{\dfrac{1143}{81} - \dfrac{1}{729}} = \sqrt{\dfrac{10286}{729}} = 3.75$$

Karl Pearson's Coefficient of Skewness $= \dfrac{3\,(\text{Mean} - \text{Median})}{\text{S.D.}}$

$$= \dfrac{3\,(62.46 - 62.5)}{3.75} = \dfrac{-0.12}{3.75} = -0.032 \qquad\qquad \textbf{Ans.}$$

Example 21. *Find mean, variance for the following table:*

Class	0 – 10	10 – 20	20 – 30	30 – 40	40 – 50	50 – 60
Frequency	5	15	25	20	10	5

(Bihar SBTE, 2014)

Solution. Assumed mean = 25

Class	Frequency	Mid value	$d = x - 25$	fd	fd^2
0 – 10	5	5	–20	– 100	– 500
10 – 20	15	15	–10	– 150	–2250
20 – 30	25	25	0	0	0
30 – 40	20	35	10	200	4000
40 – 50	10	45	20	200	2000
50 – 60	5	55	30	150	750
	$\Sigma f = 80$			300	4000

$$\text{Mean} = a + \frac{\Sigma fd}{\Sigma f} = 25 + \frac{300}{80} = 28.75$$

$$\text{Variance} = \frac{\Sigma fd^2}{\Sigma f} - \left(\frac{\Sigma fd}{\Sigma f}\right)^2 = \frac{400}{80} - \left(\frac{300}{80}\right)^2 = 50 - 14.06 = 35.94$$ **Ans.**

Example 22: *Find the standard deviation of the following distribution:*

Class interval	4 – 6	6 – 8	8 – 10	10 – 12	12 – 14	14 – 16
Frequency	13	10	9	5	8	5

(Bihar SBTE, 2010)

Solution.

Let assumed mean = 9

Cass	x	f	d	fd	fd^2
4 – 6	5	13	–4	–52	208
6 – 8	7	10	–2	–20	40
8 – 10	9	9	0	0	0
10 – 12	11	5	2	10	20
12 – 14	13	8	4	32	128
14 – 16	15	5	6	30	180
		$\Sigma f = 50$		$\Sigma fd = 0$	$\Sigma fd^2 = 576$

$$\text{S.D.} = \sigma = \sqrt{\frac{\Sigma fd^2}{\Sigma f} - \left(\frac{\Sigma fd}{\Sigma f}\right)^2} = \sqrt{\frac{576}{50} - 0}$$

$$= \frac{24}{5\sqrt{2}} = \frac{12\sqrt{2}}{5}$$ **Ans.**

Example 23. *Calculate Karl Pearson's Coefficient of Skewness from the given data :*

Life time in months	30-40	40-50	50-60	60-70	70-80	80-90	90-100	100-110	110-120
No. of mobile	4	6	8	26	28	12	8	5	3

Solution. Let $a = 75$

Life time in months	No. of mobile (f)	Mid-value (x)	$d = x - 75$	fd	fd^2	Cumulative frequency
30 – 40	4	35	– 40	– 160	6400	4
40 – 50	6	45	– 30	– 180	5400	10
50 – 60	8	55	– 20	– 160	3200	18
60 – 70	26	65	– 10	– 260	2600	44
70 – 80	28	75	0	0	0	72
80 – 90	12	85	10	120	1200	84
90 – 100	8	95	20	160	3200	92
100 – 110	5	105	30	150	4500	97
110 – 120	3	115	40	120	4800	100
	$\Sigma f = 100$			$\Sigma fd = -210$	$\Sigma fd^2 = 31300$	

$$\text{Mean} \quad = a + \frac{\Sigma fd}{\Sigma f} = 75 + \frac{-210}{100} = 72.9$$

$$\text{Median} \quad = l + \frac{\frac{N}{2} - c.f.}{f} i \ = 70 + \frac{\frac{100}{2} - 44}{28} (10)$$

$$= 70 + 2.143 = 72.143$$

$l = 70$
$N = 100$
$cf = 40$
$\Sigma f = 100$
$i = 10$

$$\text{S.D.} \quad = \sqrt{\frac{\Sigma fd^2}{\Sigma f} - \left(\frac{\Sigma fd}{\Sigma f}\right)^2}$$

$$= \sqrt{\frac{31300}{100} - \left(\frac{-210}{100}\right)^2} = \sqrt{313 - 4.41}$$

$$= \sqrt{308.59} = 17.57$$

$$\text{Karl Pearson's Coefficient of Skewness} = \frac{3\,(\text{Mean} - \text{Median})}{\text{S.D.}} = \frac{3\,(72.9 - 72.143)}{17.57}$$

$$= \frac{3\,(0.757)}{17.57} = \frac{2.271}{17.57} = 0.1293 \qquad \textbf{Ans.}$$

EXERCISE 9.1

Calculate Karl Pearson's Coefficient of Skewness from the data given below:

1. S.D. = 6.5, AM = 29.6, mode = 27.52

            Ans. $S_k = 0.32$

2. Mean = 100, Variance = 35, Median = 99.61.

            Ans. $S_k = 0.2$

3. AM = 45, Median = 48, S.D. = 22.5

            Ans. $S_k = -0.4$

4. The sum of the 20 observation is 300 and sum of the squares of the observation is 5000, Median = 15.

            Ans. $S_k = 0$

5. Find the mean, median and mode of the following frequency distribution:

Class	6 – 10	11 – 15	16 – 20	21 – 25	26 – 30
Frequency	20	30	50	40	10

      (*Bihar SBTE, 2009*) **Ans.** 18.33, 18, 17.34

6. Find the value of median from the following data:

No. of days for which absent (less than)	5	10	20	25	30	35	40	45
No. of student	29	224	582	634	644	650	653	655

      Ans. 12.87

7. The mean of 10 numbers is 15, the mean of first six numbers is 17 and the last five is 10. Find the sixth number.

      (*Bihar SBTE, 2004*) **Ans.** 6

8. Find mode for the following data:

Age	0 – 6	6 – 12	12 – 18	18 – 24	24 – 30	30 – 36	36 – 42
Frequency	6	11	25	35	35	12	6

      (*Bihar SBTE, 2010*) **Ans.** 20.22

9. Find the mean, median and mode of the following data

Class	11 – 15	16 – 20	21 – 25	26 – 30	31 – 35	36 – 40	41 – 45
Frequency	7	10	13	26	55	22	11

      (*Bihar SBTE, 2010*) **Ans.** Mean = 30.7, Median = 31.95, Mode = 32.5

10. Calculate the mean deviation from the mean and the standard deviation of the following distribution:

Class	10 - 30	30 – 50	50 – 70	70 – 90	90 – 110	110 – 130
Frequency	4	7	15	12	7	5

      (*Bihar SBTE, 2003*)

11. Calculate mean deviation of numbers 5, 7, 8, 11, 14.   (*Bihar SBTE, 2008*) **Ans.** 0

12. Compute the standard deviation for the following frequency distribution by assumed mean method.

Class interval :	0 – 4	4 – 8	8 – 12	12 – 16
Frequency	4	8	2	1

      (*Bihar SBTE, 2010*) **Ans.** 3.26

13. From the following frequency distribution, compute the standard deviation of 100 students by assumed mean method:

Mass (in kg)	60 – 62	63 – 65	66 – 68	69 – 71	72 – 74
No. of Student	5	18	42	27	8

      (*Bihar SBTE, 2009*) **Ans.** 2.56

14. Find the mean, S.D. of the following data:

Class	10 – 20	20 – 30	30 – 40	40 – 50	50 – 60	60 – 70	70 – 80
Frequency	6	8	15	7	3	0	1

(Bihar SBTE, 2004) **Ans.** 12.92

15. Find the Karl Pearson's Coefficient of Skewness for the following

Years under	10	20	30	40	50	60
No. of persons	15	32	51	78	97	109

Ans. – 0.32

16. Calculate Karl Pearson's Coefficient of Skewness from the following data :

Cost per item (in Rs.)	4.5	5.5	6.5	7.5	8.5	9.5	10.5	11.5
No. of items	35	40	48	100	125	87	43	22

Ans. – 0.2445

17. From the following data calculate Karl Pearson's Coefficient of Skewness.

Scores	0	10	20	30	40	50	60	70	80
No. of players	150	140	100	80	80	70	30	14	0

Ans. – 0.462

18. Find the Pearson's Coefficient of Skewness for the following data:

Class	10 - 19	20 - 29	30 - 39	40 - 49	50 - 59	60 - 69	70 - 79	80 - 89
Frequency	5	9	14	20	25	15	8	4

Ans. – 0.2064

19. At Lucknow students appeared in math– III and got the marks as given in the following table. Calculate Karl Pearson's Coefficient of Skewness from the said data :

Marks	0 - 10	10 - 20	20 - 30	30 - 40	40 - 50	50 - 60	60 - 70	70 - 80
No. of students	10	40	20	0	10	40	16	14

Ans. 0.754

20. Find the mean and s.d. of the following series:

Expenditure	Below Rs. 5	Below Rs. 10	Below Rs. 15	Below Rs. 20	Below Rs. 25
No. of students	6	16	28	38	46

Ans. Mean = Rs. 12.93, S.D. = Rs. 6.41

21. Calculate the arithmetic mean and the standard deviation of the following values of the world's annual gold output (in millions of pounds) for 10 different years:

Year	94	95	96	93	87	79	73	69	68	67
Gold output	78	82	83	89	95	103	108	117	130	97

Ans. Mean = 80.50, standard deviation = 9.72

22. The following table shows the marks obtained by students in an examination. Calculate the mean median, mode and standard devaition of the distribution.

Marks	0 – 10	10 – 20	20 – 30	30 – 40	40 – 50	50 – 60
No. of Students	05	10	15	25	20	15

(*Bihar SBTE, 2012*) **Ans.** $\bar{x} = 35$, Median = 30.6, Mode = 21.8, $\sigma = 14$

23. For a frequency distribution of marks in History of 200 candidates (grouped in intervals 0–5, 5–10, ...) the mean and standard deviation (s.d.) were found to be 40 and 15. Later it was discovered that the score 43 was misread as 53 in obtaining the frequency distribution. Find the corrected mean and s.d. corresponding to the corrected frequency distribution. **Ans.** Mean = 39.95, Standard deviation = 14.975 approx.

24. A student while calculating the mean and the standard deviation on 25 observations, obtained the following values:

mean = 56 cms: standard deviation = 2 cms.

It was later discovered at the time of checking that he had wrongly copied down an observation as 64. What is the mean and s.d. if correct value is omitted? **Ans.** Mean = 55.67 cms., S.D. = 1.18 cms. approx

25. The weekly wages in Rs. of the workers in a shoe factory are given below :

Weekly wages (in Rs.)	500 - 600	600 - 700	700 - 800	800 - 900	900 - 1000	1000-1100
No. of workers	8	12	4	2	1	1

Calculate Karl Pearson's Coefficient of Skewness. **Ans.** 0.34

26. Which of the following two series is symmetrical :

Series (*a*) : Mean = 32, Median = 34, S.D. = 20

Series (*b*) : Mean = 32, Median = 36, S.D. = 25 **Ans.** series (*a*) is more symmetrical than series (*b*)

27. Karl Pearson's Coefficient of Skewness of a distribution = 0.32, Standard deviation = 6.5

A.M. = 29.6.

From the above data find the mode and the median for the distribution. **Ans.** Mode = 27.52, Median = 28.91

28. Find the first four moments about mean for the following frequency distribution :

Marks	0 – 10	10 – 20	20 – 30	30 – 40	40 – 50
No. of students	5	10	40	20	25

Ans. $\mu_1 = 0$, $\mu_2 = 125$, $\mu_3 = -300$, $\mu_4 = 37625$

29. The following table gives the monthly wages of workers in a factory. Compute the standard deviation, and skewness.

Monthly wages (in Rs.)	No. of workers	Monthly wages (in Rs.)	No. of workers
125 – 175	2	375 – 425	4
175 – 225	22	425 – 475	6
225 – 275	19	475 – 525	1
275 – 325	14	525 – 575	1
325 – 375	3		

Ans. S.D. = Rs. 88.52, Skewness = 0.7

30. Find Mean, Median, Mode and standard Devision for the following table:

Class interval	0 – 5	5 – 10	10 – 15	15 – 20	20 – 25
Frequency	5	7	6	10	2

(*Bihar SBTE, 2011*) **Ans.** Mean = 12.4, Median = 13.82, Mode = 16.67, S.D. = 6.11

31. **Choose the correct alternative:**

(*i*) The Median of first eleven natural numbers is:

(*a*) 9   (*b*) 11   (*c*) 13   (*d*) 15   (*Bihar SBTE, 2009*) **Ans.** (*b*) (*ii*) Imperical formula connecting mean, median and mode is:

(*a*) 2 mean = 3 median – mode   (*b*) mean = median – mode

(*c*) 2 mode = mean – median   (*d*) mode = 2 median – mean   (*Bihar SBTE, 2009*) Ans. (*a*)

(*iii*) The square of standard deviation is known as ....

(*a*) variance   (*b*) mean deviation   (*c*) quartile   (*d*) none of these

(*Bihar SBTE, 2014, 2011*) Ans. (*a*)

(*iv*) Ratio of standard deviation to the mean is known as:

(*a*) Co-efficient of standard deviation   (*b*) Variance

(*c*) Quality control   (*d*) None of these   (*Bihar SBTE, 2009*) Ans. (*a*)

(*v*) The mean of first six prime numbers is ....

(*a*) 6   (*b*) 6.83   (*c*) 7.83   (*d*) None of these

(*Bihar SBTE, 2011*) Ans. (*b*)

(*vi*) The median of the numbers 6, 8, 9, 10, 11, 12, 13 is:

(*a*) 0   (*b*) 10   (*c*) 11   (*d*) None of these

(*Bihar SBTE, 2009*) Ans. (*b*)

(*vii*) The median for the data 4, 6, 6, 7, 9, 10 is:

(*a*) 6   (*b*) 6.5   (*c*) 7   (*d*) 8   (*Bihar SBTE, 2010*) Ans. (*b*)

(*viii*) The variance of the set of numbers 1, 4, 5, 7, 8 is ....

(*a*) 5   (*b*) $\sqrt{5}$ (*c*) 6   (*d*) $\sqrt{6}$ (*Bihar SBTE, 2008*) Ans. (*d*)

(*ix*) The mean of 20 numbers is 43. If 6 is subtracted from each of the numbers, then the mean of new numbers is:

(*a*) 37   (*b*) 35   (*c*) 38   (*d*) None of these

(*Bihar SBTE, 2009*) Ans. (*a*)

(*x*) The ages of nine students in a group were found to be 6, 11, 16, 6, 16, 17, 11, 6 and 8 the mode age is ....

(*a*) 16   (*b*) 11   (*c*) 9   (*d*) 6   (*Bihar SBTE, 2008*) Ans. (*d*)

(*xi*) In a given table

Numbers	4	10	15	20
Frequency	5	8	8	4

(*a*) 10   (*b*) 12   (*c*) 15   (*d*) 18   (*Bihar SBTE, 2008*) Ans. (*b*)

(*xii*) In a given table find the value of mean

Numbers	5	10	15	20	25
Frequency	2	2	3	2	1

(*a*) 12   (*b*) 13   (*c*) 14   (*d*) 15   (*Bihar SBTE, 2008*) Ans. (*c*)

(*xiii*) The mean of 20 numbers is 43. If 6 is subsracted from each of the numbers, then the mean of new numbers is:

(*a*) 37   (*b*) 35   (*c*) 38   (*d*) None of these

(*Bihar SBTE, 2009*) Ans. (*a*)

(*xiv*) Mode of sample:

(*a*) Observed most frequently   (*b*) Occurs most frequently

(*c*) Both (a) and (b)   (*d*) None of these   (*Bihar SBTE, 2012*) Ans. (*d*)

(*xv*) Expected value E (x) of a discrete probability distribution is ....

(*a*) Σ p.x   (*b*) Σ p   (*c*) Σ x   (*d*) None of these

(*Bihar SBTE, 2010*) **Ans.** (*a*)

(*xvi*) If mean of two numbers be 5 and that of another 3 numbers be 10, then combined mean of all five numbers is:

(*a*) 5   (*b*) 8   (*c*) 10   (*d*) 15   (*Bihar SBTE, 2010*) **Ans.** (*b*)

(*xvii*) The arithmetic mean (A.M.) of 10 numbers is 20. If the numbers are doubled, which of the following is the A.M. of the new numbers?

(*a*) 10   (*b*) 20   (*c*) 30   (*d*) 40 (*Bihar SBTE, 2010, 2004*) **Ans.** (*d*)

Fill up the blanks:

(*xviii*) The mean of two numbers be 3 and that of another three numbers be 3. The combined mean of all the five numbers will be ....

(*Bihar SBTE, 2005*) **Ans.** (3)

(*xix*) If 5 is the A.M. of the numbers 8, 3, 5, 6, x, ... is the new mean if 2 is added to each number.

(*Bihar SBTE, 2005*) **Ans.** 7

9.25 MOMENT ABOUT MEAN

Let $\overline{x}$ be the arithmetic mean, then

$$\mu_r = \frac{1}{N} \sum_{i=1}^{n} f_i (x_i - \overline{x})^r, \quad r = 0, 1, 3.......$$

where, $N = \sum_{i=1}^{n} f_i$

If $r = 0$, $\mu_0 = \frac{1}{N} \sum_{i=1}^{n} f_i (x_i - \overline{x})^0 = 1$

If $r = 1$, $\mu_1 = \frac{1}{N} \sum_{i=1}^{n} f_i (x_i - \overline{x}) = 0$

If $r = 2$, $\mu_2 = \frac{1}{N} \sum_{i=1}^{n} f_i (x_i - \overline{x})^2$ [μ_2 = variance]

If $r = 3$, $\mu_3 = \frac{1}{N} \sum_{i=1}^{n} f_i (x_i - \overline{x})^3$

If $r = 4$, $\mu_4 = \frac{1}{N} \sum_{i=1}^{n} f_i (x_i - \overline{x})^4$

9.26 MEASURE OF SKEWNESS BASED ON MOMENT

1. Measure of skewness is given by β_1 where

$$\beta_1 = \frac{\mu_3^2}{\mu_2^3}$$

The sign of Karl Pearson's coefficient of skewness is determined from the sign of μ_3.

2. Measure of Kurtosis is given by β_2, where

$$\beta_2 = \frac{\mu_4}{\mu_2^2}$$

3. Gamma Coefficients

$$\boxed{\gamma_1 = \pm \sqrt{\beta_1}}$$

$$\boxed{\gamma_2 = \beta_2 - 3}$$

9.27 KURTOSIS
(*U.P. III Semester Dec. 2006*)

It measures the degree of peakedness of a distribution and is given by measure of kurtosis

$$\beta_2 = \frac{\mu_4}{\mu_2^2}, \quad \mu_2 = \frac{\Sigma (x - \overline{x})^2}{N}, \quad \mu_4 = \frac{\Sigma (x - \overline{x})^4}{N}$$

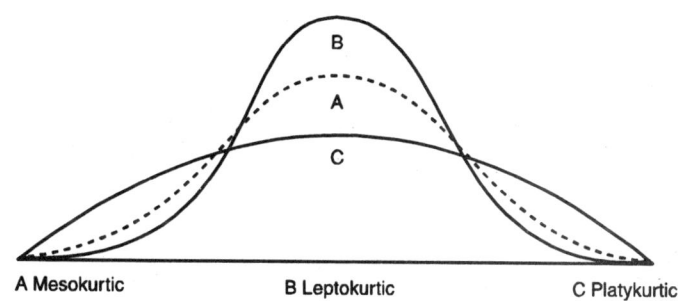

A Mesokurtic   B Leptokurtic   C Platykurtic

If $\beta_2 = 3$, the curve is normal or mesokurtic.

If $\beta_2 > 3$, the curve is peaked or leptokurtic.

If $\beta_2 < 3$, the curve is flat topped or platykurtic.

$$\gamma_2 = \beta_2 - 3$$

CHAPTER 10 Probability

10.1 PROBABILITY

(Bihar SBTE, 2010)

Probability is a concept which numerically measure the degree of uncertainty and therefore, of certainty of the occurrence of events.

If an event A can happen in m ways, and fail in n ways, all these ways being equally likely to occur, then the probability of the happening of A is

$$= \frac{\text{Number of favourable cases}}{\text{Total number of mutually exclusive and equally likely cases}} = \frac{m}{m+n}$$

and that of its failing is defined as $\dfrac{n}{m+n}$

If the probability of the happening $= p$
and the probability of not happening $= q$

then
$$p+q = \frac{m}{m+n} + \frac{n}{m+n} = \frac{m+n}{m+n} = 1 \text{ or } p+q = 1$$

For instance, on tossing a coin, the probability of getting a head is $\dfrac{1}{2}$.

10.2 DEFINITIONS

1. **Die :** It is a small cube. Dots are . .. ... :: ::. ::: marked on its faces. Plural of the die is dice. On throwing a die, the outcome is the number of dots on its upper face.

2. **Cards :** A pack of cards consists of four suits *i.e.* Spades, Hearts, Diamonds and Clubs. Each suit consists of 13 cards, nine cards numbered 2, 3, 4, ..., 10, and Ace, a King, a Queen and a Jack or Knave. Colour of Spades and Clubs is black and that of Hearts and Diamonds is red. Aces, Kings, Queens, and Jacks are known as *face* cards.

3. **Exhaustive Events or Sample Space :** The set of all possible outcomes of a single performance of an experiment is exhaustive events or sample space. Each outcome is called a sample point. In case of tossing a coin once, $S = (H,T)$ is the *sample space*. Two outcomes Head and Tail constitute an exhaustive event because no other outcome is possible.

4. **Random Experiment :** There are experiments, in which results may be altogether different, even though they are performed under identical conditions. They are known as random experiments. Tossing a coin or throwing a die is random experiment.

5. **Continuous Random Variables:** A *continuous random variable* is one which can assume any value within a number. *i.e.,* all values of continuous scale. For example (*i*) the weights (in kg) of a group of individuals, (*ii*) the heights of a group of individuals.

6. **A discrete random variable** is one which can assume only isolated values. For example, (*i*) the number of heads in 4 tosses of a coin is a discrete random variable as it cannot assume values other than 0, 1, 2, 3, 4.

7. **Trail and Event :** Performing a random experiment is called a trial and outcome is termed as event. Tossing of a coin is a trial and the turning up of head or tail is an event.

8. **Equally likely events:** Two events are said to be '*equally likely*', if one of them cannot be expected in preference to the other. For instance, if we draw a card from well-shuffled pack, we may get any card, then the 52 different cases are equally likely.

9. **Independent event :** Two events may be *independent*, when the actual happening of one does not influence in any way the probability of the happening of the other.

   Example. The event of getting head on first coin and the event of getting tail on the second coin in a simultaneous throw of two coins are independent.

10. **Mutually Exclusive events:** Two events are known as *mutually exclusive*, when the occurrence of one of them excludes the occurrence of the other. For example, on tossing of a coin, either we get head or tail, but not both.

11. **Compound Event :** When two or more events occur in composition with each other, the simultaneous occurrence is called a compound event. When a die is thrown, getting a 5 or 6 is a compound event.

12. **Favorable Events :** The events, which ensure the required happening, are said to be favourable events. For example, in throwing a die, to have the even numbers, 2, 4 and 6 are favourable cases.

13. **Conditional Probability :** The probability of happening an event A, such that event B has already happened, is called the conditional probability of happening of A on the condition that B has already happened. It is usually denoted by $P(A/B)$.

14. **Odds in favour of an event and odds against an event**

    If number of favourable ways = m, number of not favourable events = n

    (*i*) Odds in favour of the event = $\dfrac{m}{n}$, Odds against the event = $\dfrac{n}{m}$.

15. **Classical Definition of Probability.** If there are N equally likely, mutually, exclusive and exhaustive events of an experiment and m of these are favourable, then the probability of the happening of the event is defined as $\dfrac{m}{N}$.

16. **Expected value.** If $p_1, p_2, p_3, ..., p_n$ are the probabilities of the events $x_1, x_2, x_3 ... x_n$ respectively the expected value

$$E(x) = p_1 x_1 + p_2 x_2 + p_3 x_3 + ... + p_n x_n$$

$$= \sum_{r=1}^{n} p_r x_r$$

Example 1. *Two dice are thrown. Find the probability that the sum of the numbers coming up on them is 9, if it is known that the number 5 always occurs on the first die.* (*Bihar SBTE, 2011*)

Solution. Sample space i.e. S = {(5, 1), (5, 2), (5, 3), (5, 4), (5, 5), (5, 6)} $\Rightarrow$ $n(S) = 6$

E = Favourable outcome = (5, 4) $\Rightarrow$ $n(E) = 1$

$$P(E) = \frac{n(E)}{n(S)} = \frac{1}{6}$$ **Ans.**

Example 2. *Find the probability of throwing*
          (a) 5, (b) an even number with an ordinary six faced die.

Solution.   (a) There are 6 possible ways in which the die can fall and there is only one way of throwing 5.

$$\text{Probability} = \frac{\text{Number of favourable ways}}{\text{Total number of equally likely ways}} = \frac{1}{6}$$ **Ans.**

(b) Total number of ways of throwing a die = 6

   Number of ways falling 2, 4, 6 = 3

$$\text{The required probability} = \frac{3}{6} = \frac{1}{2}$$ **Ans.**

Example 3.   *Find the probability of throwing 9 with two dice.*

Solution. Total number of possible ways of throwing two dice

$$= 6 \times 6 = 36$$

Number of ways getting 9. *i.e.,* (3 + 6), (4 + 5), (5 + 4), (6 + 3) = 4.

$\therefore$ The required probability $= \dfrac{4}{36} = \dfrac{1}{9}$.   **Ans.**

Example 4. *From a pack of 52 cards, one is drawn at random. Find the probability of getting a king.*

Solution. A king can be chosen in 4 ways.

   But a card can be drawn in 52 ways.

$\therefore$ The required probability $= \dfrac{4}{52} = \dfrac{1}{13}$ **Ans.**

Example 5. *One card is drawn from a pack of cards. Find the probability that the card is either red or a king or both.*
                    (Bihar SBTE, 2009)

Solution.   The probability of drawing a red card $= \dfrac{26}{52} = \dfrac{1}{2}$

          The probability of drawing a king card $= \dfrac{4}{52} = \dfrac{1}{13}$

          The probability of drawing a red king (both) $= \dfrac{1}{2} \times \dfrac{1}{13} = \dfrac{1}{26}$

          Required probability $= \dfrac{1}{2} + \dfrac{1}{13} - \dfrac{1}{26} = \dfrac{7}{13}$ **Ans.**

EXERCISE 10.1

1.   In a class of 12 students, 5 are boys and the rest are girls. Find the probability that a student selected will be a girl.   **Ans.** $\dfrac{7}{12}$

2. A bag contains 7 red and 8 black balls. Find the probability of drawing a red ball.   **Ans.** $\dfrac{7}{15}$

3.   Three of the six vertices of a regular hexagon are chosen at random. Find the probability that the triangle with three vertices is equilateral.   **Ans.** $\dfrac{1}{10}$

4. What is the probability that a leap year, selected at random, will contain 53 Sundays.

(*A.M.I.E., Summer 2001*) **Ans.** $\dfrac{2}{7}$

Fill in the blanks with appropriate correct answer

5. Chance of throwing 6 at least once in four throws with single dice is ....

(*A.M.I.E., Summer 2000*) **Ans.** $\dfrac{671}{1296}$

6. A pair of fair dice is thrown and one die shows a four. The probability that the other die shows 5 is ....

(*A.M.I.E., Summer 2000*) **Ans.** $\dfrac{1}{36}$

Choose the correct answer:

7. In a given race, the odds in favour of horses *A, B, C, D* are 1 : 3, 1 : 4, 1 : 5, 1 : 6 respectively. The probability that horse C wins the race is

(*i*) $\dfrac{1}{4}$ (*ii*) $\dfrac{1}{5}$ (*iii*) $\dfrac{1}{6}$ (*iv*) $\dfrac{1}{7}$ **Ans.** (*iii*)

8. In tossing a fair die, the probability of getting on odd number or a number less than 4 is

(*i*) 2   (*ii*) 1/2   (*iii*) 2/3   (*iv*) 3/4   **Ans.** (*iii*)

10.3 ADDITION LAW OF PROBABILITY

(*Bihar SBTE, 2008, 2010*)

If $p_1, p_2, \ldots p_n$ be separate probabilities of mutually exclusive events, then the probability P, that any of these events will happen is given by $P = p_1 + p_2 + p_3 + \ldots + p_n$ (*Bihar SBTE, 2011, 2010*)

Proof. Let A, B, C,...... be the events, where probabilities are respectively $p_1, p_2, \ldots p_n$.
Let *n* be the total number of favourable cases to either A or B or C or....

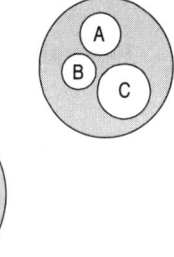

$$= m_1 + m_2 + m_3 + \ldots + m_n$$

Hence, $P(A + B + C + \ldots) = \dfrac{m_1 + m_2 + m_3 + \ldots + m_n}{n}$

$$= \dfrac{m_1}{n} + \dfrac{m_2}{n} + \dfrac{m_3}{n} + \ldots + \dfrac{m_n}{n}$$

$$= P(A) + P(B) + P(C) + \ldots$$

$$P = p_1 + p_2 + p_3 + \ldots + p_n \qquad \textbf{Proved.}$$

Note Mutually Exclusive Events

Consider the case where two events A and B are not mutually exclusive. The probability of the event that either A or B or both occur is given as

$$P(A \cup B) = P(A) + P(B) - P(A \cap B)$$

Example 6. *A bag contains four white and two black balls and a second bag contains three of each colour. A bag is selected at random, and a ball is then drawn at random from the bag chosen. What is the probability that the ball drawn is white ?.*

Solution. There are two mutually exclusive cases,

(*i*) when the first bag is chosen, (*ii*) when the second bag is chosen.

Now the chance of choosing the first bag is $\dfrac{1}{2}$ and if this bag is chosen, the probability of drawing a white ball is 4/6. Hence the probability of drawing a white ball from first bag is

$$\dfrac{1}{2} \times \dfrac{4}{6} = \dfrac{1}{3}$$

Similarly the probability of drawing a white ball from second bag is

$$\frac{1}{2} \times \frac{3}{6} = \frac{1}{4}$$

Since the events are mutually exclusive the required probability

$$= \frac{1}{3} + \frac{1}{4} = \frac{7}{12}$$

Ans.

Example 7. *Three machines I, II and III manufacture respectively 0.4, 0.5 and 0.1 of the total production. The percentage of defective items produced by I, II and III is 2, 4 and 1 per cent respectively. For an item chosen at random, what is the probability it is defective ?*

Solution. The defective item produced by machine I

$$= \frac{0.4 \times 2}{100} = \frac{0.8}{100}$$

The defective item produced by machine II $= \dfrac{0.5 \times 4}{100} = \dfrac{2}{100}$

The defective item produced by machine III $= \dfrac{0.1 \times 1}{100} = \dfrac{0.1}{100}$

The total defective items produced by machines I, II, III

$$= \frac{0.8}{100} + \frac{2}{100} + \frac{0.1}{100} = \frac{2.9}{100} = 0.029$$

The required probability $= \dfrac{0.029}{1} = 0.029$

Ans.

Example 8. *An urn contains 10 black and 10 white balls. Find the probability of drawing two balls of the same colour.* *(Bihar SBTE, 2009)*

Solution. Probability of drawing two black balls $= \dfrac{^{10}C_2}{^{20}C_2}$

$\therefore$ Probability of drawing two white balls $= \dfrac{^{10}C_2}{^{20}C_2}$

$\therefore$ Probability of drawing two balls of the same colour

$$= \frac{^{10}C_2}{^{20}C_2} + \frac{^{10}C_2}{^{20}C_2} = 2. \frac{^{10}C_2}{^{20}C_2} = 2. \frac{\dfrac{10 \times 9}{2 \times 1}}{\dfrac{20 \times 19}{2 \times 1}} = \frac{9}{19}$$

Ans.

10.4 MULTIPLICATION LAW OF PROBABILITY
(Bihar SBTE, 2008, 2009, 2010)

If there are two independent events the respective probabilities of which are known, then the probability that both will happen is the product of the probabilities of their happening respectively.

$$P(AB) = P(A) \times P(B) \qquad \textit{(Bihar, SBTE, 2010, 2009)}$$

Proof. Suppose A and B are two independent events. Let A happen in m_1 ways and fail in n_1 ways.

$\therefore$
$$P(A) = \frac{m_1}{m_1 + n_1}$$

Also let B happen in m_2 ways and fail in n_2 ways.

∴ $P(B) = \dfrac{m_2}{m_2 + n_2}$

Now there are four possibilities

A and B both may happen, then the number of ways $= m_1 \cdot m_2$.

A may happen and B may fail, then the number of ways $= m_1 \cdot n_2$

A may fail and B may happen, then the number of ways $= n_1 \cdot m_2$

A and B both may fail, then the number of ways $= n_1 \cdot n_2$

Thus, the total number of ways $= m_1 m_2 + m_1 n_2 + n_1 m_2 + n_1 n_2 = (m_1 + n_1)(m_2 + n_2)$

Hence the probabilities of the happening of both A and B

$$P(AB) = \dfrac{m_1 m_2}{(m_1 + n_1)(m_2 + n_2)} = \dfrac{m_1}{m_1 + n_1} \cdot \dfrac{m_2}{m_2 + n_2} = P(A) \cdot P(B) \textbf{ Proved.}$$

Example 9. *Find the probability of drawing an ace, a king and a queen in three successive draws from a pack of cards if the cards are not replaced after each draw.*   *(Bihar SBTE, 2003)*

Solution.   The probability of drawing an ace $= \dfrac{4}{52} = \dfrac{1}{13}$

If the card is not replaced, the pack will now have 51 cards.

Thus the probability of drawing a king $= \dfrac{4}{51}$.

The probability of drawing a queen $= \dfrac{4}{50} = \dfrac{2}{25}$

∴   The probability of drawing three cards $= \dfrac{1}{13} \times \dfrac{4}{51} \times \dfrac{2}{25} = \dfrac{8}{16575}$ **Ans.**

Example 10. Aa card is drawn from an ordinary pack and a gambler bets that it is a spade or a king. What the probability of his not winning the bet?   *(Bihar SBTE, 2012)*

Solution.   P: (king) $= \dfrac{4}{52} = \dfrac{1}{13}$

P (spade) $= \dfrac{13}{52} = \dfrac{1}{4}$,   P (king of spade) $= \dfrac{1}{52}$

Required P i.e. $= \dfrac{1}{13} + \dfrac{1}{4} - \dfrac{1}{52} = \dfrac{4}{13}$

P (of his not winging the bet) $= 1 - \dfrac{4}{13} = \dfrac{9}{13}$ **Ans.**

Example 11. *An article manufactured by a company consists of two parts A and B. In the process of manufacture of part A, 9 out of 100 are likely to be defective. Similarly, 5 out of 100 are likely to be defective in the manufacture of part B. Calculate the probability that the assembled article will not be defective (assuming that the events of finding the part A non-defective and that of B are independent).*

Solution. Probability that part A will be defective $= \dfrac{9}{100}$

Probability that part A will not be defective $= \left(1 - \dfrac{9}{100}\right) = \dfrac{91}{100}$

Probability that part B will be defective $= \dfrac{5}{100}$

Probability that part B will not be defective $= \left(1 - \dfrac{5}{100}\right) = \dfrac{95}{100}$

Probability that the assembled article will not be defective = (Probability that part A will not be defective)× (Probability that part B will not be defective)

$$= \left(\dfrac{91}{100}\right) \times \left(\dfrac{95}{100}\right) = 0.8645 \qquad \textbf{Ans.}$$

Example 12. *The probability that machine A will be performing an usual function in 5 years' time is* $\dfrac{1}{4}$, *while the probability that machine B will still be operating usefully at the end of the same period, is* $\dfrac{1}{3}$

Find the probability in the following cases that in 5 years time :

    (*i*) *Both machines will be performing an usual function.*

   (*ii*) *Neither will be operating.*

  (*iii*) *Only machine B will be operating.*

  (*iv*) *At least one of the machines will be operating.*

Solution. $P\,(A \text{ operating usefully}) = \dfrac{1}{4}$, so $q\,(A) = 1 - \dfrac{1}{4} = \dfrac{3}{4}$

$$P\,(B \text{ operating usefully}) = \dfrac{1}{3}, \text{ so } q\,(B) = 1 - \dfrac{1}{3} = \dfrac{2}{3}$$

(*i*) P (Both A and B will operate usefully) $= P\,(A).\,P\,(B) = \left(\dfrac{1}{4}\right) \times \left(\dfrac{1}{3}\right) = \dfrac{1}{12}$

(*ii*) P (Neither will be operating) $= q\,(A)\,.\,q\,(B) = \left(\dfrac{3}{4}\right)\left(\dfrac{2}{3}\right) = \dfrac{1}{2}$

(*iii*) P (Only B will be operating) $= P\,(B) \times q\,(A) = \left(\dfrac{1}{3}\right) \times \left(\dfrac{3}{4}\right) = \dfrac{1}{4}$

(*iv*) P (At least one of the machines will be operating)

$$= 1 - P \text{ (none of them operates)} = 1 - \dfrac{1}{2} = \dfrac{1}{2} \qquad \textbf{Ans.}$$

Example 13. *There are two groups of subjects one of which consists of 5 science and 3 engineering subjects and the other consists of 3 science and 5 engineering subjects. An unbiased die is cast. If number 3 or number 5 turns up, a subject is selected at random from the first group, otherwise the subject is selected at random from the second group. Find the probability that an engineering subject is selected ultimately.*

Solution. Probability of turning up 3 or 5 $= \dfrac{2}{6} = \dfrac{1}{3}$

Probability of selecting engineering subject from first group $= \dfrac{3}{8}$

Now the probability of selecting engineering subject from first group on turning up 3 or 5

$$= \left(\frac{1}{3}\right) \times \left(\frac{3}{8}\right) = \frac{1}{8} \qquad \qquad ...(1)$$

Probability of not turning up 3 or 5 $= 1 - \frac{1}{3} = \frac{2}{3}$

Probability of selecting engineering subject from second group $= \frac{5}{8}$

Now probability of selecting engineering subject from second group on not turning up 3 or 5

$$= \frac{2}{3} \times \frac{5}{8} = \frac{5}{12} \qquad \qquad ...(2)$$

Probability of the selection of engineering subject $= \frac{1}{8} + \frac{5}{12}$ $\qquad$ [From (1) and (2)]

$$= \frac{13}{24} \qquad \qquad \textbf{Ans.}$$

Example 14. *An urn A contains 2 white and 4 black balls. Another urn B contains 5 white and 7 black balls. A ball is transferred from the urn A to the urn B, then a ball is drawn from urn B. Find the probability that it is white.*

Solution. Urn A contains 2 white and 4 black balls.

Urn B contains 5 white and 7 black balls.

Now there are two cases of transferring a ball from A to B.

Case I. When a white ball is transferred from A to B

$$P \text{ (Transfer of a white ball)} = \frac{2}{2+4} = \frac{1}{3}$$

After transfer of a white ball, urn B contains 6 white balls and 7 black balls.

P (Drawing a white ball from urn B after transfer)

$$= P \text{ (Transfer of a white ball)} \times P \text{ (Drawing of a white ball)}$$

$$= \left(\frac{1}{3}\right)\left(\frac{6}{6+7}\right) = \frac{1}{3} \times \frac{6}{13} = \frac{2}{13}$$

Case II. When a black ball is transferred from A to B.

$$P \text{ (Transfer of a black ball)} = \frac{4}{2+4} = \frac{2}{3}$$

After transfer of a black ball, urn B contains 5 white and 8 black balls.

P (Drawing a white ball from urn B after transfer)

$$= P \text{ (Transfer of a black ball)} \times P \text{ (Drawing of a white ball)}$$

$$= \frac{2}{3}\left(\frac{5}{5+8}\right) = \frac{10}{39}$$

Required probability $= \frac{2}{13} + \frac{10}{39} = \frac{16}{39}$ $\qquad \qquad$ **Ans.**

Example 15. *A bag contains 5 white and 7 black balls. Two balls are drawn in succession.. What is the probability that first is white and second is black ?* $\qquad$ *(Bihar SBTE, 2010)*

Solution. $\qquad$ Let A be the event of a white ball and a black ball $n(A) = {}^5C_1 \times {}^7C_1$

$$\text{Reqd. probability} = \frac{{}^5C_1 \times {}^7C_1}{12_2} = \frac{5 \times 7}{66} \qquad \qquad \textbf{Ans.}$$

Example 16. *A bag contains 10 white and 15 black balls. Two balls are drawn in succession. What is the probability that both are of different colour?*   **(Bihar SBTE, 2009)**

Solution.   Probability of drawing a white ball $= \dfrac{10}{25}$

After a white ball has been drawn there remains 15 black and 9 white balls.

Now, probability of drawing black ball $= \dfrac{15}{24}$

$\therefore$ The probability of drawing first a white and 2nd black $= \dfrac{10}{25} \times \dfrac{15}{24} = \dfrac{1}{4}$

Similarly the probability of drawing first a black and then a white ball $= \dfrac{15}{25} \times \dfrac{10}{24} = \dfrac{1}{4}$

Thus the required probability $= \dfrac{1}{4} + \dfrac{1}{4} = \dfrac{1}{2}$ **Ans.**

Example 17. *Three groups of children contain respectively 3 girls and 1 boy; 2 girls and 2 boys; 1 girl and 3 boys. One child is selected at random from each group. Find the chance of selecting 1 girl and 2 boys.*

Solution. There are three ways of selecting 1 girl and two boys.

I way : Girl is selected from first group, boy from second group and second boy from third group.

Probability of the selection of (Girl + Boy + Boy)

$$= \frac{3}{4} \times \frac{2}{4} \times \frac{3}{4} = \frac{18}{64}$$

II way : Boy is selected from first group, girl from second group and second boy from third group.

Probability of the selection of (Boy + Girl + Boy)

$$= \frac{1}{4} \times \frac{2}{4} \times \frac{3}{4} = \frac{6}{64}$$

III way : Boy is selected from first group, second boy from second group and the girl from the third group.

Probability of selection of (Boy + Boy + Girl)

$$= \frac{1}{4} \times \frac{2}{4} \times \frac{1}{4} = \frac{2}{64}$$

Total probability $= \dfrac{18}{64} + \dfrac{6}{64} + \dfrac{2}{64} = \dfrac{26}{64} = \dfrac{13}{32}$ **Ans.**

Example 18. *The number of children in a family in a region are either 0, 1 or 2 with probability 0.2, 0.3 and 0.5 respectively. The probability of each child being a boy or girl 0.5. Find the probability that a family has no boy.*

Solution. Here there are three types of families

(*i*) Probability of zero child (boys) = 0.2

(*ii*)

Boy	Girl
0	1
1	0

Probability of zero boy in case II

$$= 0.3 \times 0.5 = 0.15$$

(iii)

Boy	Girl
0	2
1	1
2	0

In this case probability of zero boy $= 0.5 \times \dfrac{1}{3} = 0.167$

Considering all the three cases, the probability of zero boy

$$= 0.2 + 0.15 + 0.167 = 0.517 \qquad \text{Ans.}$$

Example 19. *A husband and wife appear in an interview for two vacancies in the same post. The probability of husband's selection is* $\dfrac{1}{7}$ *and that of wife's selection is* $\dfrac{1}{5}$. *What is the probability that*

(i)   *both of them will be selected.*   (ii)   *only one of them will be selected, and*

(iii) *none of them will be selected ?*

Solution. P (husband's selection) $= \dfrac{1}{7}$, P (wife's selection) $= \dfrac{1}{5}$

(i) P (both selected) $= \dfrac{1}{7} \times \dfrac{1}{5} = \dfrac{1}{35}$

(ii) P (only one selected) $= P$ (only husband's selection) $+ P$ (only wife's selection)\

$$= \dfrac{1}{7} \times \dfrac{4}{5} + \dfrac{1}{5} \times \dfrac{6}{7} = \dfrac{10}{35} = \dfrac{2}{7}$$

(iii) P (none of them will be selected) $= \dfrac{6}{7} \times \dfrac{4}{5} = \dfrac{24}{35}$ \qquad **Ans.**

Example 20. *A problem of statistics is given to three students A, B and C whose chances of solving it are* $\dfrac{1}{2}, \dfrac{1}{3}$ *and* $\dfrac{1}{4}$ *respectively. What is the probability that the problem will be solved ?*

(Bihar SBTE, 2011, 2010)

Solution.   The probability that A can solve the problem $= \dfrac{1}{2}$

The probability that A cannot solve the problem $= 1 - \dfrac{1}{2}$.

Similarly the probability that B and C cannot solve the problem are $\left(1 - \dfrac{1}{3}\right)$ and $\left(1 - \dfrac{1}{4}\right)$

$\therefore$ The probability that A, B, C cannot solve the problem $= \left(1 - \dfrac{1}{2}\right) \times \left(1 - \dfrac{1}{3}\right) \times \left(1 - \dfrac{1}{4}\right) = \dfrac{1}{2} \times \dfrac{2}{4} \times \dfrac{3}{4} = \dfrac{1}{4}$

Hence, the probability that the problem can be solved $= 1 - \dfrac{1}{4} = \dfrac{3}{4}$ \qquad **Ans.**

Example 21. *A student takes his examination in four subjects* α, β, γ, δ. *He estimates his chances of passing in* α *as* $\dfrac{4}{5}$, *in* β *as* $\dfrac{3}{4}$, *in* γ *as* $\dfrac{5}{6}$ *and in* δ *as* $\dfrac{2}{3}$. *To qualify, he must pass in* α *and at least two other subjects. What is the probability that he qualifies?*

Solution. $P(\alpha) = \dfrac{4}{5}$, $\quad P(\beta) = \dfrac{3}{4}$, $\quad P(\gamma) = \dfrac{5}{6}$, $\quad P(\delta) = \dfrac{2}{3}$

There are four possibilities of passing at least two subjects.

(*i*) Probability of passing β, γ and failing δ $= \dfrac{3}{4} \times \dfrac{5}{6} \times \left(1 - \dfrac{2}{3}\right) = \dfrac{3}{4} \times \dfrac{5}{6} \times \dfrac{1}{3} = \dfrac{5}{24}$

(*ii*) Probability of passing γ, δ and failing β $= \dfrac{5}{6} \times \dfrac{2}{3} \times \left(1 - \dfrac{3}{4}\right) = \dfrac{5}{6} \times \dfrac{2}{3} \times \dfrac{1}{4} = \dfrac{5}{36}$

(*iii*) Probability of passing δ, β and failing γ $= \dfrac{2}{3} \times \dfrac{3}{4} \times \left(1 - \dfrac{5}{6}\right) = \dfrac{2}{3} \times \dfrac{3}{4} \times \dfrac{1}{6} = \dfrac{1}{12}$

(*iv*) Probability of passing β, γ , δ $= \dfrac{3}{4} \times \dfrac{5}{6} \times \dfrac{2}{3} = \dfrac{5}{12}$

Probability of passing at least two subjects. $= \dfrac{5}{24} + \dfrac{5}{36} + \dfrac{1}{12} + \dfrac{5}{12} = \dfrac{61}{72}$

Probability of passing α and at least two subjects. $= \dfrac{4}{5} \times \dfrac{61}{72} = \dfrac{61}{90}$ **Ans.**

Example 22. *A box contains 9 tickets numbered 1 to 9 inclusive. If 3 tickets are drawn from the box one at a time, find the probability they are alternatively either odd, even, odd or even, odd, even.*
                                                                                                  (MDU. Dec. 2009)

Solution.   Total number of tickets = 9

                    Number of odd tickets = 5

                    Number of even tickets = 4

$P(\text{odd, even, odd}) = P(\text{odd}) \cdot P(\text{even}) \cdot P(\text{odd}) = \dfrac{5}{9} \times \dfrac{4}{8} \times \dfrac{4}{7} = \dfrac{10}{63}$

$P(\text{even, odd, even}) = P(\text{even}) \cdot P(\text{odd}) \cdot P(\text{even}) = \dfrac{4}{9} \times \dfrac{5}{8} \times \dfrac{3}{7} = \dfrac{5}{42}$

Required probability $= P(\text{odd, even, odd or even, odd, even})$

$$= P(\text{odd, even, odd}) + P(\text{even, odd, even}) = \dfrac{10}{63} + \dfrac{5}{42} = \dfrac{5}{8}$$ **Ans.**

Example 23. *Bag A contains 10 red and 5 white balls. Bag B contains 8 red and 7 white balls. If any one ball (red or white) is transferred from A to bag B, find the probability of drawing one white ball from bag B.*
                                                                                                  (Bihar SBTE, 2012)

Solution.   Now there are two case of (transferring a ball from A to B).
                    Case 1. When a white ball is transferred from A + B)

$$P \text{ (transfer of a white ball)} = \dfrac{5}{10 + 5} = \dfrac{1}{3}$$

P (drawing a white ball from bag. B after transferring

$$= p \text{ (transfer of white ball)} \times p \text{ (drawing of a white ball} \left(\dfrac{1}{3}\right)\left(\dfrac{8}{8 + 8}\right) = \dfrac{1}{3} \times \dfrac{1}{2} = \dfrac{1}{6}$$

Case 2. When a red ball is transfer from A to B

$$P \text{ (transfer of R Red ball)} = \dfrac{10}{10 + 5} = \dfrac{10}{15} = \dfrac{2}{3}$$

P (drawing a white ball from bag B after transfer)

= P (transfer of a Red ball) × P (Drawing of a white II)

$$= \frac{2}{3}\left(\frac{7}{7+9}\right) = \frac{2}{3} \times \frac{7}{16} = \frac{14}{48}$$

Required Productivity $= \frac{1}{6} + \frac{14}{40} = \frac{31}{60}$ **Ans.**

Example 24. *A bag contains 10 black and 10 white balls. Find the probability of drawing two balls of the same colour* (*Bihar SBTE, 2009*)

Solution. Here p (s) = $^{20}C_2$ = 190

n(W) = $^{10}C_2$ = 45

$$P\ (B) = \frac{45}{190} = p(w)$$

The probability of drawing two balls of the same colour $= \frac{45}{190} = \frac{45}{190} + \frac{45}{190}$

$$\frac{90}{190} = \frac{9}{19}$$ **Ans.**

Example 25. *A committee is to be formed by choosing two boys and four girls out of a group of five boys and six girls. What is the probability that a particular boy named A and a particular girl named B are selected in the committee?*

Solution. Two boys are to be selected out of 5 boys. A particular boy A is to be included in the committee. It means that only 1 boy is to be selected out of 4 boys.

Number of ways of selection = $^4C_1$

Similarly a girl B is to be included in the committee.

Then only 3 girls are to be selected out of 5 girls.

Number of ways of selection = $^5C_3$

Required probability $= \frac{^4C_1 \times {}^5C_3}{^5C_2 \times {}^6C_4} = \frac{4 \times 10}{10 \times 15} = \frac{4}{15}$ **Ans.**

Example 26. *There are 6 positive and 8 negative numbers. Four numbers are chosen at random, without replacement, and multiplied. What is the probability that the product is a positive number?*

Solution. To get from the product of four numbers, a positive number, the possible combinations are as follows :

S. No.	Out of 6 Positive Numbers	Out of 8 Negative Numbers	Positive Numbers
1.	4	0	$^6C_4 \times {}^8C_0 = \frac{6 \times 5}{1 \times 2} \times 1 = 15$
2.	2	2	$^6C_2 \times {}^8C_2 = \frac{6 \times 5}{1 \times 2} \times \frac{8 \times 7}{1 \times 2} = 420$
3.	0	4	$^6C_0 \times {}^8C_4 = 1 \times \frac{8 \times 7 \times 6 \times 5}{1 \times 2 \times 3 \times 4} = 70$
			Total = 505

$$\text{Required Probability} = \frac{{}^6C_4 \times {}^8 C_0 + {}^6 C_2 \times {}^8 C_2 + {}^6 C_0 \times {}^8 C_4}{{}^{14}C_4}$$

$$= \frac{15 + 420 + 70}{\dfrac{14 \times 13 \times 12 \times 11}{1 \times 2 \times 3 \times 4}} = \frac{505 \times 4 \times 3 \times 2 \times 1}{14 \times 13 \times 12 \times 11} = \frac{505}{1001} \qquad \textbf{Ans.}$$

Example 27. *A six-faced die is so biased that, when thrown, it is twice as likely to show an even number than an odd number. If it is thrown twice, what is the probability that the sum of two numbers thrown is odd.*

Solution. A biased die, when thrown, shows even number twice than an odd number.

Probability of showing even number $= \dfrac{2}{2+1} = \dfrac{2}{3}$

Probability of showing odd number $= \dfrac{1}{1+2} = \dfrac{1}{3}$

Sum of two numbers is odd if the first is even and the second is odd or vice versa.
Probability of sum to be odd = Probability of an even number × Probability of an odd number + Probability of an odd number × Probability of an even number.

$$= \frac{2}{3} \times \frac{1}{3} + \frac{1}{3} \times \frac{2}{3} = \frac{2}{9} + \frac{2}{9} = \frac{4}{9} \qquad \textbf{Ans.}$$

Example 28. *A can hit a target 3 times in 5 shots, B 2 times in 5 shots and C three times in 4 shots. All of them fire one shot each simultaneously at the target. What is the probability that*

(i) *2 shots hit*   (ii) *At least two shots hit?*

Solution.   Probability of A hitting the target $= \dfrac{3}{5}$

Probability of B hitting the target $= \dfrac{2}{5}$

Probability of C hitting the target $= \dfrac{3}{4}$

(i) Probability that 2 shots hit the target

$$= P(A)\, P(B)\, q(C) + P(A)\, P(C)\, q(B) + P(B)\, P(C)\, q(A)$$

$$= \frac{3}{5} \times \frac{2}{5} \times \left(1 - \frac{3}{4}\right) + \frac{3}{5} \times \frac{3}{4} \times \left(1 - \frac{2}{5}\right) + \frac{2}{5} \times \frac{3}{4} \times \left(1 - \frac{3}{5}\right)$$

$$= \frac{6}{25} \times \frac{1}{4} + \frac{9}{20} \times \frac{3}{5} + \frac{6}{20} \times \frac{2}{5}$$

$$= \frac{6 + 27 + 12}{100} = \frac{45}{100} = \frac{9}{20} \qquad \textbf{Ans.}$$

(ii) Probability of at least two shots hitting the target

= Probability of 2 shots + probability of 3 shots hitting the target

$$= \frac{9}{20} + P(A)\, P(B)\ P(C) = \frac{9}{20} + \frac{3}{5} \times \frac{2}{5} \times \frac{3}{4} = \frac{63}{100} \qquad \textbf{Ans.}$$

Example 29. *A factory, manufacturing televisions has four units A, B, C, D. The units A, B, C, D manufacture 15%, 20%, 30% and 35% of the total output respectively. It was found that out of their outputs 1%, 2%, 2% and 3% are defective. A television is chosen at random from the total output and found to be defective. What is the probability that it came from unit D ?*   *(Bihar SBTE, 2010, 2009)*

Solution. Let the factory manufacture 100 televisions.

Unit A manufactures = 15 TV,   Unit B manufactures = 20 TV

Unit C manufactures = 30 TV,   Unit D manufactures = 35 TV

Defective TV manufactured by unit $A = 15 \times \dfrac{1}{100} = 0.15$

Defective TV manufactured by unit $B = 20 \times \dfrac{2}{100} = 0.4$

Defective TV manufactured by unit $C = 30 \times \dfrac{2}{100} = 0.6$

Defective TV manufactured by unit $D = 35 \times \dfrac{3}{100} = 1.05$

Total defective TV = 0.15 + 0.4 + 0.6 + 1.05 = 2.20.

Probability of defective TV from unit $D = \dfrac{1.05}{2.20} = \dfrac{21}{44}$ **Ans.**

Example 30. *A and B take turns in throwing two dice, the first to throw 10 being awarded the prize. Show that if A has the first throw, their chances of winning are in the ratio 12:11.*

Solution. The combinations of throwing 10 from two dice can be

$$(6 + 4), (4 + 6), (5 + 5).$$

The number of combinations is 3.

Total combinations from two dice = 6 × 6 = 36.

$\therefore$ The probability of throwing 10 = $p = \dfrac{3}{36} = \dfrac{1}{12}$

The probability of not getting 10 = $q = 1 - \left(\dfrac{1}{12}\right) = \dfrac{11}{12}$

If A is to win, he should throw 10 in either the first, the third, the fifth, ... throws.

Their respective probabilities are = $p, q^2 p, q^4 p, \ldots = \dfrac{1}{12}, \left(\dfrac{11}{12}\right)^2 \dfrac{1}{12}, \left(\dfrac{11}{12}\right)^4 \dfrac{1}{12} \ldots$

A's total probability of winning = $\dfrac{1}{12} + \left(\dfrac{11}{12}\right)^2 \cdot \dfrac{1}{12} + \left(\dfrac{11}{12}\right)^4 \cdot \dfrac{1}{12} + \ldots$

$$= \dfrac{\dfrac{1}{12}}{1 - \left(\dfrac{11}{12}\right)^2} = \dfrac{12}{23} \qquad \left[\text{This is infinite G.P. Its sum} = \dfrac{a}{1-r}\right]$$

B can win in either 2nd, 4th, 6th ... throws.

So B's total chance of winning = $qp + q^3 p + q^5 p + \ldots\ldots\ldots$

$$= \left(\frac{11}{12}\right)\left(\frac{1}{12}\right) + \left(\frac{11}{12}\right)^3\left(\frac{1}{12}\right) + \left(\frac{11}{12}\right)^5\left(\frac{1}{12}\right) + ... = \frac{\left(\frac{11}{12}\right)\left(\frac{1}{12}\right)}{1 - \left(\frac{11}{12}\right)^2} = \frac{11}{23}$$

Hence A's chance to B's chance $= \frac{12}{23} : \frac{11}{23} = 12 : 11$ **Proved.**

Example 31. *A and B throw alternatively a pair of dice. A wins if he throws 6 before B throws 7 and B wins if he throws 7 before A throws 6. Find their respective chances of winning, if A begins.*

Solution. Number of ways of throwing 6

i.e. $(1 + 5), (2 + 4), (3 + 3), (4 + 2), (5 + 1) = 5.$

Probability of throwing $6 = \frac{5}{36} = p_1, \quad q_1 = \frac{31}{36}$

Number of ways of throwing 7

i.e.; $(1 + 6), (2 + 5), (3 + 4), (4 + 3), (5 + 2), (6 + 1) = 6$

Probability of throwing $6 = \frac{6}{36} = \frac{1}{6} = P_2, \quad q_2 = \frac{5}{6}$

$$P(A) = p_1 + q_1 q_2 p_1 + q_1^2 q_2^2 p_1 + ...$$

$$P(B) = q_1 p_2 + q_1^2 q_2 p_2 + q_1^3 q_2^2 p_2 + ...$$

Probability of A's winning $= p_1 + q_1 q_2 p_1 + q_1^2 q_2^2 p_1 + ...$

$$= \frac{p_1}{1 - q_1 q_2} = \frac{\frac{5}{36}}{1 - \frac{31}{36} \times \frac{5}{6}} = \frac{5}{36} \times \frac{36 \times 6}{61} = \frac{31}{61}$$

Probability of B's winning $= q_1 p_2 + q_1^2 q_2 p_2 + q_1^3 q_2^2 p_2 + ...$

$$= \frac{q_1 p_2}{1 - q_1 q_2} = \frac{\frac{31}{36} \times \frac{1}{6}}{1 - \left(\frac{31}{36}\right)\left(\frac{5}{6}\right)} = \frac{31}{36 \times 6} \times \frac{36 \times 6}{61} = \frac{30}{61}$$ **Ans.**

EXERCISE 10.2

1. The probability that Nirmal will solve a problem is $\frac{2}{3}$ and the probability that Satyajit will solve it is $\frac{3}{4}$. What is

    the probability that (*a*) the problem will be solved (*b*) neither can solve it.   **Ans.** (*a*) $\frac{11}{12}$, (*b*) $\frac{1}{12}$

2. An urn contains 13 balls numbering 1 to 13. Find the probability that a ball selected at random is a ball with

    number that is a multiple of 3 or 4.   **Ans.** $\frac{6}{13}$

3. Four persons are chosen at random from a group containing 3 men, 2 women, and 4 children. Show that the

    probability that exactly two of them will be children is $\frac{10}{21}$.

4. A five digit number is formed by using the digits 0, 1, 2, 3, 4 and 5 without repetition. Find the probability that the number is divisible by 6.

Ans. $\dfrac{4}{25}$

5. The chances that doctor A will diagnose a disease X correctly is 60%. The chances that a patient will die by his treatment after correct diagnosis is 40% and the chances of death by wrong diagnosis is 70%. A patient of doctor A, who had disease X, died, what is the chance that his disease was diagnosed correctly.

Ans. $\dfrac{6}{13}$

6. An anti-aircraft gun can take a maximum of four shots on enemy's plane moving from it. The probabilities of hitting the plane at first, second, third and fourth shots are 0.4, 0.3, 0.2 and 0.1 respectively. Find the probability that the gun hits the plane.
Ans. 0.6976.

7. An electronic component consists of three parts. Each part has probability 0.99 of performing satisfactorily. The component fails if two or more parts do not perform satisfactorily. Assuming that the parts perform independently, determine the probability that the component does not perform satisfactorily.
Ans. 0.000298

8. The face cards are removed from a full pack. Out of the remaining 40 cards, 4 are drawn at random. What is the probability that they belong to different suits?

Ans. $\dfrac{1000}{9139}$

9. Of the cigarette smoking population, 70% are men and 30% women, 10% of these men and 20% of these women smoke 'WILLS.' What is the probability that a person seen smoking a 'WILLS' will be a man.
Ans. $\dfrac{7}{13}$

10. A machine contains a component C that is vital to its operation. The reliability of component C is 80%. To improve the reliability of a machine, a similar component is used in parallel to form a system S. The machine will work provided that one of these components functions correctly. Calculate the reliability of the system S.
Ans. 96%

11. In a bolt factory, machines A, B and C manufacture 25%, 35% and 40% of the total output respectively. Of their outputs, 5%, 4% and 2% are defective bolts. A bolt is chosen at random and found to be defective. What is the probability that the bolt came from machine A ? B ? C ?

Ans. $\dfrac{25}{69}, \dfrac{28}{69}, \dfrac{16}{69}$

12. One bag contains four white and two black beads and another contains three of each colour. A bead is drawn from each bag. What is the probability that one is white and one is black ?

Ans. $\dfrac{1}{2}$

13. The odds that a book will be favourably reviewed by three independent critics are 5 to 2, 4 to 3, 3 to 4 respectively. What is the probability that of the three reviews, a majority will be favourable?

Ans. $\dfrac{209}{343}$

14. Let E and F be independent events. The probability that both E and F happen is $\dfrac{1}{12}$ and the probability that neither E nor F happen is $\dfrac{1}{2}$. Then find $P(E)$ and $P(F)$.

Ans. $P(E) = \dfrac{1}{3}, P(F) = \dfrac{1}{4}$

15. Given a random variable whose range set is (1, 2) and whose probability is $f(1) = \dfrac{1}{4}$ and $f(2) = \dfrac{3}{4}$. Find the mean and variance of the distribution.

Ans. Mean $= \dfrac{7}{4}$, Var $= \dfrac{3}{16}$

16. A man takes a step forward with probability 0.4 and backward with probability 0.6. Find the probability that at the end of 11 steps, he is just one step away from the starting point. **Ans.** 0.5263

17. What would be the expectation of the number of failures preceding the first success in an infinite series of independent trials with the constant probability of success p ?

      Solution. The probabilities of success in 1st, 2nd, 3rd trials respectively are p , qp , $q^2 p$, $q^3 p$,....
The expected number of failures preceding the first success

$$E(x) = (0 \cdot p) + (1 \cdot qp) + (2 \cdot q^2 p) + ...\infty$$
$$= qp[1 + 2q + 3q^2 + ...\infty] \quad \text{where} \quad q < 1.$$
$$= \frac{qp}{(1-q)^2} = \frac{qp}{p^2} = \frac{q}{p} \qquad\qquad \textbf{Ans.}$$

18. A candidate is selected for interview for three posts. For the first post there are three candidates, for the second there are 4, and for the third are 2. What is the chance of getting at least one post?

                                               (A.M.I.E., Summer 2001) **Ans.** $\dfrac{3}{4}$

19. The chance of hitting a target by a bomb is 50% when 4 bombs are dropped, what is the probability of destroying the target, if one bomb is just sufficient to destroy it.   *(A.M.I.E., Winter 2003)* **Ans.** $\dfrac{15}{16}$

20. A pair of dice is tossed twice. Find the probability of scoring 7 points (*i*) once, (*ii*) at least once (*iii*) twice.

                    (K.U.K. Dec. 2009)   **Ans.** (*i*) $\dfrac{5}{8}$ (*ii*) $\dfrac{11}{36}$ (*iii*) $\dfrac{1}{36}$

21. Fill in the blanks :

    (*a*) If the probabilities of n independent events are p_1 , p_2 , p_3 ,..., p_n , then the probability that at least one of the event will happen is ....

    (*b*) For a biased die, the probabilities for the different faces to turn up are given below :

Face	1	2	3	4	5	6
Prob.	0.1	0.32	0.21	0.15	0.05	0.17

       The die is tossed and you are told that either face 1 or face 2 has turned up. Then the probability that it is face 1, is ....

    (*c*) The probability of getting a ticket of number of multiple of 5 in a random draw from a bag containing tickets of even numbers from 1 to 100, is ....

    (*d*) A town has two doctors X and Y operating independently. If the prob. the doctor X is available, is 0.9 and that for Y is 0.8, then the prob. that at least one doctor is available, when needed is ....

    (*e*) From a pack of well shuffled cards, one card is drawn randomly. A gambler bets it as a diamond or a king. The odds in favour of his winning the bet are ....

    (*f*) From a pack of cards, 2 cards are drawn, the first being replaced before the second is drawn. The probability that the first is a diamond and the second is a king will be ....

    (*g*) From an urn containing 12 white and 8 black balls two balls are drawn at random. The probability that both the balls will turn to be black is ....

    (*h*) A ball is taken out of a pot containing 6 white and 12 red balls. The probability that the ball is white is ....

    (*i*) A speaks truth in 75% and B in 80% of the cases. The percentage of cases in which they likely to contradict each other narrating the same incident is ....

Ans. (*a*) $1 - (1 - p_1)(1 - p_2) ..... (1 - p_n)$,   (*b*) $\dfrac{5}{21}$,   (*c*) $\dfrac{1}{5}$,   (*d*) 0.98,   (*e*) 4 : 9,

    (*f*) $\dfrac{1}{52}$,   (*g*) $\dfrac{14}{95}$,   (*h*) $\dfrac{1}{3}$,   (*i*) 35%

22. Tick √ the correct answer :

(i) The probability that at least one of the events A and B occurs is 0.8 and the probability that both the events occur simultaneously is 0.25. The probability $P(A) + P(B)$ is
    (i) 0.65   (ii) 0.75   (iii) 0.85   (iv) 0.95

(ii) A, B, C are independent events such that $P(A) = P(B)$ and probability that at least one of them happens is 1/2. The probability that A or B happens given that at least one of A, B, or C happens is $\frac{2}{9}$. Find $P(A)$ and $P(C)$.

                                                  Ans. $P(A) = 1 - \frac{\sqrt{7}}{3}$, $P(C) = \frac{5}{14}$.

(iii) An unbiased coin is tossed five times. Given that heads were obtained in two of the tosses, the probability that these were obtained in the first two tosses is
    (a) 1/10   (b) 1/4   (c) 1/32   (d) None of these.

(iv) Groups are formed of 4 persons out of 12 persons. The probability that one particular person is never included is
    (a) 2/3   (b) 1/3   (c) 1/4   (d) none of these

(v) 50 tickets are serially numbered 1 to 50. One ticket is drawn from these at random. The probability of its being a multiple of 3 or 4 is
    (a) 12/25   (b) 14/25   (c) 2/5   (d) none of these

(vi) The probabilities of occurring of two events E, F are 0.25 and 0.5 respectively and of occurring both simultaneously is 0.14. Then the probability of the occurrence of the neither event is
    (a) 0.61   (b) 0.39   (c) 0.89   (d) none of these

(vii) A bag contains 5 black and 4 white balls. Two balls are drawn at random. The probability that they match, is
    (a) 7/12   (b) 5/8   (c) 5/9   (d) 4/9

(viii) A, B, C in order toss a coin, the first to throw a head wins. Assuming the game continues indefinitely their respective chances of

    (a) $\frac{4}{7}, \frac{2}{7}, \frac{1}{7}$ (b) $\frac{1}{7}, \frac{4}{7}, \frac{2}{7}$ (c) $\frac{2}{7}, \frac{4}{7}, \frac{1}{7}$ (d) None of these (*A.M.I.E., winter 2000*)

(ix) A purse contains 4 copper coins, 3 silver coins, the second purse contains 6 copper coins and 2 silver coins. A coin is taken out of any purse, the probability that it is a copper coin is
    (a) 4/7   (b) 3/4   (c) 3/7   (d) 37/56

(x) In rolling two fair dice, the probability of getting equal numbers or numbers with an even product is
    (a) 6/36   (b) 30/36   (c) 27/36   (d) 3/36

(xi) One of the two events must occur. If the chance of one is 2/3 of the other, then odds in favour of the other are
    (a) 1 : 3   (b) 2 : 3   (c) 3 : 1   (d) none of these

(xii) The probability that a certain beginner at golf gets a good shot if he uses the correct club is 1/3, and the probability of a good shot with an incorrect club is 1/4. In his bag are 5 different clubs, only one of which is correct for the shot in question. If he chooses a club at random and takes a stroke, the probability that he gets a good shot is

    (a) $\frac{1}{3}$ (b) $\frac{1}{12}$ (c) $\frac{4}{15}$ (d) $\frac{7}{12}$

(xiii) India plays two matches each with West Indies and Australia. In any match, the probabilities of India getting points 0, 1 and 2, are 0.45, 0.05 and 0.50 respectively. Assuming that the outcomes are independent, the probability of India getting at least 7 points is   (*AMIETE., Summer 2001*)
    (a) 0.8750   (b) 0.0875   (c) 0.625   (d) 0.0250.

(xiv) A bag contains 10 bolts, 3 of which are defective. Two bolts are drawn without replacement. The probability that both the bolts drawn are not defective is

    (a) $\frac{49}{100}$ (b) $\frac{7}{15}$ (c) $\frac{4}{9}$ (d) $\frac{3}{10}$

(xv) The probability that a family has k children is $(0.5)^{k+1}$, $k = 0, 1, 2, ...$ If four families are chosen at random, the probability that each family has at least one child is
    (a) 1/16   (b) 1/256   (c) 3/16   (d) 3/256

(*xvi*) The random variable X has $N(1, 4)$ distribution, then

(*a*) $P(x > 3) > P(x > 1)$ (*b*) $P(x > 3) < P(x < 1)$

(*c*) $P(x < 3) < P(x > 1)$ (*d*) $P(x < 3) < P(x < 1)$

(*xvii*) Two distinguishable dice are tossed simultaneously. The probability that multiple of 2 does not occur on the first die or multiple of 3 does not occur on the second die is

(*a*) $\dfrac{5}{36}$ (*b*) $\dfrac{10}{36}$ (*c*) $\dfrac{20}{36}$ (*d*) $\dfrac{30}{36}$

(*xviii*) An unbiased die with faces marked 1, 2, 3, 4, 5, 6 is rolled 4 times, out of four face values obtained, the probability that the minimum face value is not less than 2 and the maximum face value is not greater than 5 is then

(*a*) $\dfrac{16}{81}$ (*b*) $\dfrac{2}{9}$ (*c*) $\dfrac{80}{81}$ (*d*) $\dfrac{8}{9}$

(A.M.I.E.T.E., Summer 2000)

(*xix*) There are q persons sitting in a row. Two of them are selected at random, the probability that the two selected persons are not together is

(*a*) $\dfrac{2}{q}$ (*b*) $1 - \dfrac{2}{q}$ (*c*) $\dfrac{q(q-1)}{(q+1)(q+2)}$ (*d*) None of these

(*xx*) Probability of any event can not be greater than .... and less than ....

(A.M.I.E. Winter 2001)

Ans. (*i*) (*b*), (*ii*) (*b*), (*iii*) (*a*), (*iv*) (*a*), (*v*) (*a*), (*vi*) (*b*), (*vii*) (*d*), (*viii*) (*a*), (*ix*) (*d*), (*x*) (*b*), (*xi*) (*d*), (*xii*) (*c*), (*xiii*) (*b*), (*xiv*) (*b*), (*xv*) (*a*), (*xvi*) (*b*), (*xvii*) (*d*), (*xviii*) (*a*), (*xix*) (*b*), (*xx*) 0, 1

10.5 CONDITIONAL PROBABILITY

Let A and B be two events of a sample space S and let $P(B) \neq 0$. Then conditional probability of the event A, given B, denoted by $P(A/B)$, is defined by

$$P(A/B) = \frac{P(A \cap B)}{P(B)}$$

... (1)

Theorem. If the events A and B defined on a sample space S of a random experiment are independent, then

$$P(A/B) = P(A) \text{ and } P(B/A) = P(B)$$

Proof. A and B are given to be independent events,

$$P(A \text{ and } B) = P(A) \cdot P(B)$$

$\Rightarrow$ $$P(A/B) = \frac{P(A \cap B)}{P(B)} = \frac{P(A).P(B)}{P(B)} = P(A)$$

$\Rightarrow$ $$P(B/A) = \frac{P(B \cap A)}{P(A)} = \frac{P(B).P(A)}{P(A)} = P(B)$$

Example 32. *Two coins are tossed. What is probability of coming up of two heads, if it is known that at least one head comes up?*

(Bihar SBTE, 2009)

Solution.   Let S be the sample space.

and   A = the event of coming up of two heads.

B = the event of coming up of least one head.

Then,   S = {(H, H), (H, T), (T, H), (T, T)}

A = {(H, H)}

and   B = {(H, H), (H, T), (T, H)}

$\therefore$ $n(S) = 4$, $n(A) = 1$, $n(B) = 3$

Also, $A \cap B = \{(H, H)\}$

$\therefore \qquad n(A \cap B) = 1$

Now, $P(B) = \dfrac{n(B)}{n(S)} = \dfrac{3}{4}$ and $P(A \cap B) = \dfrac{n(A \cap B)}{n(S)} = \dfrac{1}{4}$

$\therefore \qquad$ Required probability $P\left(\dfrac{A}{B}\right) = \dfrac{P(A \cap B)}{P(B)} = \dfrac{\frac{1}{4}}{\frac{3}{4}} = \dfrac{1}{3}.$ **Ans.**

10.6 BAYES THEOREM

If $B_1, B_2, B_3, \ldots\ldots B_n$ are mutually exclusive events with $P(B_i) \neq 0$, $(i = 1, 2, \ldots n)$ of a random experiment then for any arbitrary event A of the sample space of the above experiment with $P(A) > 0$, we have ·

$$P(B_i / A) = \frac{P(B_i) P(A / B_i)}{\displaystyle\sum_{i=1}^{n} P(B_i) P(A / B_i)} \qquad\qquad \text{(for } n = 3)$$

$$P(B_2 / A) = \frac{P(B_2) P(A / B_2)}{P(B_1) P(A / B_1) + P(B_2) P(A / B_2) + P(B_3) P(A / B_3)}$$

Proof. Let S be the sample space of the random experiment.

The events $B_1, B_2, \ldots, B_n$ being exhaustive

$\qquad\qquad S = B_1 \cup B_2 \cup \ldots \cup B_n \qquad\qquad\qquad\qquad\qquad [\because A \subset S]$

$\therefore \qquad A = A \cap S = A \cap (B_1 \cup B_2 \cup \ldots \cup B_n)$

$\qquad\qquad = (A \cap B_1) \cup (A \cap B_2) \cup \ldots \cup (A \cap B_n) \qquad\qquad$ [Distributive Law]

$\Rightarrow \qquad P(A) = P(A \cap B_1) + P(A \cap B_2) + \ldots + P(A \cap B_n)$

$\qquad\qquad = P(B_1) P(A / B_1) + P(B_2) P(A / B_2) + \ldots + P(B_n) P(A / B_n)$

$$= \sum_{i=1}^{n} P(B_i) P(A / B_i) \qquad\qquad\qquad\qquad \ldots (1)$$

Now, $\qquad P(A \cap B_i) = P(A) P(B_i / A)$

$$P(B_i / A) = \frac{P(A \cap B_i)}{P(A)} = \frac{P(B_i) P(A / B_i)}{\displaystyle\sum_{i=1}^{n} P(B_i) P(A / B_i)} \qquad\qquad \text{[Using (1)]}$$

$\Rightarrow$

Note. $P(B)$ is the probability of occurrence B. If we are told that the event A has already occurred. On knowing about the event A, $P(B)$ is changed to $P(B/A)$. With the help of Baye's theorem we can calculate $P(B / A)$.

Example 33. *An urn I contains 3 white and 4 red balls and an urn II contains 5 white and 6 red balls. One ball is drawn at random from one of the urns and is found to be white. Find the probability that it was drawn from urn I.*

Solution. Let $\qquad U_1$: the ball is drawn from urn I

$\qquad\qquad\qquad U_2$: the ball is drawn from urn II

$\qquad\qquad\qquad W$: the ball is white.

We have to find $P(U_1/W)$

By Baye's Theorem

$$P(U_1/W) = \frac{P(U_1)P(W/U_1)}{P(U_1)P(W/U_1) + P(U_2)P(W/U_2)} \qquad \ldots (1)$$

Since two urns are equally likely to be selected, $P(U_1) = P(U_2) = \dfrac{1}{2}$

$P(W/U_1) = P$ (a white ball is drawn from urn I) $= \dfrac{3}{7}$

$P(W/U_2) = P$ (a white ball is drawn from urn II) $= \dfrac{5}{11}$

∴   From (1), $P(U_1/W) = \dfrac{\dfrac{1}{2} \times \dfrac{3}{7}}{\dfrac{1}{2} \times \dfrac{3}{7} + \dfrac{1}{2} \times \dfrac{5}{11}} = \dfrac{33}{68}$ **Ans.**

Example 34. *Three urns contains 6 red, 4 black; 4 red, 6 black; 5 red, 5 black balls respectively. One of the urns is selected at random and a ball is drawn from it. If the ball drawn is red find the probability that it is drawn from the first urn.*

Solution. Let U_1: the ball is drawn from U_1.

U_2 : the ball is drawn from U_2.

U_3 : the ball is drawn from U_3.

R : the ball is red.

We have to find $P(U_1/R)$.

By Baye's Theorem, $P(U_1/R) = \dfrac{P(U_1)P(R/U_1)}{P(U_1)P(R/U_1) + P(U_2)P(R/U_2) + P(U_3)P(R/U_3)}$ \ldots (1)

Since the three urns are equally likely to be selected $P(U_1) = P(U_2) = P(U_3) = \dfrac{1}{3}$

Also $P(R/U_1) = P$ (a red ball is drawn from urn I) $= \dfrac{6}{10}$

$P(R/U_2) = P$ (a red ball is drawn from urn II) $= \dfrac{4}{10}$

$P(R/U_3) = P$ (a red ball is drawn from urn III) $= \dfrac{5}{10}$

∴ From (1), we have $P(U_1/R) = \dfrac{\dfrac{1}{3} \times \dfrac{6}{10}}{\dfrac{1}{3} \times \dfrac{6}{10} + \dfrac{1}{3} \times \dfrac{4}{10} + \dfrac{1}{3} \times \dfrac{5}{10}} = \dfrac{2}{5}$ **Ans.**

Example 35. *The contents of urns I, II and III are as follows:*

1 white, 2 black and 3 red balls,

2 white 1 black and 1 red balls, and

4 white, 5 black and 3 red balls.

One urn is chosen at random and two balls drawn. They happen to be white and red. What is the probability that they come from urns I, II and III?   *(KUK. 2005)*

Solution. Let E_1: urn I is chosen; E_2: urn II is chosen; E_3: urn III is chosen

and , A: the two balls are white and red.

We have to find $P(E_1/A)$, $P(E_2/A)$ and $P(E_3/A)$.

Now $\qquad P(E_1) = P(E_2) = P(E_3) = \dfrac{1}{3}$

$\qquad\qquad P(A/E_1) = P$ (a white and a red ball are drawn from urn I) $= \dfrac{{}^1C_1 \times {}^3C_1}{{}^6C_2} = \dfrac{1}{5}$

$\qquad P(A/E_2) = \dfrac{{}^2C_1 \times {}^1C_1}{{}^4C_2} = \dfrac{1}{3}; \quad P(A/E_3) = \dfrac{{}^4C_1 \times {}^3C_1}{{}^{12}C_2} = \dfrac{2}{11}$

By Baye's Theorem, we have

$P(E_1/A) = \dfrac{P(E_1)P(A/E_1)}{P(E_1)P(A/E_1) + P(E_2)P(A/E_2) + P(E_3)P(A/E_3)} = \dfrac{\dfrac{1}{3} \times \dfrac{1}{5}}{\dfrac{1}{3} \times \dfrac{1}{5} + \dfrac{1}{3} \times \dfrac{1}{3} + \dfrac{1}{3} \times \dfrac{2}{11}} = \dfrac{33}{118}$

Similarly $P(E_2/A) = \dfrac{55}{118}$, $P(E_3/A) = \dfrac{15}{59}$. **Ans.**

Example 36. *There bags A, B, C contain 4 red, 3 black, 2 white; 3 red, 4 black, 4 white; and 5 red, 2 black, 6 white balls respectively. If a bag is selected at random and a ball is drawn from it, find the probability that the ball drawn is red.* (M.D.U. May 2006, Dec. 2007)

Solution. Let E_1, E_2, E_3 denote the events of choosing bags A, B, C respectively,

then $\qquad P(E_1) = P(E_2) = P(E_3) = \dfrac{1}{3}$

Let R denote the event of drawing a red ball, then

$\qquad P(R/E_1) = \dfrac{4}{4+3+2} = \dfrac{4}{9},$

$\qquad P(R/E_2) = \dfrac{3}{3+4+4} = \dfrac{3}{11},$

$\qquad P(R/E_3) = \dfrac{5}{5+2+6} = \dfrac{5}{13}$

$\qquad P(R) = P(E_1)\, P(R/E_1) + P(E_2)\, P(R/E_2) + P(E_3)\, P(R/E_3)$

$\qquad\qquad = \dfrac{1}{3} \times \dfrac{4}{9} + \dfrac{1}{3} \times \dfrac{3}{11} + \dfrac{1}{3} \times \dfrac{5}{13} = \dfrac{4}{27} + \dfrac{1}{11} + \dfrac{5}{39}$

$\qquad\qquad = 0.148 + 0.090 + 0.128 = 0.366.$ **Ans.**

Example 37. *In a bolt factory, machines A, B and C manufacture respectively 25%, 35% and 40% of the total. If their output 5, 4 and 2 per cent are defective bolts. A bolt is drawn at random from the product and is found to be defective. What is the probability that it was manufactured by machine B?*

Solution. $\qquad A :$ bolt is manufactured by machine A.

$\qquad\qquad B :$ bolt is manufactured by machine B.

$\qquad\qquad C :$ bolt is manufactured by machine C.

$\qquad\qquad P(A) = 0.25, P(B) = 0.35, P(C) = 0.40$

The probability of drawing a defective bolt manufactured by machine A is $P(D/A) = 0.05$

Similarly, $P(D/B) = 0.04$ and $P(D/C) = 0.02$

By Baye's theorem

$$P(B/D) = \frac{P(B)P(D/B)}{P(A)P(D/A) + P(B)P(D/B) + P(C)P(D/C)}$$

$$= \frac{0.35 \times 0.04}{0.25 \times 0.05 + 0.35 \times 0.04 + 0.40 \times 0.02} = 0.41$$ **Ans.**

EXERCISE 10.3

1. There are three bags: first containing 1 white, 2 red 3 green balls; second 2 white, 3 red, 1 green balls and third 3 white, 1 red, 2 green balls. Two balls are drawn from a bag chosen at random. They are found to be 1 red and 1 white. Find the

    probability that balls so drawn came from the second bag.   (*M.D.U. Dec. 2008*) **Ans.** $\frac{6}{11}$

2. An insurance company insured 2000 scooter drivers, 4000 car drivers and 6000 truck drivers. The probability of accident is 0.01, 0.03 and 0.15 respectively. One of the insured persons meets an accident. What is the probability that he is a

    scooter driver?   (*M.D.U. 2006*)   **Ans.** $\frac{1}{52}$

3. In a bolt factory, there are four machines A, B, C, D manufacturing 20%, 15%, 25% and 40% of the total output respectively. Of their outputs 5%, 4%, 3% and 2%, in the same order, are defective bolts. A bolt is chosen at random from the factory's production and is found defective. What is the probability that the bolt was manufactured by machine A or machine D?   (*M.D.U., Dec. 2006*) **Ans.** 0.3175, 0.254

4. A survey was conducted to find the supplied of the consumer durables for the market. It was found that the three major companies A, B and C have market share of 35%, 25% and 40% respectively out of which 2%, 1% and 3% are not upto the satisfaction. A consumer buys a product and is dissatisfied with it. What is the probability that it might be from the

    company C?   (*M.D.U. May 2006*) **Ans.** $\frac{24}{43}$

Choose the correct alternatives:

(*i*) A dies is thrown. The probability that the digit coming up is greater than 4 is:

(a) $\frac{1}{4}$ (b) $\frac{1}{3}$ (c) $\frac{1}{5}$ (d) 0   (*Bihar SBTE, 2009*) **Ans.** (b)

(*ii*) A fair die is rolled once. What is the probability that an odd number turns up?

(a) $\frac{1}{6}$ (b) $\frac{1}{3}$ (c) $\frac{1}{2}$ (d) 1   (*Bihar SBTE, 2003*) **Ans.** (c)

(*iii*) Three fair dice are rolled together. The probability that the same number will appear on each of them is

(a) $\frac{1}{6}$ (b) $\frac{1}{36}$ (c) $\frac{1}{18}$ (d) none   (*Bihar SBTE, 2003*) **Ans.** (b)

(*iv*) A card is drawn from a pack of cards. The probability that the card is an ace or a queen is ....

(a) $\frac{1}{4}$ (b) $\frac{1}{13}$ (c) $\frac{1}{52}$ (d) $\frac{2}{13}$ (*Bihar SBTE,2011*) **Ans. ??**

(*v*) What is the probability of drawing an ace from a pack of 52 cards?

(a) $\frac{1}{13}$ (b) $\frac{4}{13}$ (c) $\frac{1}{52}$ (d) $\frac{1}{4}$ (*Bihar SBTE, 2004*) **Ans.** (a)

(*vi*) A bag contains 5 white and 7 red balls. What will be the probability of drawing a red ball?

(a) $\frac{5}{12}$ (b) $\frac{7}{12}$ (c) $\frac{5}{7}$ (d) $\frac{7}{5}$ (*Bihar SBTE, 2011*) **Ans.** (b)

(*vii*) A bag contains 4 white, 3 red and 5 blue balls. The probability of drawing a red ball is:

(a) $\frac{1}{12}$ (b) $\frac{1}{3}$ (c) $\frac{1}{4}$ (d) $\frac{1}{5}$ (*Bihar SBTE, 2010*) **Ans.** (c) (*viii*) If p is probability of an event A and q is the probability of complementry of A, then

(*a*) $p = 1 + q$ (*b*) $P = \dfrac{1}{q}$ (*c*) $p = 1 - q$ (*d*) $p = q - 1$ (*Bihar SBTE, 2009*) **Ans.** (*c*)

(*ix*) If A and B are two mutually exclusive events with probabilities p and q respectively, then the probability of either of them is ....

(*a*) $p - q$ (*b*) $p + q$ (*c*) $p \cdot q$ (*d*) p/q (*Bihar SBTE, 2008*) **Ans.** (*b*)

(*x*) The probability that A will solve the problem is $\dfrac{2}{3}$ and probability that B will solve the problem is $\dfrac{3}{4}$. The probability that neither will solve the problem is:

(*a*) $\dfrac{1}{12}$ (*b*) $\dfrac{5}{12}$ (*c*) $\dfrac{7}{12}$ (*e*) $\dfrac{11}{12}$ (*Bihar SBTE, 2008*) **Ans.** (*a*)

(*xi*) What is the probability of getting two heads in tossing a coin twice?

(*a*) 2×2 (*b*) $\dfrac{1}{2}$ (*c*) $\dfrac{1}{4}$ (*d*) 1 (*Bihar SBTE, 2005*) **Ans.** (*c*)

(*xii*) A urn contains 2 white and 3 red balls. One ball is drawn at random. What is the probability that it is drawn red?

(*a*) $\dfrac{1}{2}$ (*b*) $\dfrac{1}{3}$ (*c*) $\dfrac{2}{5}$ (*d*) $\dfrac{3}{5}$ (*Bihar SBTE, 2005*) **Ans.** (*d*)

Fill up the blanks:

(*xiii*) If $P(A) = \dfrac{4}{9}$, then $P(\overline{A})$ = ...   (*Patna 2005*) **Ans.** $\left(\dfrac{5}{9} \right)$

CHAPTER 11 Probability Distribution

11.1 RANDOM VARIABLE

(*Bihar SBTE, 2009*)

If we toss a coin three times, then the sample space is

$$S = \{HHH, HHT, HTH, THH, HTT, THT, TTH, TTT\}$$

This is a random experiment.

Let X represent the number of heads, which can come up. With each sample point, we can associate a number for x. Sample points are given below with the number.

Sample point (X)	Number of heads
HHH	3
HHT	
HTH	2
THH	
HTT	
THT	1
TTH	
TTT	0

Heads (X)	0 head	1 head	2 heads	3 heads	Total outcomes
Number of occurence	1	3	3	1	8

In the above experiment a number is associated with each outcome. This single real number varies with different outcomes of the experiment. Therefore this number is a variable and this variable is called as random variable.

A random variable is denoted by X.

Definition. A random variable is a real valued function whose domain is the sample space of a random experiment.

There are two types of random variables

1. discrete random variable

2. continuous random variable

Here, we will discuss the discrete variable.

11.2 PROBABILITY DISTRIBUTION

(*Bihar SBTE, 2009*)

Example 1. *Find the probability distribution of the number of heads in the simultaneous toss of four coins.*

Solution. On tossing four coins simultaneously, all the possible outcomes are :

{*HHHH, HHHT, HHTH, HTHH, THHH, HHTT, HTHT, HTTH, THHT, THTH, TTHH, TTTH, TTHT, THTT,*
HTTT, TTTT} $\Rightarrow$ $n(S) = 16$

The possible values of X are 0, 1, 2, 3, 4.

Number of heads (X)	0 Head	1 Head	2 Heads	3 Heads	4 Heads	Total outcomes
Number of occurrence	1	4	6	4	1	16

Let X be the random variable "number of heads" on tossing four coins simultaneously.

Let E_0 be the event of getting no head

$\Rightarrow$ $\qquad\qquad\qquad\qquad E_0 = \{TTTT\} \Rightarrow n(E_0) = 1$

Now, $P(X = 0)$ = Probability of getting no head

$$= P \text{ (no head)} = \frac{n(E_0)}{n(S)} = \frac{1}{16}$$

Let E_1 be the event of getting one head

$$E_1 = \{TTTH, TTHT, THTT, HTTT\}$$

$\Rightarrow$ $\qquad\qquad n(E_1) = 4$

$\Rightarrow$ $\qquad\qquad P(x = 1) = $ Probability of getting one head

$$P \text{ (one head)} = \frac{n(E_1)}{n(S)} = \frac{4}{16} = \frac{1}{4}$$

Let E_2 be the event of getting two heads

$\Rightarrow$ $\qquad\qquad E_2 = \{HHTT, HTHT, HTTH, THHT, THTH, TTHH\}$

$\Rightarrow$ $\qquad\qquad n(E_2) = 6$

$\Rightarrow$ $\qquad P(X = 2) = P \text{ (2 heads)} = \frac{n(E_2)}{n(S)} = \frac{6}{16} = \frac{3}{8}$

Let E_3 be the event of getting three heads

$\Rightarrow$ $\qquad\qquad E_3 = \{HHHT, HHTH, HTHH, THHH\}$

$\Rightarrow$ $\qquad\qquad n(E_3) = 4$

$\Rightarrow$ $\qquad P(X = 3) = P \text{ (3 heads)} = \frac{n(E_3)}{n(S)} = \frac{4}{16} = \frac{1}{4}$

Let E_4 be the event of getting four heads.

$\Rightarrow$ $\qquad\qquad E_4 = \{HHHH\}$

$\Rightarrow$ $\qquad\qquad n(E_4) = 1$

$\Rightarrow$ $\qquad P(X = 4) = P \text{ (4 heads)} = \frac{n(E_4)}{n(S)} = \frac{1}{16}$

$\therefore$ The probability distribution of the random variable X is :

Number of heads (X)	0	1	2	3	4
Probability P(X)	$\frac{1}{16}$	$\frac{1}{4}$	$\frac{3}{8}$	$\frac{1}{4}$	$\frac{1}{16}$

Ans.

Example 2. *Four defective oranges are accidentally mixed with sixteen good ones. Three oranges are drawn at random from the mixed lot. Find the probability distribution of X, the number of defective oranges.*

Solution. Number of good oranges = 16

Number of defective oranges = 4.

Let X denote the random variable showing the number of defective oranges drawn. Then X can take values 0, 1, 2, 3.

Now,
$$P(X = 0) = \frac{{}^4C_0 \cdot {}^{16}C_3}{{}^{20}C_3} = \frac{28}{57}$$

$$P(X = 1) = \frac{{}^4C_1 \cdot {}^{16}C_2}{{}^{20}C_3} = \frac{24}{57}$$

$$P(X = 2) = \frac{{}^4C_2 \cdot {}^{16}C_1}{{}^{20}C_3} = \frac{24}{285}$$

$$P(X = 3) = \frac{{}^4C_3 \cdot {}^{16}C_0}{{}^{20}C_3} = \frac{1}{285}$$

∴ The reqduired probability distribution is

X	0	1	2	3
$P(X)$	$\dfrac{28}{57}$	$\dfrac{24}{57}$	$\dfrac{24}{285}$	$\dfrac{1}{285}$

Ans.

Example 3. *From a lot of 30 bulbs which include 6 defectives, a sample of 4 bulbs is drawn at random with replacement. Find the probability distribution of the number of defective bulbs.*

Solution. Let X be the random variable which denote the values of X are 0, 1, 2, 3 or 4.

$$\text{Total number of bulbs} = 30$$
$$\text{Defective bulbs} = 6$$
$$\text{Non-defective bulbs} = 30 - 6 = 24$$

∴
$$P(\text{Defective bulbs}) = \frac{6}{30} = \frac{1}{5}$$

$$P(\text{Non defective bulbs}) = \frac{24}{30} = \frac{4}{5}$$

Let D stands for Defective and N stands for non-defective, then $P(D) = \dfrac{1}{5}$ and $P(N) = \dfrac{4}{5}$.

$\Rightarrow P(X = 0) = P$ (0 defective bulb or all non-defective bulbs)

$$= P(NNNN) = P(N) \cdot P(N) \cdot P(N) \cdot P(N) = \frac{4}{5} \cdot \frac{4}{5} \cdot \frac{4}{5} \cdot \frac{4}{5} = \frac{256}{625}$$

$P(X = 1) = P$ (one defective aud 3 non-defectives) $= P(DNNN) + P(NDNN) + P(NNDN) + P(NNND)$

$$= 4 \times P(D) \times P(N) \times P(N) \times P(N) = 4 \times \frac{1}{5} \times \frac{4}{5} \times \frac{4}{5} \times \frac{4}{5} = \frac{256}{625}$$

$P(X = 2) = P$ (two defectives and two non-defectives)
$$= P(DDNN) + P(NDDN) + P(NNDD) + P(NDND) + P(DNDN) + P(DNND)$$

$$P(X = 2) = 6 \times P(D) \cdot P(D) \cdot P(N) \cdot P(N) = 6 \times \frac{1}{5} \times \frac{1}{5} \times \frac{4}{5} \times \frac{4}{5} = \frac{96}{625}$$

$P(X = 3) = P$ (three defectives and one non-defective)

$$= P(DDDN) + P(DDND) + P(DNDD) + P(NDDD) = 4 \times P(D) \times P(D) \times P(D) \times P(N)$$

$$= 4 \times \frac{1}{5} \times \frac{1}{5} \times \frac{1}{5} \times \frac{4}{5} = \frac{16}{625}$$

$P(X = 4) = P(\text{all the four defectives}) = P(DDDD) = P(D) \cdot P(D) \cdot P(D) \cdot P(D)$

$$= \frac{1}{5} \times \frac{1}{5} \times \frac{1}{5} \times \frac{1}{5} = \frac{1}{625}$$

Therefore, the required probability is given below :

Number of Defective bulbs (X)	0	1	2	3	4
Probability $P(X)$	$\dfrac{256}{625}$	$\dfrac{256}{625}$	$\dfrac{96}{625}$	$\dfrac{16}{625}$	$\dfrac{1}{625}$

Ans.

Example 4. *From a well shuffled pack of 52 cards, 3 cards are drawn one-by-one with replacement. Find the probability distribution of number of queens.*

Solution. Let E_i ($i = 1, 2, 3$) be the event of drawing a queen. Let X denote the discrete random variable "Number of Queens" in 3 draws one by one with replacement. Here

$X = 0, 1, 2, 3$

Now $\qquad\qquad P(E_i) = \dfrac{4}{52} : \ P(\bar{E}_i) = \dfrac{48}{52} \qquad (i = 1, 2, 3).$

$\therefore \qquad P(X = 0) = P(\bar{E}_1 \bar{E}_2 \bar{E}_3) = P(\bar{E}_1) P(\bar{E}_2) P(\bar{E}_3) = \left(\dfrac{48}{52}\right)^3 = \left(\dfrac{12}{13}\right)^3 = \dfrac{1728}{2197}.$

$P(X = 1) = P(E_1 \bar{E}_2 \bar{E}_3 \ \text{or} \ \bar{E}_1 E_2 \bar{E}_3 \ \text{or} \ \bar{E}_1 \bar{E}_2 E_3)$

$\qquad\qquad = P(E_1 \bar{E}_2 \bar{E}_3) + P(\bar{E}_1 E_2 \bar{E}_3) + P(\bar{E}_1 \bar{E}_2 E_3)$

$\qquad\qquad = P(E_1) P(\bar{E}_2) P(\bar{E}_3) + P(\bar{E}_1) P(E_2) P(\bar{E}_3) + P(\bar{E}_1) P(\bar{E}_2) P(E_3)$

$\qquad\qquad = 3 \times \dfrac{4}{52} \times \left(\dfrac{48}{52}\right)^2 = 3 \times \dfrac{1}{13} \times \left(\dfrac{12}{13}\right)^2 = \dfrac{432}{2197}$

$P(X = 2) = P(E_1 E_2 \bar{E}_3 \ \text{or} \ E_1 \bar{E}_2 E_3, \ \text{or} \ \bar{E}_1 E_2 E_3)$

$\qquad\qquad = P(E_1 E_2 \bar{E}_3) + P(E_1 \bar{E}_2 E_3) + P(\bar{E}_1 E_2 E_3)$

$\qquad\qquad = P(E_1) P(E_2) P(\bar{E}_3) + P(E_1) P(\bar{E}_2) P(E_3) + P(\bar{E}_1) P(E_2) P(E_3)$

$\qquad\qquad = 3 \times \left(\dfrac{4}{52}\right)^2 \times \left(\dfrac{48}{52}\right) = 3 \times \left(\dfrac{1}{13}\right)^2 \times \dfrac{12}{13} = \dfrac{36}{2197}$

$P(X = 3) = P(E_1 E_2 E_3) = P(E_1) P(E_2) P(E_3) = \left(\dfrac{4}{52}\right)^3 = \dfrac{1}{2197}$

$\therefore$ The requdired probability distribution is

X	0	1	2	3
$P(X)$	$\dfrac{1728}{2197}$	$\dfrac{432}{2197}$	$\dfrac{36}{2197}$	$\dfrac{1}{2197}$

Ans.

Example 5. *Two cards are drawn one by one without replacement from a well shuffled pack of 52 cards. Find the probability distribution of the number of aces.*

Solution. Let $A \equiv$ Event of drawing an Ace and

$X \equiv$ Random Variable of "Number of Aces" in two draws

$\Rightarrow \qquad\qquad\qquad\qquad\qquad X = 0, 1, 2.$

Probability of an ace in first toss $= P(A) = \dfrac{4}{52}$

Probability of non ace in second toss (without replacement) $= P(\bar{A}) = \dfrac{48}{51}$

Probability of non ace in first toss $= P(\bar{A}) = \dfrac{48}{52}$

Probability of an ace in the second toss (without replacement) $= P(A) = \dfrac{47}{51}$

Now
$$P(X = 0) = P(\bar{A}_1 \bar{A}_2) = P(\bar{A}_1)\, P(\bar{A}_2) = \frac{48}{52} \times \frac{47}{51} = \frac{188}{221}$$

$$P(X = 1) = P(A_1 \bar{A}_2 \text{ or } \bar{A}_1 A_2) = P(A_1 \bar{A}_2) + P(\bar{A}_1 A_2) = P(A_1)\, P(\bar{A}_2) + P(\bar{A}_1)\, P(A_2)$$

$$= \frac{4}{52} \times \frac{48}{51} + \frac{48}{52} \times \frac{4}{51} = \frac{32}{221}$$

$$P(X = 2) = P(A_1 A_2) = P(A_1)\, P(A_2) = \frac{4}{52} \times \frac{3}{51} = \frac{1}{221}.$$

∴ The probability distribution of the random variable X is

X	0	1	2
$P(X)$	$\dfrac{188}{221}$	$\dfrac{32}{221}$	$\dfrac{1}{221}$

Clearly, $\{P(X = 0) + P(X = 1) + P(X = 2)\} = \left(\dfrac{188}{221} + \dfrac{32}{221} + \dfrac{1}{221}\right) = 1$ **Ans.**

EXERCISE 11.1

1. Two cards are drawn one by one without replacement from a well shuffled pack of 52 cards. Find the probability distribution of the number of kings. **Ans.**

X	0	1	2
$P(X)$	$\dfrac{188}{221}$	$\dfrac{32}{221}$	$\dfrac{1}{221}$

2. A box contains 16 bulbs out of which 4 bulbs are defective. 3 bulbs are drawn one by one from the box without replacement. Find the probability distribution of the number of defective bulbs drawn. **Ans.**

X	0	1	2	3
$P(X)$	$\dfrac{11}{28}$	$\dfrac{33}{70}$	$\dfrac{9}{70}$	$\dfrac{1}{140}$

3. Determine the probability distribution of the number of jack cards when drawn, two cards are drawn one by one without replacement from a pack of 52 playing cards. **Ans.**

X	0	1	2
$P(X)$	$\dfrac{188}{221}$	$\dfrac{32}{221}$	$\dfrac{1}{221}$

4. Three balls are drawn without replacement from a bag containing 5 white and 4 red balls. Find the probability distribution of the number of red balls drawn. **Ans.**

X	0	1	2	3
$P(X)$	$\dfrac{5}{42}$	$\dfrac{10}{21}$	$\dfrac{5}{14}$	$\dfrac{1}{21}$

5. Four bad eggs are mixed with 10 good ones. If 3 eggs are drawn one by one without replacement, find the probability distribution of the number of bad eggs drawn. **Ans.**

X	0	1	2	3
$P(X)$	$\dfrac{30}{91}$	$\dfrac{45}{91}$	$\dfrac{15}{91}$	$\dfrac{1}{91}$

6. A coin is tossed 5 times. What is the probability distribution that head appears an even number of times? (Take 0 as an even number). **Ans.**

X	0	2	4
$P(X)$	$\dfrac{1}{32}$	$\dfrac{10}{32}$	$\dfrac{1}{32}$

7. Find the probability distribution of the number of sixes in three tosses of a die.

Ans.

X	0	1	2	3
$P(X)$	$\dfrac{125}{216}$	$\dfrac{75}{216}$	$\dfrac{15}{216}$	$\dfrac{1}{216}$

8. Let X be a random variable, which assumes the values x_1, x_2, x_3, x_4 such that

$$2P(X = x_1) = 3P(X = x_2) = P(X = x_3) = 5P(X = x_4)$$

Find the probability distribution of X.   **Ans.**

X	x_1	x_2	x_3	x_4
$P(X)$	$\dfrac{15}{61}$	$\dfrac{10}{61}$	$\dfrac{30}{61}$	$\dfrac{6}{61}$

9. Let X represent the difference between the number of heads and the number of tails obtained when a coin is tossed 6 times. What are possible values of X ?   **Ans.** X = 6, 4, 2, 0

10. An urn contains 5 red and 2 black balls. Two balls are randomly drawn. Let X represent the number of black balls. What are the possible values of X? Is X a random variable?   **Ans.** X = 0, 1, 2, yes.

11.3 MEAN OF A RANDOM VARIABLE

Some times we have to describe the random variable by means of a single number. This number is calculated from probability distribution.

This number can be

1. Mean   **2.** Median   **3.** Mode

In this article we will discuss the mean only.

Definition. *Let* $x_1, x_2, x_3, \ldots, x_n$ *be a random variables with probabilities* $p_1, p_2, p_3, \ldots, p_n.$

$$Mean \quad = x_1 p_1 + x_2 p_2 + x_3 p_3 + \ldots + x_n p_n \quad = \sum_{i=1}^{n} x_i p_i$$

The mean of a random variable is also called the expectation of the random variable and it is denoted as :
Expectation of X = E (X)

$$E(X) = \mu = x_1 p_1 + x_2 p_2 + x_3 p_3 + \ldots + x_n p_n.$$

Example 6. *Two dice are thrown simultaneously. If X denotes the number of sixes, find the expectation of X.*

Solution. Total possible outcomes (= 36) are as follows :

(1, 1),	(1, 2),	(1, 3),	(1, 4),	(1, 5),	(1, 6),
(2, 1),	(2, 2),	(2, 3),	(2, 4),	(2, 5),	(2, 6),
(3, 1),	(3, 2),	(3, 3),	(3, 4),	(3, 5),	(3, 6),
(4, 1),	(4, 2),	(4, 3),	(4, 4),	(4, 5),	(4, 6),
(5, 1),	(5, 2),	(5, 3),	(5, 4),	(5, 5),	(5, 6),
(6, 1),	(6, 2),	(6, 3),	(6, 4),	(6, 5),	(6, 6),

Number of ordered pairs which does not contain 6 = 25
Number of ordered pairs which contains 6 only once = 10
Number of ordered pairs which contains 6 twice = 1
Clearly, X contains the values 0, 1 and 2.
Thus, the probability distribution of X is given by

X	0	1	2
$P(X)$	$\dfrac{25}{36}$	$\dfrac{10}{36}$	$\dfrac{1}{36}$

Computation of mean and variance

x_i	$p_i = P(X = x_i)$	$p_i x_i$
0	$\dfrac{25}{36}$	0
1	$\dfrac{10}{36}$	$\dfrac{10}{36}$
2	$\dfrac{1}{36}$	$\dfrac{2}{36}$
		$\Sigma p_i x_i = \dfrac{12}{36}$

We have,

$$\Sigma p_i x_i = \frac{12}{36} = \frac{1}{3}$$

$\therefore$
$$E(X) = \Sigma p_i x_i = \frac{1}{3}$$

Hence, Expectation of $X = \dfrac{1}{3}$. **Ans.**

Example 7. *In roulette, (see figure) the wheel has 13 numbers 0, 1, 2, ..., 12 marked on equally spaced slots. A player sets ₹ 10 on a given number. He receives ₹ 100 from the organiser of the game if the ball comes to rest in this slot; otherwise he gets nothing. If X denotes the player's net gain/loss, find E(X).*

Solution. If the player sets ₹ 10 on a given number but the ball does not come to rest on the slot of the given number then he looses ₹ 10.

$\therefore \qquad X = -10 \qquad \Rightarrow \qquad P(X = -10) = \dfrac{12}{13}$

Or if the player sets ₹ 10 on a given number and the ball comes to rest on the slot of the given number then he wins ₹ 100.

$\therefore \quad X = -10 + 100 = 90 \qquad \Rightarrow \qquad P(X = 90) = \dfrac{1}{13}$

Thus, the probability distribution of X is as follows :

X	-10	90
$P(X)$	$\dfrac{12}{13}$	$\dfrac{1}{13}$

$\therefore \qquad E(X) = (-10)\left(\dfrac{12}{13}\right) + (90)\left(\dfrac{1}{13}\right)$

$\Rightarrow \qquad E(X) = \dfrac{-120}{13} + \dfrac{90}{13} \qquad \Rightarrow \quad E(X) = \dfrac{-30}{13}$

Hence, $\qquad E(X) = \dfrac{-30}{13}$ **Ans.**

Example 8. *In a game a man wins a rupee for a six and looses a rupee for any other number when a fair die is thrown. The man decided to throw a die thrice but to quit as and when he gets a six. Find the expected value of the amount he wins/looses :*

Solution. The man may get six in the first throw and then he quits the game.

Thus, he wins = Re 1

i.e., $\qquad X = 1 \quad \Rightarrow \qquad P[X = 1] = \dfrac{1}{6}$

Again in the first throw, he may get a number other than six and in the second throw he may get six and quits the game.

Here, in the first throw he looses Re 1 and in the second throw he wins Re 1.

Thus $\qquad\qquad\qquad\qquad X = -1 + 1 = 0$

$$P(X = 0) = \frac{5}{6} \times \frac{1}{6} = \frac{5}{36}$$

Again, in the first two throws he gets a number other than six and in third throw he may get a six.

So, $\qquad\qquad\qquad\qquad X = -1 -1 + 1 = -1$

$$P(X = -1) = \frac{5}{6} \times \frac{5}{6} \times \frac{1}{6} = \frac{25}{216}$$

Or he may not get six in any one of three throws.

So, $\qquad\qquad\qquad\qquad X = -1 -1 -1 = -3.$

$$P(X = -3) = \frac{5}{6} \times \frac{5}{6} \times \frac{5}{6} = \frac{125}{216}$$

Thus, the probability distribution of X is as follows :

X	1	0	-1	-3
$P(X)$	$\dfrac{1}{6}$	$\dfrac{5}{36}$	$\dfrac{25}{216}$	$\dfrac{125}{216}$

$$E(X) = \frac{1}{6} \times 1 + \frac{5}{36} \times 0 + \frac{25}{216} \times (-1) + \frac{125}{216} \times (-3)$$

$\Rightarrow \qquad\qquad E(X) = \dfrac{36 + 0 - 25 - 375}{216} = -\dfrac{364}{216} = \dfrac{91}{54}$ **Ans.**

11.4 VARIANCE

The mean of a random variable does not indicate about the variability in the values of the random variable. *The variance is the most important measure of the dispersion. The variance of the probability distribution of a random variable X is the average square deviation measured from the expected value of a probability distribution of X.*

$$\text{Variance} = \sum_{i=1}^{n} [X_i - E(X)]^2 \, P_i = \sum_{i=1}^{n} \left[(X_i)^2 + \{E(X)\}^2 - 2X_i E(X) \right] P_i$$

$$= \sum_{i=1}^{n} P_i (X_i)^2 + \{E(X)\}^2 \sum_{i=1}^{n} P_i - 2E(X) \sum_{i=1}^{n} P_i X_i = \Sigma P_i X_i^2 + [E(X)^2] - 2E(X).E(X)$$

$$= E(X^2) - [E(X)]^2 \qquad\qquad \left[\sum_{i=1}^{n} P_i = 1, \ E(X) = \sum_{i=1}^{n} P_i X_i \right]$$

$$\boxed{\text{Variance} = E(X^2) - [E(X)]^2}$$

11.5 STANDARD DEVIATION

Standard deviation of $X = \sqrt{\text{Var}(X)}$

$$\boxed{\text{S.D.}(X) = \sqrt{\text{Var}(X)}}$$

Example 9. *In a meeting 70% of the members favour a certain proposal, 30% being opposed. A member is selected at random and we let X = 0 if he opposed, and X = 1 if he is in favour. Find E(X) and Var (X).*

Solution. The probability distribution of X is

X	0	1
$P(X)$	$\dfrac{30}{100}$	$\dfrac{70}{100}$

∴ $$E(X) = \frac{30}{100} \times 0 + \frac{70}{100} \times 1 = \frac{7}{10} = 0.7$$

and, $$E(X^2) = \frac{30}{100} \times 0^2 + \frac{70}{100} \times 1^2 = \frac{7}{10} = 0.7$$

∴ $$\text{Var (X)} = E(X^2) - [E(X)]^2 = \frac{7}{10} - \frac{49}{100} = \frac{21}{100} = 0.21$$

Hence, $E(X) = 0.7$ and Var $(X) = 0.21$ **Ans.**

Example 10. *A class has 15 students whose ages are 14, 17, 15, 14, 21, 17, 19, 20, 16, 18, 20, 17, 16, 19 and 20 years. One student is selected in such a manner that each has the same chance of being chosen and the age X of the selected student is recorded. What is the probability distribution of the random variable X ? Find mean, variance and standard deviation of X.*

Solution. We observe that X takes the values 14, 15, 16, 17, 18, 19, 20 and 21 such that

X	14	15	16	17	18	19	20	21	Total
$f(X)$	2	1	2	3	1	2	3	1	15
$P(X)$	$\dfrac{2}{15}$	$\dfrac{1}{15}$	$\dfrac{2}{15}$	$\dfrac{3}{15}$	$\dfrac{1}{15}$	$\dfrac{2}{15}$	$\dfrac{3}{15}$	$\dfrac{1}{15}$	1

Computation of mean and variance

x_i	$p_i = P(X = x_i)$	$p_i x_i$	$p_i x_i^2$
14	$\dfrac{2}{15}$	$\dfrac{28}{15}$	$\dfrac{392}{15}$
15	$\dfrac{1}{15}$	$\dfrac{15}{15}$	$\dfrac{225}{15}$
16	$\dfrac{2}{15}$	$\dfrac{32}{15}$	$\dfrac{512}{15}$
17	$\dfrac{3}{15}$	$\dfrac{51}{15}$	$\dfrac{867}{15}$
18	$\dfrac{1}{15}$	$\dfrac{18}{15}$	$\dfrac{324}{15}$
19	$\dfrac{2}{15}$	$\dfrac{38}{15}$	$\dfrac{722}{15}$
20	$\dfrac{3}{15}$	$\dfrac{60}{15}$	$\dfrac{1200}{15}$
21	$\dfrac{1}{15}$	$\dfrac{21}{15}$	$\dfrac{441}{15}$
		$\Sigma p_i x_i = \dfrac{263}{15}$	$\Sigma p_i x_i^2 = \dfrac{4683}{15}$

We have,

$$\Sigma p_i x_i = \frac{263}{15} \quad \text{and} \quad \Sigma p_i x_i^2 = \frac{4683}{15}$$

$$\therefore \quad \text{Mean} = \Sigma p_i x_i = \frac{263}{15} = 17.53$$

$$\text{Variance} = \Sigma p_i x_i^2 - (\Sigma p_i x_i)^2 = \frac{4683}{15} - \left(\frac{263}{15}\right)^2 = \frac{70245 - 69169}{225} = \frac{1076}{225} = 4.78$$

$$\therefore \quad \text{Standard Deviation} = \sqrt{\text{Variance}} = \sqrt{4.78} = 2.186$$

Hence, Mean = 17.53, Variance = 4.78, Standard Deviation = 2.186 **Ans.**

Example 11. *Find the mean, standard deviation of the number of tails in two tosses of a coin.*

(Bihar SBTE, 2009)

Solution. In two tosses of a coin, the sample space is given by S = {HH, HT, TH, TT) $\Rightarrow$ n(S) = 4

So, every single outcome has a probability $= \frac{1}{4}$

Let x = number of tails in two tosses.
In two tosses, we may have no tail, 1 tail or 2 tails.
So, the possible values of x are 0, 1, 2.

$$P(x = 0) = P \text{ (getting no tail)} = p(HH) = \frac{1}{4}$$

$$P(x = 1) = P \text{ (getting 1 tail)} = P (HT, TH) = \frac{2}{4} = \frac{1}{2}$$

$$P(x = 2) = P \text{ (getting 2 tails)} = P (TT) = \frac{1}{4}$$

Hence, the probability distribution of x is given by

$X = x_i$	0	1	2
P_i	$\frac{1}{4}$	$\frac{1}{2}$	$\frac{1}{4}$

$$\text{Mean, } \mu = \Sigma x_i p_i = \left(0 \times \frac{1}{4}\right) + \left(1 \times \frac{1}{2}\right) + \left(2 \times \frac{1}{4}\right) = 1$$

$$\text{Variance of } \sigma^2 = \Sigma x_i^2 p_i - \mu^2 = \left[\left(0 + \frac{1}{4}\right) + \left(1 \times \frac{1}{2}\right) + \left(4 \times \frac{1}{4}\right)\right] - 1^2 = \frac{1}{2}$$

$$\text{Standard deviation, } \sigma = \frac{1}{\sqrt{2}} \times \frac{\sqrt{2}}{\sqrt{2}} = \frac{\sqrt{2}}{2} = \frac{1.414}{2} = 0.707 \quad \textbf{Ans.}$$

Example 12. *Two cards are drawn simultaneously (or successively without replacement) from a well shuffled pack of 52 cards. Find the mean, variance and standard deviation of the number of kings.*

Solution. Let X denote the number of kings in a draw of two cards. X is a random variables which can assume the values 0, 1 or 2.

Now, $$P(X = 0) = P \text{ (no king)} = \frac{^{48}C_2}{^{52}C_2} = \frac{\frac{48!}{2!(48-2)!}}{\frac{52!}{2!(52-2)!}} = \frac{48 \times 47}{52 \times 51} = \frac{188}{221}$$

$$P(X = 1) = P \text{ (one king and one other card)} = \frac{^4C_1 \, ^{48}C_1}{^{52}C_2} = \frac{4 \times 48 \times 2}{52 \times 51} = \frac{32}{221}$$

and $$P(X = 2) = P \text{ (two kings)} = \frac{^4C_2}{^{52}C_2} = \frac{4 \times 3}{52 \times 51} = \frac{1}{221}$$

Thus, the probability distribution of X is

X	0	1	2
$P(X)$	$\dfrac{188}{221}$	$\dfrac{32}{221}$	$\dfrac{1}{221}$

Now   Mean of $X = E(X) = \sum\limits_{i=1}^{n} x_i p(x_i) = 0 \times \dfrac{188}{221} + 1 \times \dfrac{32}{221} + 2 \times \dfrac{1}{221} = \dfrac{34}{221}$

Also, $E(X^2) = \sum\limits_{i=1}^{n} x_i^2 p(x_i) = 0^2 \times \dfrac{188}{221} + 1^2 \times \dfrac{32}{221} + 2^2 \times \dfrac{1}{221} = \dfrac{36}{221}$

Now, $\text{Var}(X) = E(X^2) - [E(X)]^2 = \dfrac{36}{221} - \left(\dfrac{34}{221}\right)^2 = \dfrac{6800}{(221)^2}$

Therefore $\sigma_x = \sqrt{\text{Var}(X)} = \dfrac{\sqrt{6800}}{221} = 0.37$ **Ans.**

Example 13. *Find the mean and standard deviation of the number of heads when three coins are tossed.*

(*Bihar SBTE, 2010*)

Solution.   Tosses of three coins

= {(H, H, H) (T, H, H), (H, T, H), (H, H, T), (H, T, T), (T, H, T), (T, T, H), (T, T, T)]

$n(s) = 8$

$P(x = 0) = P(\text{getting no head}) = \dfrac{1}{8}$

$P(X = 1) = P(\text{getting 1 head}) = \dfrac{3}{8}$

$P(x = 2) = P(\text{getting 2 head}) = \dfrac{3}{8}$

$P(x = 3) = P(\text{getting 3 head}) = \dfrac{1}{8}$

Hence the probability distribution of x is given by

$X = X$	0	1	2	3
P	$\dfrac{1}{8}$	$\dfrac{3}{8}$	$\dfrac{3}{8}$	$\dfrac{1}{8}$

$\text{Mean} = \Sigma x_1 P_1 = 0, \dfrac{3}{8} + \dfrac{6}{8} + \dfrac{3}{8} = 1.5$

$\text{Variance} = \Sigma x_1^2 P_i - \mu^2 = \left(0 + \dfrac{3}{8} + 4 \times \dfrac{3}{8} + \dfrac{9}{8}\right) - \left(\dfrac{3}{2}\right)^2 = \dfrac{3}{4}$

$\text{S.D.} = \sqrt{\text{Variance}} = \sqrt{\dfrac{3}{4}} = \dfrac{\sqrt{3}}{2}$ **Ans.**

EXERCISE 11.2

1. A die is thrown at random. What is the expectation of the number on it ?   **Ans.** $\dfrac{7}{2}$

2. Find the mean number of heads in three tosses of a fair coin.   **Ans.** 1.5

3. A fair coin is tossed until a head or five tails occur. If X denotes the number of tosses of the coin, find the mean of X.   **Ans.** 1.9

4. In four tosses of a coin, let X be the number of heads. Tabulate the 16 possible outcomes with the corresponding values of X. By simple counting, derive the distribution of X and hence calculate the expected value of X.   **Ans.** 2

5. Three coins whose faces are marked 1 and 2, are tossed. What is the expectation of the total value of numbers on their faces ? **Ans.** 4.5

6. In a game, a person is paid ₹ 5 if he gets all heads or all tails when three coins are tossed, and he will pay ₹ 3 if either one or two heads show. What can he expect to win on the average per game ? **Ans.** Loss of ₹ 1 per toss.

7. Find the mean and variance of the number of heads in the two tosses of a coin. **Ans.** $1, \dfrac{1}{2}$

8. Two dice are thrown. Find the probability that the sum of the numbers coming up on them is 9, if it is known that the number 5 always occurs on the first die. *(Bihar SBTE, 2010)* **Ans.** $\dfrac{1}{36}$

9. Two coins are tossed. What is probability of coming up of two heads, if it is known that at least one head comes up? **Ans.** $\dfrac{1}{3}$

10. If $P(A) = \dfrac{3}{8}, P = \dfrac{1}{2}$ and $P(A \cap B) = \dfrac{1}{4}$ then find: $P(A/B)$ and (ii) $P(A^1 B^1)$ *(Bihar SBTE, 2012)* **Ans.** $\dfrac{1}{3}$

11. Bag 'A' contains 05 red balls and 07 black balls. Bag 'B' contains 06 red balls and 08 black balls. It any one ball (Red or Black) is transferred from bag A to bag B. Find the probability of drawing one black bal from Bag. *(Bihar SBTE, 2014)*

Choose the correct answer of the following:

12. The mean of the numbers obtained on throwing a die having written 1 on three faces, 2 on two faces and 5 on one face is

(a) 1   (b) 2   (c) 5   (d) $\dfrac{8}{3}$ **Ans.** (b)

13. Suppose that two cards are drawn at random from a deck of cards. Let X be the number of aces obtained. What is the value of E (X)?

(a) $\dfrac{37}{221}$ (b) $\dfrac{5}{13}$ (c) $\dfrac{1}{13}$ (d) $\dfrac{2}{13}$ **Ans.** (d)

14. Expected value of E (x) of a discrete probability distribution is given by .... *(Bihar SBTE, 2004, 2005, 2011)*

Ans. $\left(\displaystyle\sum_i x_i\, p(x_i) \right)$

15. What is the probability of drawing one spade from a pack of 52 cards?

(a) $\dfrac{1}{52}$ (b) $\dfrac{1}{4}$ (c) $\dfrac{4}{52}$ (d) $\dfrac{13}{51}$ *(Bihar SBTE, 2011)* **Ans.** (c)

16. A card is drawn from pack of 52 cards. What is the probability of getting a king

(a) $\dfrac{1}{52}$ (b) $\dfrac{1}{3}$ (c) $\dfrac{4}{13}$ (d) None of these *(Bihar SBTE, 2014)* **Ans.** (d)

17. A card is drawn from pack of 52 cards, what is the probability of getting a black card?

(a) $\dfrac{1}{4}$ (b) $\dfrac{1}{2}$ (c) $\dfrac{1}{3}$ (d) $\dfrac{3}{52}$ *(Bihar, SBTE, 2014)* **Ans.** (b)

18. A card is drawn from a peak of 52 cards. What is the probability of getting queen or king?

(a) $\dfrac{1}{13}$ (b) $\dfrac{2}{13}$ (c) $\dfrac{3}{13}$ (d) $\dfrac{4}{13}$ *(Bihar, SBTE, 2012)* **Ans.** (b)

19. A card is drawn from a pack of cards. The probability that the card is an ace or a queen is ....

(a) $\dfrac{1}{4}$ (b) $\dfrac{1}{13}$ (c) $\dfrac{1}{52}$ (d) $\dfrac{2}{13}$ *(Bihar, SBTE, 2010)* **Ans.** (d)

20. A bag contains 3 white and 2 black balls. Find the probability of drawing white ball.

(a) $\dfrac{3}{5}$ (b) $\dfrac{2}{5}$ (c) $\dfrac{1}{5}$ (d) $\dfrac{4}{5}$ *(Bihar, SBTE, 2011)* **Ans.** (a)

21. A bag contains 4 white, 3 red and 5 blue balls. The probability of drawing a red ball is:

(a) $\frac{1}{12}$ (b) $\frac{1}{3}$ (c) $\frac{1}{4}$ (d) $\frac{1}{5}$ *(Bihar, SBTE, 2010)* **Ans.** *(c)*

22. A bag contains 10 black and 10 white balls. The probability of drawing two balls of the same colour is:

(a) $\frac{1}{20}$ (b) $\frac{9}{19}$ (c) $\frac{8}{19}$ (d) None of these   *(Bihar, SBTE, 2010)* **Ans.** *(b)*

23. The probability of event lies between ....

(a) $(-\infty$ and $+ \infty)$ (b) $(-1$ and $+ 1)$ (c) $(0$ and $-1)$ (d) None of these   *(Bihar, SBTE, 2010)* **Ans.** *(d)*

24. A die is thrown. The probability that the digit coming up is greater than 4 is:

(a) $\frac{1}{4}$ (b) $\frac{1}{3}$ (c) $\frac{1}{5}$ (d) 0   *(Bihar, SBTE, 2009)* **Ans.** *(b)*

25. A die is tossed t times. A success is getting 1 or 6 on a toss. Find the mean and variance of the number of success.

(a) $\left(\frac{t}{3}, \frac{2t}{9}\right)$ (b) $\left(\frac{t}{2}, \frac{2t}{3}\right)$ (c) $\left(\frac{t}{4}, \frac{2t}{7}\right)$ (d) None of these   *(Bihar, SBTE, 2014)* **Ans.** *(a)*

26. What is the productivity that a leap your, selecting at random, will contain 53 Monda?

(a) $\frac{1}{7}$ (b) $\frac{2}{53}$ (c) $\frac{1}{53}$ (d) $\frac{2}{7}$ *(Bihar, SBTE, 2011)* **Ans.** *(d)*

27. The probability that A will solve the problem is $\frac{2}{3}$ and probability that B will solve the problem is $\frac{3}{4}$. The probability that neither will solve the problem is

(a) $\frac{1}{12}$ (b) $\frac{5}{12}$ (c) $\frac{7}{12}$ (d) $\frac{11}{12}$ *(Bihar, SBTE, 2008)* **Ans.** *(a)*

28. What is the chance of getting 7 or 11 with 2 dice?   *(Bihar, SBTE, 2012)* **Ans.** *(a)*

(a) $\frac{2}{9}$ (b) $\frac{2}{7}$ (c) $\frac{2}{5}$ (d) None of these   *(Bihar, SBTE, 2008)* **Ans.** *(a)*

29. Mutualy Exclusive events with probabilities P and q respectively then the probability of either of them

(a) $p - q$ (b) $p + q$ (c) $p \cdot q$ (d) $\frac{p}{q}$ *(Bihar, SBTE, 2008)* **Ans.** *(a)*

30. Which is correct relation?

(a) $PA + P(A^1) = \infty$ (b) $P(A) + P(A^1) = 0$

(c) $P(A) + P(A^1) = 1$ (d) $P(A) + P(A^1) = -1$ *(Bihar, SBTE, 2012)* **Ans.** *(c)*

31. If A and B are two independent events then.........

(a) $P(A \cap B) = P(A) \cdot P(B/A)$ (b) $P(A \cap B) = P(A) \cdot P(A/B)$

(c) $P(A \cap B) = P(A) \cdot P(B)$ (d) $P(A \cap B) = P(A) + P(B)$ *(Bihar, SBTE, 2011)* **Ans.** *(d)*

32. If p is probability of an event A and q is the probability of complementary event of A, then

(a) $p = 1 + q$ (b) $p = \frac{1}{q}$ (c) $p = 1 - q$ (d) $p = q - 1$ *(Bihar, SBTE, 2009)* **Ans.** *(c)*

33. Mutually exclusive events with probabilities p and q respectively, then the probability of either of them is....................

(a) $p - q$ (b) $p + q$ (c) $p \cdot q$ (d) p/q *(Bihar, SBTE, 2008)* **Ans.** *(b)*

12 Binomial Distribution

12.1 Descrete random variable:
(*Bihar SBTE, 20009*)

If a random variable takes a finite set of values, is called descrete random variable.

12.2 BINOMIAL DISTRIBUTION $P(r) = {}^nC_r \ P^r \cdot q^{n-r}$

To find the probability of the happening of an event once, twice, thrice, ... r times ...... exactly in n trials.

Let the probability of the happening of an event A in one trial be p and

its probability of not happening be $1 - p = q$.

We assume that there are n trials and the happening of the event A is r times and its not happening is $n - r$ times.

$$A \ A....... A \qquad\qquad \overline{A}.\overline{A}......\overline{A}$$

$$r \ \text{times} \qquad\qquad n - r \ \text{times} \qquad\qquad\qquad ...(1)$$

A indicates its happening, $\overline{A}$ its failure and $P(A) = p$ and $P(\overline{A}) = q$.

We see that (1) has the probability

$$p p \ .. \ p \quad q \cdot q .... q \ = p^r \cdot q^{n-r} = p^r q^{n-r} \qquad\qquad ...(2)$$

$$r \ \text{times} \quad n - r \ \text{times}$$

Clearly (1) is merely one order of arranging r A's.

The probability of (1) $= p^r q^{n-r} \times$ Number of different arrangements of

$$r \ A\text{'s and} \ (n - r)\overline{A}\text{'s}.$$

The number of different arrangements of r A's and $(n - r) \overline{A}$'s $= {}^nC_r$

$\therefore$ Probability of the happening of an event r times ${}^nC_r \ P^r \cdot q^{n-r}$.

$$p(r) = {}^nC_r \ P^r \cdot q^{n-r} \ (r = 0, 1, 2, ...,n).$$

$$= (r + 1)\text{th term of} \ (q + p)^n$$

If $r = 0$, probability of happening of an event 0 times

$$= {}^nC_0 \ q^n p^0 = q^n$$

If $r = 1$, probability of happening of an event 1 time $= {}^nC_1 q^{n-1} p$

If $r = 2$, probability of happening of an event 2 times $= {}^nC_2 \ q^{n-2} p^2$

If $r = 3$, probability of happening of an event 3 times $= {}^nC_3 \, q^{n-3} \, p^3$

and so on.

These terms are clearly the successive terms in the expansion of $(q + p)^n$.

Hence it is called Binomial Distribution.

Example 1. *Find the probability of getting 4 heads in 6 tosses of a fair coin.*   (*Bihar SBTE, 2011*)

Solution. $p = \dfrac{1}{2}, q = \dfrac{1}{2}, \; n = 10, \; r = 4$

We know that $P(r) = {}^nC_r \, q^{n-r} \, p^r$

$P(4) = {}^{10}C_4 \, q^{10-4} p^4$

$$= \frac{10 \times 9 \times 8 \times 7}{4 \times 3 \times 2} \cdot \left(\frac{1}{2}\right)^6 \left(\frac{1}{2}\right)^4 = 210\left(\frac{1}{2}\right)^{10} = \frac{210}{1024} = \frac{105}{512} \qquad \textbf{Ans.}$$

Example 2. *If on an average one ship in every ten is wrecked, find the probability that out of 5 ships expected to arrive, 4 at least will arrive safely.*

Solution. Out of 10 ships, one ship is wrecked.

i.e., Nine ships out of ten ships are safe.

$$P(\text{safety}) = \frac{9}{10}$$

$P(\text{At least 4 ships out of 5 are safe}) = P(4 \text{ or } 5) = P(4) + P(5)$

$$= {}^5C_4 \, P^4 q^{5-4} + {}^5C_5 P^5 q^0 = 5\left(\frac{9}{10}\right)^4\left(\frac{1}{10}\right) + \left(\frac{9}{10}\right)^5 = \left(\frac{9}{10}\right)^4\left(\frac{5}{10} + \frac{9}{10}\right) = \frac{7}{5}\left(\frac{9}{10}\right)^4 \qquad \textbf{Ans.}$$

Example 3. *The overall percentage of failures in a certain examination is 20. If six candidates appear in the examination, what is the probability that at least five pass the examination ?*

Solution. Probability of failures $= 20\% = \dfrac{20}{100} = \dfrac{1}{5}$

Probability of pass $(P) = 1 - \dfrac{1}{5} = \dfrac{4}{5}$

Probability of at least five pass $= P(5 \text{ or } 6)$

$$= P(5) + P(6) = {}^6C_5 \, P^5 q + {}^6C_6 P^6 q^0 = 6\left(\frac{4}{5}\right)^5\left(\frac{1}{5}\right) + \left(\frac{4}{5}\right)^6 = \left(\frac{4}{5}\right)^5\left[\frac{6}{5} + \frac{4}{5}\right] = 2\left(\frac{4}{5}\right)^5 = \frac{2048}{3125} = 0.65536 \; \textbf{Ans.}$$

Example 4. *Ten percent of screws produced in a certain factory turn out to be defective. Find the probability that in a sample of 10 screws chosen at random, exactly two will be defective.*

Solution. $P = \dfrac{1}{10}, \qquad q = \dfrac{9}{10}, \; n = 10, \; r = 2$

$$P(r) = {}^nC_r \, P^r \, q^{n-r}$$

$$P(2) = {}^{10}C_2\left(\frac{1}{10}\right)^2\left(\frac{9}{10}\right)^{10-2} = \frac{10 \times 9}{1 \times 2}\left(\frac{1}{10}\right)^2\left(\frac{9}{10}\right)^8$$

$$= \frac{1}{2} \cdot \left(\frac{9}{10}\right)^9 = 0.1937 \qquad \textbf{Ans.}$$

Example 5. *The probability that a man aged 60 will live to be 70 is 0.65. What is the probability that out of 10 men, now 60, at least 7 will live to be 70 ?*

Solution. The probability that a man aged 60 will live to be $70 = p = 0.65$

$$q = 1 - p = 1 - 0.65 = 0.35$$

Number of men $= n = 10$

Probability that at least 7 men (7 or 8 or 9 or 10) will live to 70

$$= P(7) + P(8) + P(9) + P(10) = {}^{10}C_7\, q^3\, p^7 + {}^{10}C_8 q^2 p^8 + {}^{10}C_9 q\, p^9 + p^{10}$$

$$= \frac{10 \times 9 \times 8}{1 \times 2 \times 3}(0.35)^3(0.65)^7 + \frac{10 \times 9}{1 \times 2}(0.35)^2(0.65)^8 + 10(0.35)(0.65)^9 + (0.65)^{10}$$

$$= (0.65)^7\,[\,120\,(0.35)^3 + 45\,(0.35)^2\,(0.65) + 10\,(0.35)\,(0.65)^2 + (0.65)^3\,]$$

$$= (0.65)^7 \times 125\,[120 \times (0.07)^3 + 45 \times (0.07)^2\,(0.13) + 10\,(0.07)\,(0.13)^2 + (0.13)^3]$$

$$= 0.04902 \times 125\,[0.04 + 0.028665 + 0.011830 + 0.002197]$$

$$= 6.1275 \times 0.082692 = 0.5067 \qquad \text{Ans.}$$

Example 6. *If 10 % of bolts produced by a machine are defective. Determine the probability that out of 10 bolts, chosen at random (i) 1 (ii) none (iii) at most 2 bolts will be defective.*

Solution. Probability of defective bolts $= p = 10\% = 0.1$

Probability of not defective bolts $= q = 1 - p = 1 - 0.1 = 0.9$

Total number of bolts $= n = 10$

(i) Probability of 1 defective bolt $\quad = {}^{10}C_1\,(0.1)^1\,(0.9)^9 = 0.3874$

(ii) Probability that none is defective $=$ Probability of 0 defective bolt

$$= P(0) = {}^{10}C_0\,(0.1)^0\,(0.9)^{10} = 0.3487$$

(iii) Probability of 2 defective $\quad = {}^{10}C_2\,(0.1)^2\,(0.9)^8 = 0.1937$

Probability of at most 2 defective $\quad = P(0 \text{ or } 1 \text{ or } 2)$

$$= P(0) + P(1) + P(2) = 0.3487 + 0.3874 + 0.1937$$

$$= 0.9298 \qquad \text{Ans.}$$

Example 7. *A die is thrown 8 times and it is required to find the probability that 3 will show (i) Exactly 2 times (ii) At least seven times (iii) At least once.*

Solution. The probability of throwing 3 in a single trial $= P = \dfrac{1}{6}$

The probability of not throwing 3 in a single trial $\quad = q = \dfrac{5}{6}$

(i) P (getting 3, exactly 2 times)

$$= {}^8C_2\, q^6\, p^2 = 28\left(\frac{5}{6}\right)^6\left(\frac{1}{6}\right)^2 = \frac{28 \times 5^6}{6^8}$$

(ii) P (getting 3, at least seven times) $= P$ (getting 3, at 7 or 8 times)

$$= P(7) + P(8) = {}^8C_7\, q^1\, p^7 + {}^8C_8\, q^0\, p^8$$

$$= 8\left(\frac{5}{6}\right)\left(\frac{1}{6}\right)^7 + \left(\frac{1}{6}\right)^8 = \frac{41}{6^8}$$

(iii) P (getting 3 at least once)

$$= P \text{ (getting 3, at 1 or 2 or 3 or 4 or 5 or 6 or 7 or 8 times)}$$

$$= P(1) + P(2) + P(3) + P(4) + P(5) + P(6) + P(7) + P(8)$$

$$= P(0) + P(1) + P(2) + P(3) + P(4) + P(5) + P(6) + P(7) + P(8) - P(0)$$
$$= 1 - P(0) = 1 - {}^8C_0 \, q^8 \, p^0$$
$$= 1 - \left(\frac{5}{6}\right)^8 \qquad\qquad\qquad\qquad\qquad \textbf{Ans.}$$

Example 8. *An underground mine has 5 pumps installed for pumping out storm water, the probability of any one of the pumps failing during the storm is $\dfrac{1}{8}$. What is the probability that (i) at least 2 pumps will be working; (ii) all the pumps will be working during a particular storm ?*

(*Bihar SBTE, 2010*)

Solution. (*i*) Probability of pump failing $= \dfrac{1}{8}$

Probability of pump working $= 1 - \dfrac{1}{8} = \dfrac{7}{8}, \quad P = \dfrac{7}{8}, \quad q = \dfrac{1}{8}, \quad n = 5$

(*i*) P (At least 2 pumps working) $= P$ (2 or 3 or 4 or 5 pumps working)

$$= P(2) + P(3) + P(4) + P(5) = {}^5C_2 \, p^2 q^3 + {}^5C_3 \, p^3 q^2 + {}^5C_4 \, p^4 q + {}^5C_5 p^5 q^0$$

$$= 10\left(\frac{7}{8}\right)^2\left(\frac{1}{8}\right)^3 + 10\left(\frac{7}{8}\right)^3\left(\frac{1}{8}\right)^2 + 5\left(\frac{7}{8}\right)^4\left(\frac{1}{8}\right) + \left(\frac{7}{8}\right)^5 = \frac{1}{8^5}[10 \times 49 + 10 \times 343 + 5 \times 2401 + 16807]$$

$$= \frac{1}{8^5}[490 + 3430 + 12005 + 16807] = \frac{32732}{8^5} = \frac{8183}{8192}$$

(*ii*) P (All the 5 pumps working) $= P(5) = {}^5C_5 p^5 q^0 = \left(\dfrac{7}{8}\right)^5 = \dfrac{16807}{32768}$ **Ans.**

Example 9. *Assuming that 20 % of the population of a city are literate, so that the chance of an individual being literate is $\dfrac{1}{5}$ and assuming that 100 investigators each take 10 individuals to see whether they are literate, how many investigators would you expect to report 3 or less were literate.*

(*A.M.I.E.T.E., Summer 2000*)

Solution. $P = \dfrac{1}{5}, n = 10$

P (3 or less) $= P$ (0 or 1 or 2 or 3) $= P(0) + P(1) + P(2) + P(3)$

$$= {}^{10}C_0\left(\frac{1}{5}\right)^0\left(\frac{4}{5}\right)^{10} + {}^{10}C_1\left(\frac{1}{5}\right)^1\left(\frac{4}{5}\right)^9 + {}^{10}C_2\left(\frac{1}{5}\right)^2\left(\frac{4}{5}\right)^8 + {}^{10}C_3\left(\frac{1}{5}\right)^3\left(\frac{4}{5}\right)^7$$

$$= \left(\frac{4}{5}\right)^{10} + \frac{10}{5}\left(\frac{4}{5}\right)^9 + \frac{45}{25}\left(\frac{4}{5}\right)^8 + \frac{120}{125}\left(\frac{4}{5}\right)^7$$

$$= \left(\frac{4}{5}\right)^7 [(0.8)^3 + 2(0.8)^2 + 1.8(0.8) + 0.96]$$

$$= 0.2097152 \, [0.512 + 1.28 + 1.44 + 0.96]$$

$$= 0.2097152 \times 4.192 = 0.879126118$$

Required number of investigators $= 0.879126118 \times 100 = 87.9126118$

$$= 88 \text{ approximate} \qquad\qquad\qquad \textbf{Ans.}$$

Example 10. *Write two-three areas where binomial distribution is applied. The probability of entering student in chartered accountant will be graduate 0.5. Determine the probability that out of 10 students (i) none (ii) one or (iii) at least one will graduate.* (*R.G.P.V., Bhopal, Dec., 2003*)

Solution.   Given, the probability of an entering student in chartered accountant will graduate is $p = 0.5$

∴   The probability of an entering student in chartered accountant will not graduate is $q = 0.5$.

Therefore

(*i*) The probability of none will graduate out of 10 students

$$P(0) = {}^{10}C_0\, p^0\, q^{10} = {}^{10}C_0\, (0.5)^0\, (0.5)^{10}$$

$$= 9.765625 \times 10^{-4} \qquad\qquad\qquad\qquad\text{Ans.}$$

(*ii*) The probability of exactly one student will be graduate out of 10 students.

$$P(1) = {}^{10}C_1\, (0.5)^1\, (0.5)^9 = 10 \times 0.5 \times (0.5)^9$$

$$= 9.765625 \times 10^{-3} \qquad\qquad\qquad\qquad\text{Ans.}$$

(*iii*) The probability of at least one will be graduate out of 10 students

$$P(\text{At least one}) = 1 - (\text{probability of none will graduate})$$

$$= 1 - 9.765625 \times 10^{-4} = 0.99 \qquad\qquad\text{Ans.}$$

Example 11. *Assuming half the population of a town consumes chocolates and that 100 investigators each take 10 individuals to see whether they are consumers, how many investigators would you expect to report that three people or less were consumers?*

Solution. The chance for an individual to be consumer is $p = \dfrac{1}{2}$

The chance of not being a consumer $= q = 1 - \dfrac{1}{2} = \dfrac{1}{2}$.

Here we have to find the probabilities of 0, 1, 2 and 3 successes.

$$P(r \le 3) = P(0) + P(1) + P(2) + P(3)$$

$$= {}^{10}C_0\, q^{10}\, p^0 + {}^{10}C_1\, q^9\, p^1 + {}^{10}C_2\, q^8\, p^2 + {}^{10}C_3\, q^7\, p^3$$

$$= \left(\frac{1}{2}\right)^{10} + 10\left(\frac{1}{2}\right)^9\left(\frac{1}{2}\right) + 45\left(\frac{1}{2}\right)^8\left(\frac{1}{2}\right)^2 + 120\left(\frac{1}{2}\right)^7\left(\frac{1}{2}\right)^3 = \left(\frac{1}{2}\right)^{10}[1+10+45+120] = \frac{176}{1024}$$

The number of investigators to report that three or less people were consumers of chocolates is given by

$$\frac{176}{1024} \times 100 = 17.2$$

Hence, 17 investigators would report that 3 or less people are consumers.   **Ans.**

Example 12. *The probability that a bomb dropped from a plane will strike the target is $\dfrac{1}{5}$. If six bombs are dropped, find the probability that:*

(i) *Exactly two will strike the target.* (M.D.U., May 2007)

(ii) *At least two will strike the target.* (R.G.P.V., Bhopal, II Semester, Feb. 2006)

Solution.   Here, $p = \dfrac{1}{5}$, $q = 1 - \dfrac{1}{5} = \dfrac{4}{5}$, $n = 6$

We know that $P(r) = {}^nC_r\, p^r\, q^{n-r}$

$$P(2) = {}^6C_2\left(\frac{1}{5}\right)^2\left(\frac{4}{5}\right)^{6-2} = 15\left(\frac{256}{15625}\right) = \frac{768}{3125} = 0.24576$$

$P(\text{at least } 2) = P(2, 3, 4, 5, 6)$

$$= P(2) + P(3) + P(4) + P(5) + P(6)$$

$$= P(0) + P(1) + P(2) + P(3) + P(4) + P(5) + P(6) - P(0) - P(1)$$

$$= 1 - [P(0) + P(1)]$$

$$= 1 - \left[{}^6C_0 \left(\frac{1}{5}\right)^0 \left(\frac{4}{5}\right)^6 + {}^6C_1 \left(\frac{1}{5}\right) \left(\frac{4}{5}\right)^5 \right] = 1 - \left[\frac{4096}{15625} + 6 \left(\frac{1024}{15625}\right) \right]$$

$$= 1 - \frac{10240}{15625} = \frac{5385}{15625} = \frac{1077}{3125} = 0.34464$$

Hence (i) $P = 0.24576$ (ii) $P = 0.34464$ **Ans.**

Example 13. *Three defective bulbs are mixed with 7 good ones. Find the probability distribution of the number of defective bulbs, if bulbs are drawn at random.*   *(Bihar SBTE, 2003)*

Solution.   Here p = Prbability of a defective bulb = $\dfrac{3}{10}$

q = Probability of non-defective bulb = $1 - \dfrac{3}{10} = \dfrac{7}{10}$

∴ The probability distribution of number of defective bulb = ${}^3C_r \left(\dfrac{3}{10}\right)^r \left(\dfrac{7}{10}\right)^{3-r}$ **Ans.**

Example 14. *Two cards are drawn successively with replacement from a well-shuffled pack of 52 cards. Find the probability distribution of the number of kings.*   *(Bihar SBTE, 2004)*

Solution.   Here

$$P_1 = \frac{{}^4C_1}{{}^{52}C_1} = \frac{4}{52} = \frac{1}{13}, \qquad P_2 = \frac{{}^4C_1}{{}^{52}C_1} = \frac{4}{52} = \frac{1}{13}$$

Let P = probability that the two cards are drawn successively with replacement of the king.

$$= \frac{1}{13} \times \frac{1}{13} = \frac{1}{169}$$

$$q = 1 - p = 1 - \frac{1}{169} = \frac{168}{169}$$

∴   Probability distribution of the number of king = ${}^2C_r \left(\dfrac{1}{169}\right)^r \left(\dfrac{168}{169}\right)^{2-r}$ **Ans.**

Example 15. *A bag contains 7 white and 3 black balls. A ball is drawn and replaced. What is the probability of 2 white and 3 black balls in five drawings ?*   *(Bihar SBTE, 2004)*

Solution.   Let S is the sample space.

W is the event of drawing a while ball and B is the event of drawing a black ball.

According to question

        n (S) = 10,   n (W) = 7,   n (B) = 3

If P = probability of drawing a white ball in one trial.

Then, $p = \dfrac{n(W)}{n(S)} = \dfrac{7}{10}$

q = probability of drawing not a white ball in one trial.

i.e., the probability of drawing a black ball.

$$= 1 - \frac{7}{10} = \frac{3}{10}$$

$\therefore$ The probability of drawing 2 white 3 black balls in five drawing $= {}^5C_2 \cdot p^2 q^3$

$$= {}^5C_2 \cdot \left(\frac{7}{10}\right)^2 \cdot \left(\frac{3}{10}\right)^3 \qquad\qquad\qquad\qquad\qquad \textbf{Ans.}$$

Example 16. *The probability that a Television manufactured by a company will defective is $\frac{1}{10}$. If 12 such*

Television are manufacture, find the proability that:
(a) Exactly two will be defective
(b) At least two will be defective
(c) None will be defective   (Bihar SBTE, 2011)

Solution. $P = \dfrac{1}{10}$

$$q = 1 - \frac{1}{10} = \frac{9}{10}$$

(a) $P(2) = {}^nC_2 p^2 g^{n-2} = {}^{12}C_2 \left(\dfrac{1}{10}\right)^2 \left(\dfrac{9}{10}\right)^8 = \dfrac{66 \times 9^8}{10^{10}}$ **Ans.**

(b)   P (At least two defective $= 1 - P(P) - P(1)$

$$= 1 - {}^{12}C_0 \left(\frac{1}{10}\right)^0 \left(\frac{9}{10}\right)^{12} - {}^{12}C_1 \left(\frac{1}{10}\right)^1 \left(\frac{9}{10}\right)^{11} = 1 - \left(\frac{9}{10}\right)^{12} - \frac{12}{100}\left(\frac{9}{10}\right)^{11} \qquad \textbf{Ans.}$$

(c)   P (None is defective) $= {}^nC_0\, p^0 q^{12}$

$$= {}^{12}C_0 \left(\frac{1}{10}\right)^0 \left(\frac{9}{10}\right)^{12} = \left(\frac{9}{10}\right)^{12} \qquad\qquad\qquad \textbf{Ans.}$$

EXERCISE 12.1

1.   If 20% of the bolts produced by a machine are defective, determine the probability that out of 4 bolts chosen at random
     (a) 1   (b) 0   (c) At most 2

     bolts will be defective.   **Ans.** (a) 0.4096,   (b) 0.4096, (c) 0.9728.

2.   Six dice are thrown 729 times. How many times do you expect at least three dice to show a five or a six ?   **Ans.** 233

3.   Find the probability of getting a total of 7 at least once in 4 tosses of a pair of fair dice.?

                                                            (A.M.I.E., Winter 2002) **Ans.** $\dfrac{671}{1296}$

4.   If the chance that any one of the 10 telephone lines is busy at any instant is 0.2, what is the chance that 5 of the lines are
     busy ? What is the probability that all the lines are busy ?   **Ans.** ${}^{10}C_5\ (0.2)^5\ (0.8)^5,\ (0.2)^{10}$

5.   An insurance salesman sells policies to 5 men, all of identical age in good health. According to the actuarial tables the

     probability that a man of this particular age will be alive 30 years hence is $\dfrac{3}{2}$. Find the probability that in 30 years.

     (a) All 5 men   (b) At least 3 men   (c) Only 2 men   (d) At least 1 man will be alive.

                                   Ans. (a) $\dfrac{32}{243}$ (b) $\dfrac{192}{243}$ (c) $\dfrac{40}{243}$ (d) $\dfrac{242}{243}$

6.   A box contains 10 screws, 3 of which are defective. Two screws are drawn at random without replacement. Find the
     probability that none of the two screws is defective.
                                                                          Ans. $\dfrac{7}{15}$

7.   Out of 800 families with four children each, how many families would be expected to have :

     (i) 2 boys and 2 girls;   (ii) at least one boy;   (iii) no girl;   (iv) at most two girls ?

Assume equal probabilities for boys and girls.   **Ans.** (*i*) 300, (*ii*) 750, (*iii*) 50, (*iv*) 550.

8.   In a hurdle race, a player has to cross 10 hurdles. The probability that he will clear each hurdle is 5/6. What is the

probability that he will knock down less than 2 hurdles ?   **Ans.** $\dfrac{8}{3}\left(\dfrac{5}{6}\right)^9$

9.   In a lot of 500 soleniods 25 are defective. Find the probability of 0, 1, 2, 3, defective soleniods in a random sample of

20 solenoids.   (*M.D.U. May 2008*) **Ans** 0.3585, 0.3774, 0.1887, 0.0596.

10. An electronic component consists of three parts. Each part has probability 0.99 of performing satisfactorily. The component
fails if 2 or more parts do not perform satisfactorily. Assuming that the parts perform independently, determine the
probability that the component does not perform satisfactorily.   **Ans.** 0.000298

11. The incidence of occupational disease in an industry is such that the workers have 20% chance of suffering from it. What
is the probability that out of 6 workers 4 or more will catch the disease ?

(*MDU* 2006, *A.M.I.E., Winter 2005*) **Ans.** $\dfrac{53}{2125}$

12. Among 10,000 random digits, find the probability p that the digit 3 appears at most 950 times.

(*A.M.I.E., Summer 2003*) **Ans.** $\displaystyle\sum_{x=0}^{950} 10{,}000 \, C_r \left(\dfrac{1}{10}\right)^r \left(\dfrac{9}{10}\right) 10{,}000 - r$

13. In a bombing action there is 50% chance that any bomb will strike the target. Two direct hits are needed to destroy the
target completely. How many bombs are required to be dropped to give a 99% chance or better of completely destroying
the target.   (*R.G.P.V., Bhopal, June 2008*) **Ans.** 11

14. **Fill in the blanks:**

(*a*) A coin is biased so that a head is twice as likely to occur as a tail. If the coin is tossed 3 times, the probability of

getting exactly 2 tails, is ....   **Ans.** $\dfrac{2}{9}$

(*b*) The probability of getting number 5 exactly two times in five throws of an unbiased die is .... **Ans.** $10, \dfrac{5^3}{6^5}$

(*c*) A die is thrown 6 times. The probability to get greater than 4 appears at least once is .... **Ans.** $\dfrac{665}{729}$

(*d*) For what, one should be?

(*i*) Obtaining 6 at least once in 4 throws of a die.

or (*ii*) Obtaining a double-six at least once in 24 throws with two dice.   **Ans.** (*i*)

(*e*) The probability of producing a defective bolt is 0.1. The probability that out of 5 bolts one will be defective is

....   **Ans.** $\dfrac{1}{2}\left(\dfrac{9}{10}\right)^4$

(*f*) If the probability of hitting a target is 5% and 5 shots are fired independently, the probability that the target will be

hit at least once is ....   **Ans.** $1 - (0.95)^5$

15. **Tick (✓) the correct answer :**

(*a*) If a coin is tossed 6 times in succession, the probability of getting at least one head is

(*i*) 1/64   (*ii*) 3/32   (*iii*) 63/64   (*iv*) 1/2   **Ans.** (*iii*)

(*b*) A coin is tossed until a tail appears or at the most five times. Given that the tail does not appear on the first two
tosses, the probability that the coin will be tossed 5 times, is

(*i*) 1/2   (*ii*) 3/5   (*iii*) 1/3   (*iv*) 1/4   **Ans.** (*iv*)

(c) In a certain manufacturing process it is known that on an average, 1 in every 100 items is defective. What is the probability that 5 items are inspected before a defective item is found?

(i) 0.0096   (ii) 0.96   (iii) 0.096   (iv) none of these   Ans. (i)

(d) The probability that a marksman will hit a target is given as $\frac{1}{5}$. Then his probability of at least one hit in 10 shots is

(i) $1-\left(\frac{4}{5}\right)^{10}$ (ii) $\frac{1}{5^{10}}$ (iii) $1-\frac{1}{5^{10}}$ (iv) None of these.   Ans. (i)

(e) The probability of having at least one tail in 4 throws with a coin is

(i) $\frac{15}{16}$,   (ii) $\frac{1}{16}$,   (iii) $\frac{1}{4}$,   (iv) 1.   Ans. (i)

(f) A coin is tossed 3 times. The probability of obtaining two heads will be

(i) $\frac{3}{8}$,   (ii) $\frac{1}{2}$,   (iii) 1,   (iv) 2.   Ans. (i)

(g) 8 coins are tossed simultaneously. The probability of getting at least 6 heads is

(i) $\frac{57}{64}$,   (ii) $\frac{229}{256}$,   (iii) $\frac{7}{64}$,   (iv) $\frac{37}{256}$.   Ans. (iv)

(h) Three unbiased coins are tossed simultaneously. This is repeated four times. The probability of getting at least one head each time is

(i) $\left(\frac{3}{4}\right)^4$ (ii) $\left(\frac{7}{8}\right)^4$ (iii) $\left(\frac{1}{8}\right)^4$ (iv) $\left(\frac{1}{4}\right)^4$ Ans. (ii)

(i) In rolling two fair dice, the probability of getting equal numbers or numbers with an even product

(i) $\frac{6}{36}$ (ii) $\frac{30}{36}$ (iii) $\frac{27}{36}$ (iv) $\frac{3}{36}$.   Ans. (ii)

12.3 MEAN OF BINOMIAL DISTRIBUTION

(Bihar SBTE, 2009, A. M. I .E.T. E., Winter 2002, Summer 2000, A.M.I.E., Winter 2002)

$$(q+p)^n = q^n + {}^nC_1\, q^{n-1}p^1 + {}^nC_2\, q^{n-2}\, p^2 + {}^nC_3\, q^{n-3}\, p^3 + \ldots + {}^nC_r\, q^{n-r}\, p^r + \ldots + p^n$$

Successes r	Frequency f	Product rf
0	q^n	0
1	$n\, q^{n-1}p$	$n\, q^{n-1}p$
2	$\dfrac{n(n-1)}{2}q^{n-2}p^2$	$n(n-1)\, p^{n-2}\, q^2$
3	$\dfrac{n(n-1)(n-2)}{6}q^{n-3}p^3$	$\dfrac{n(n-1)(n-2)}{2}q^{n-3}p^3$
....	......	....
n	p^n	np^n

$$\Sigma f r = n q^{n-1}\, p + n(n-1)q^{n-2}p^2 + \frac{n(n-1)(n-2)}{2}q^{n-3}p^3 + \ldots + n\, p^n$$

$$= np\left[q^{n-1} + \frac{(n-1)}{1!}q^{n-2}p + \frac{(n-1)(n-2)}{2}q^{n-3}p^2 + \ldots + p^{n-1}\right]$$

$$= n p (q + p)^{n-1} = np \qquad \text{(since } q + p = 1)$$

$$\Sigma f = q^n + n q^{n-1} p + \frac{n(n-1)}{2} q^{n-2} p^2 + \ldots + p^n$$

$$= (q + p)^n = 1 \qquad \text{(since } q + p = 1)$$

Hence, $\qquad\qquad\qquad\qquad$ Mean $= \dfrac{\Sigma f\, r}{\Sigma f} = \dfrac{np}{1}$ $\qquad\qquad\qquad\qquad$ **Ans.**

12.4 STANDARD DEVIATION OF BINOMIAL DISTRIBUTION

(Bihar SBTE, 2009, 2010, 2011, 2014, A. M. I. E.T. E., Winter 2002, A.M.I.E., Winter 2002)

Successes r	Frequency f	Product $r^2 f$
0	q^n	0
1	$n q^{n-1} p$	$n q^{n-1} p$
2	$\dfrac{n(n-1)}{2} q^{n-2} p^2$	$2 n (n - 1) q^{n-2} p^2$
3	$\dfrac{n(n-1)(n-2)}{6} q^{n-3} p^3$	$\dfrac{3n(n-1)(n-2)}{2} q^{n-3} p^3$
.....	.....	.....
n	p^n	$n^2 p^n$

We know that $\qquad\qquad \sigma^2 = \dfrac{\Sigma f r^2}{\Sigma f} - \left(\dfrac{\Sigma f r}{\Sigma f}\right)^2$ $\qquad\qquad\qquad\qquad$ ...(1)

r is the deviation of items (successes) from 0.

$$\Sigma f = 1, \ \Sigma f r = np$$

$$\Sigma f r^2 = 0 + n q^{n-1} p + 2n(n-1) q^{n-2} p^2 + \frac{3n(n-1)(n-2)}{2} q^{n-3} p^3 + \ldots + n^2 p^n$$

$$= np\left[q^{n-1} + \frac{2(n-1)}{1!} q^{n-2} p + \frac{3(n-1)(n-2)}{2!} q^{n-3} p^2 + \ldots + np^{n-1} \right]$$

$$= np\left[q^{n-1} + \frac{(n-1) q^{n-2} p}{1!} + \frac{(n-1)(n-2)}{2!} q^{n-3} p^2 + \ldots + p^{n-1} \right.$$

$$\left. + \frac{(n-1) q^{n-2} p}{1!} + \frac{2(n-1)(n-2)}{2!} q^{n-3} p^2 + \ldots + (n-1) p^{n-1} \right]$$

$$= n p\left[q^{n-1} + (n-1) q^{n-2} p + \frac{(n-1)(n-2)}{2!} q^{n-3} p^2 + \ldots + p^{n-1} \right.$$

$$\left. + (n-1) p\left\{ q^{n-2} + (n-2) q^{n-3} p + \frac{(n-2)(n-3)}{2!} q^{n-4} p^2 + \ldots + p^{n-2} \right\} \right]$$

$$= np\left[(q+p)^{n-1} + (n-1) p (q+p)^{n-2} \right]$$

$$= np\left[1 + (n-1) p \right]$$

$$= np\left[np + (1-p) \right] = np[np + q] = n^2 p^2 + npq$$

Putting these values in (1), we have

$$\text{Variance} = \sigma^2 = \frac{n^2 p^2 + n\,pq}{1} - \left(\frac{np}{1}\right)^2 = npq,$$

$$S.D. = \sigma = \sqrt{n\,p\,q}$$

Hence for the binomial distribution,

$$\text{Mean} = np, \ \mu_2 = \sigma^2 = n\,p\,q$$

Example 17. *If the probability of a defective bolt is 0.1, find*

(a) the mean (b) the standard deviation for the distribution bolts in a total of 400.

Solution. $n = 400, \ p = 0.1, \ \text{Mean} = np = 400 \times 0.1 = 40$

$$\text{Standard deviation} = \sqrt{npq} = \sqrt{400 \times 0.1(1 - 0.1)} = \sqrt{400 \times 0.1 \times 0.9} = 20 \times 0.3 = 6 \qquad \textbf{Ans.}$$

Example 18. *A die is tossed thrice. A success is getting 1 or 6 on a toss. Find the mean and variance of the number of successes.* \hfill (*AMIETE, Dec. 2010*)

Solution. $$n = 3, \ p = \frac{1}{3}, \ q = \frac{2}{3}$$

$$\text{Mean} = np = 3 \times \frac{1}{3} = 1 \qquad \text{Variance} = npq = 3 \times \frac{1}{3} \times \frac{2}{3} = \frac{2}{3} \qquad \textbf{Ans.}$$

Example 19. *If mean and variance of a binomial distribution are 4 and 2 respectively, find the probability of*
(i) exactly 2 successes (ii) less than 2 successes (iii) at least 2 successes.

\hfill (*R.G.P.V., Bhopal, II Semester, June 2005*)

Solution. $\quad$ Mean = 4 $\qquad \Rightarrow \qquad np = 4 \qquad \qquad$ ... (1)

$\qquad$ Variance = 2 $\qquad \Rightarrow \qquad npq = 2 \qquad \qquad$ ... (2)

Dividing (2) by (1), we get

$$\frac{npq}{np} = \frac{2}{4} \qquad \Rightarrow \qquad q = \frac{1}{2}$$

$$p = 1 - q = 1 - \frac{1}{2} = \frac{1}{2}$$

Putting the value of p in (1), we get $\ n\left(\dfrac{1}{2}\right) = 4 \qquad \Rightarrow \qquad n = 8$

(*i*) Probability of r successes $= {}^nC_r\, p^r\, q^{n-r}$

$$P(2) = {}^8C_2 \left(\frac{1}{2}\right)^2 \left(\frac{1}{2}\right)^{8-2} = {}^8C_2 \left(\frac{1}{2}\right)^8 = \frac{8 \times 7}{2} \frac{1}{256} = \frac{7}{64}$$

(*ii*) P (less than 2 successes) $= P(0) + P(1) = {}^8C_0\, p^0 q^8 + {}^8C_1\, p^1 q^7 = \dfrac{1}{256} + 8 \dfrac{1}{2}\left(\dfrac{1}{2}\right)^7 = \dfrac{9}{256}$

(*iii*) P (at least 2 successes) $= P(2) + P(3) + \dots + P(8)$

$$= P(0) + P(1) + P(2) + P(3) + \dots + P(8) - P(0) - P(1)$$

$$= 1 - P(0) - P(1) = 1 - [P(0) + P(1)] = 1 - \frac{9}{256} = \frac{247}{256} \qquad \textbf{Ans.}$$

Example 20. *Fit a binomial distribution to the following data:*

x	0	1	2	3	4
f	30	62	46	10	2

\hfill (*M.D.U. Dec. 2009*)

Solution.   We have,

x	f	fx
0	30	0
1	62	62
2	46	92
3	10	30
4	2	8
	$\Sigma f = 150$	$\Sigma fx = 192$

Mean of observation $= \dfrac{\Sigma f x}{\Sigma f} = \dfrac{192}{150} = 1.28$

$\Rightarrow$ $np = 1.28$ $(\because n = 4)$

$\Rightarrow$ $4p = 1.28$ $\Rightarrow$ $p = 0.32$ and $q = 1 - p = 1 - 0.32 = 0.68$

Also, $N = 150$ $[N = \Sigma f]$

Hence, the binmial distributions is $N (q + p)^n = 150 (0.,68 + 0.32)^4$ **Ans.**

Example 21. *Fit a Binomial distribution for the following data and compare the theoretical frequencies with actual ones :*

x	0	1	2	3	4	5
y	2	14	20	34	22	8

(R.G.P.V., Bhopal, II Semester, June 2006)

Solution.

x	$y = f$	fx	$P = {}^5C_r\, p^r\, q^{5-r}$	*Theoretical Frequency*
0	2	0	${}^5C_0\, (0.568)^0\, (0.432)^5 = 0.015$	$100 \times 0.015 = 1.5$
1	14	14	${}^5C_1\, (0.568)^1\, (0.432)^4 = 0.099$	$100 \times 0.099 = 9.9$
2	20	40	${}^5C_2\, (0.568)^2\, (0.432)^3 = 0.260$	$100 \times 0.260 = 26.0$
3	34	102	${}^5C_3\, (0.568)^3\, (0.432)^2 = 0.342$	$100 \times 0.342 = 34.2$
4	22	88	${}^5C_4\, (0.568)^4\, (0.432)^1 = 0.225$	$100 \times 0.225 = 22.5$
5	8	40	${}^5C_5\, (0.568)^5\, (0.432)^0 = 0.0591$	$100 \times 0.0591 = 5.91$
	100	284		

$\Sigma f = 100,$ $\Sigma f x = 284$

Mean $= \dfrac{\Sigma f x}{\Sigma f} = \dfrac{284}{100} = 2.84$

Mean $= np = 2.84$

$\Rightarrow$ $5p = 2.84$ $\Rightarrow$ $p = \dfrac{2.84}{5} = 0.568$

$\Rightarrow$ $q = 1 - p = 1 - 0.568 = 0.432$

Binomial Distribution $= 100 (0.432 + 0.568)^5$ **Ans.**

12.5 RECURRENCE RELATION FOR THE BINOMIAL DISTRIBUTION

By Binomial distribution, $\quad P(r) = {}^nC_r \, p^r \, q^{n-r}$ $\qquad ...(1)$ $\qquad$ (*A.M.I.E., Summer 2002*)

$$P(r+1) = {}^nC_{r+1} \, p^{r+1} \, q^{n-r-1} \qquad ...(2)$$

On dividing (2) by (1), we get

$$\frac{P(r+1)}{P(r)} = \frac{{}^nC_{r+1}}{{}^nC_r} \frac{p^{r+1}q^{n-r-1}}{p^r q^{n-r}} = \frac{n(n-1)(n-2).....(n-r)}{(r+1)!} \frac{r!}{n(n-1)(n-2).....(n-r+1)} \frac{p}{q}$$

$$\frac{P(r+1)}{P(r)} = \frac{n-r}{r+1}\frac{p}{q} \quad \text{or} \quad P(r+1) = \frac{n-r}{r+1}\frac{p}{q} P(r) \qquad \textbf{Ans.}$$

EXERCISE 12.2

1. Fit a binomial distribution to the following frequency data:

x	0	1	3	4
f	2 8	6 2	10	4

(*U. P III Sem. Dec. 2004*)

Ans. $P(r) = {}^{104}C_r \, (0.00999)^r \, (0.99111)^{104-r}$

2. Fit a Binomial distribution to the following frequency distribution:

x	0	1	2	3	4	5	6
f	13	25	52	58	32	16	4

(*K.U.K. Dec. 2009*) **Ans.** $200 \, (0.554 + 0.446)^6$

3. Fill in the blanks :

(*a*) If three persons selected at random are stopped on a street, then the probability that all of them were born on

Sunday is........... (*A.M.I.E., Winter 2001*) **Ans.** $\dfrac{1}{343}$

(*b*) The mean, standard deviation and skewness of binomial distribution are _____ , ____ and_____ .

(*A.M.I.E., Summer 2001*) **Ans.** $n\,p, \sqrt{n\,p\,q}$

(*c*) If n and p are the parameters of a binomial distribution the standard deviation is .... **Ans.** $\sqrt{n\,p\,q}$

(*d*) The Binomial distribution of mean 5 and variance $\dfrac{10}{3}$ is .... **Ans.** ${}^{15}C_r \left(\dfrac{1}{3}\right)^r \left(\dfrac{2}{3}\right)^{15-r}$

4. Tick (✓) the correct answer :

(*a*) The variance for a Binomial distribution is :

(*i*) $n\,p$ $\qquad$ (*ii*) $\sqrt{n\,p}$ $\qquad$ (*iii*) $n\,p\,q$ $\qquad$ (*iv*) $\sqrt{n\,p\,q}$

(*R.G.P.V., Bhopal, II Semester, June 2007*) **Ans.** (*iii*)

(*b*) For the Binomial distribution $(p + q)^n$, the relation of mean and variance is :

(*i*) means = variance $\qquad$ (*ii*) means < variance

(*iii*) mean > variance $\qquad$ (*iv*) (mean)² = variance

(*R.G.P.V., Bhopal, II Semester, June 2006*) **Ans.** (*iii*)

(*c*) In usual notation, for Binomial distribution, $n\,p\,q$, is

(*i*) < $n\,p$ $\qquad$ (*ii*) = $n\,p$ $\qquad$ (*iii*) > $n\,p$ $\qquad$ (*iv*) None of the above (*A.M.I.E., Winter 2005*) **Ans.** (*i*)

(*d*) The standard deviation of Binomial distribution ${}^nC_r \, p^r \, q^{n-r} \, (r = 0, 1, 2, ..., n)$ is

(*i*) $\sqrt{np}$ $\qquad$ (*ii*) $\sqrt{nq}$ $\qquad$ (*iii*) $\sqrt{pq}$ $\qquad$ (*iv*) $\sqrt{npq}$

(*Bihar SBTE, 2014, 2012, 2008, 2004*) **Ans.** (*iv*)

13 Poisson Distribution

13.1 POISSON DISTRIBUTION

(*Bihar SBTE, 2011, 2010*)

Poisson distribution is a particular limiting form of the Binomial distribution when p (or q) is very small and n is large enough.

Poisson distribution is

$$\boxed{P(r) = \frac{m^r e^{-m}}{r!}}$$

where m is the mean of the distribution.

Proof. In Binomial distribution.

$$P(r) = {}^n C_r q^{n-r} p^r = {}^n C_r (1-p)^{n-r} p^r \qquad \left(\text{since mean} = m = np, \ p = \frac{m}{n}\right)$$

$$= {}^n C_r \left(1 - \frac{m}{n}\right)^{n-r} \left(\frac{m}{n}\right)^r \qquad (m \text{ is constant})$$

$$= \frac{n(n-1)(n-2)...(n-\overline{r-1})}{r!} \left(\frac{m}{n}\right)^r \left(1 - \frac{m}{n}\right)^{n-r}$$

$$= \frac{n\left(\frac{n}{n} - \frac{1}{n}\right)\left(\frac{n}{n} - \frac{2}{n}\right) ... \left(\frac{n}{n} - \frac{r-1}{n}\right) m^r \left(1 - \frac{m}{n}\right)^n}{r! \ \left(1 - \frac{m}{n}\right)^r} = \frac{1\left(1 - \frac{1}{n}\right)\left(1 - \frac{2}{n}\right) ... \left(1 - \frac{r-1}{n}\right) m^r \left(1 - \frac{m}{n}\right)^n}{r! \ \left(1 - \frac{m}{n}\right)^r}$$

Taking limits, when n tends to infinity

$$\lim_{n \to \infty} \left(1 - \frac{m}{n}\right)^n = \lim_{n \to \infty} \left[\left(1 - \frac{m}{n}\right)^{-\frac{n}{m}}\right]^{-m} = e^{-m}$$

$$P(r) = \frac{m^r}{r!} e^{-m}$$

$$P(r) = \frac{e^{-m} . m^r}{r!}$$

Example 1. *Prove that the variance of a poisson distribution is equal to its mean. Also, if x is a poisson variate so that $P(0) = P(1) = k$, then prove that $k = \dfrac{1}{e}$.* (*Bihar SBTE, 2004*)

Solution. Poisson distribution is

$$P(r) = \frac{e^{-m} m^r}{r!}$$

$$P(0) = \frac{e^{-m} m^0}{0!} = e^{-m}$$

$$P(1) = \frac{e^{-m} m^1}{1!} = me^{-m}$$

Given P (0) = P (1)

$\Rightarrow \qquad e^{-m} = me^{-m} \qquad \Rightarrow \qquad m = 1$

$\qquad\qquad P(0) = P(1) = k$

$e^{-1} = 1.e^{-1} = k \qquad\qquad \Rightarrow \qquad k = \frac{1}{e}$ **Proved.**

13.2 MEAN OF POISSON DISTRIBUTION

$$P(r) = \frac{e^{-m}.m^r}{r!} \qquad \text{(Bihar, SBTE, 2012, 2010, 2008, A.M.I.E.T.E., Summer 2004, 2002)}$$

Successes r	Frequency f	f.r
0	$\dfrac{e^{-m} m^0}{0!}$	0
1	$\dfrac{e^{-m} m^1}{1!}$	$e^{-m}. m$
2	$\dfrac{e^{-m} m^2}{2!}$	$e^{-m}. m^2$
3	$\dfrac{e^{-m} m^3}{3!}$	$\dfrac{e^{-m}. m^3}{2!}$
...	...	...
r	$\dfrac{e^{-m} m^r}{r!}$	$\dfrac{e^{-m}. m^r}{(r-1)!}$
...	...	...

$$\sum f r = 0 + e^{-m}.m + e^{-m}.m^2 + e^{-m}.\frac{m^3}{2!} + ... + e^{-m}\frac{m^r}{(r-1)!} + ...$$

$$= e^{-m}.m\left[1 + \frac{m}{1!} + \frac{m^2}{2!} + ... + \frac{m^{r-1}}{(r-1)!} + ...\right] = m . e^{m}. [e^m] = m$$

$$\text{Mean} = \frac{\sum fr}{\sum f} = \frac{m}{1},$$

$$\boxed{\text{Mean} = m.}$$ **Ans.**

13.3 STANDARD DEVIATION OF POISSON DISTRIBUTION

$$P(r) = \frac{e^{-m} m^r}{r!} \qquad \text{(Bihar, SBTE, 2012, 2010, A.M.I.E.T.E., Summer 2002)}$$

Successes r	Frequency f	Product rf	Product r^2f
0	$\dfrac{e^{-m}m^0}{0!}$	0	0
1	$\dfrac{e^{-m}m^1}{1!}$	$e^{-m}.m$	$e^{-m}.m$
2	$\dfrac{e^{-m}m^2}{2!}$	$e^{-m}.m^2$	$2e^{-m}.m^2$
3	$\dfrac{e^{-m}m^3}{3!}$	$\dfrac{e^{-m}.m^3}{2!}$	$3e^{-m}.\dfrac{m^3}{2!}$
r	$\dfrac{e^{-m}m^r}{r!}$	$\dfrac{e^{-m}.m^r}{(r-1)!}$	$\dfrac{re^{-m}.m^r}{(r-1)!}$
....	....	....	....
	$\Sigma f = 1$	$\Sigma fr = m$	

$$\Sigma f r^2 = 0 + e^{-m}.m + 2e^{-m}.m^2 + 3.e^{-m}.\frac{m^3}{2} + ... + \frac{re^{-m}.m^r}{(r-1)!} + ..$$

$$= m.e^{-m}\left[1 + 2m + \frac{3m^2}{2!} + \frac{4m^3}{3!} + ... \frac{r.m^{r-1}}{(r-1)!} + ..\right]$$

$$= m.e^{-m}\left[1 + m + \frac{m^2}{2!} + \frac{m^3}{3!} + ... \frac{m^{r-1}}{(r-1)!} + ... + m + 2\frac{m^2}{2!} + \frac{3m^3}{3!} + ... + \frac{(r-1)m^{r-1}}{(r-1)!} + ...\right]$$

$$= m.e^{-m}\left[\left\{1 + m + \frac{m^2}{2!} + \frac{m^3}{3!} + ... + \frac{m^{r-1}}{(r-1)!} + ...\right\} + m\left\{1 + \frac{m}{1!} + \frac{m^2}{2!} + ... + \frac{m^{r-2}}{(r-2)!} + ...\right\}\right]$$

$$= m.e^{-m}[e^m + m\,e^m] = m + m^2$$

$$\sigma^2 = \frac{\Sigma fr^2}{\Sigma f} - \left(\frac{\Sigma fr}{\Sigma f}\right)^2 = \frac{m+m^2}{1} - (m)^2 = m \quad \text{or} \quad \sigma = \sqrt{m}$$

$$\boxed{\text{S. D.} = \sqrt{m}}$$

Hence mean and variance of a Poisson distribution are each equal to m. Similarly, we can obtain

$$\mu_3 = m, \quad \mu_4 = 3m^2 + m$$

(Bihar SBTE, 2004)

$$\beta_1 = \frac{1}{m}, \quad \beta_2 = 3 + \frac{1}{m}$$

$$\gamma_1 = \frac{1}{\sqrt{m}}, \quad \gamma_2 = \frac{1}{m}$$

13.4 MEASURES OF DISPERSION

Measures of dipersion– It is the mean of the absolute values of the deviation of a given set of numbers from their arithmetic mean. If $x_1, x_2, x_3 ...... x_n$ be a set of numbers with frequences $f_1, f_2.. f_n$ resketively. Let $\bar{x}$ be the arithmetic mean of the numbers $x_1, x_2, ... x_n$, then

$$\text{Mean deviation} = \frac{\Sigma f_{ij}\,|x - \bar{x}|}{\Sigma f_i}$$

(Bihar SBTE, 2012)

$$\text{Standard deviation} = \text{S.D.} \quad \sigma = \sqrt{\frac{\Sigma f\,(x - \bar{x})^2}{\Sigma f}}$$

$$\sigma = \frac{\Sigma f r^2}{\Sigma f^2} - \left(\frac{\Sigma f r}{\Sigma r}\right) = \frac{m + m^2}{21} - (m^2) = m$$

$$\boxed{\sigma = \sqrt{m}}$$

$$\boxed{\text{S.D.} = \sqrt{m}}$$

13.5 MEAN DEVIATION

Show that in a Poisson distribution with unit mean, the mean deviation about the mean is $\left(\dfrac{2}{e}\right)$ times the standard deviation.

Solution. $P(r) = \dfrac{m^r}{r!} e^{-m}$ But mean = 1 $\qquad i.e. \qquad$ $m = 1$ and S.D. $= \sqrt{m} = 1$

Hence, $\qquad\qquad P(r) = \dfrac{e^{-1}}{r!} = \dfrac{1}{e} \cdot \dfrac{1}{r!}$

r	$P(r)$	$\lvert r - 1 \rvert$	$P(r)\,\lvert r - 1 \rvert$
0	$\dfrac{1}{e}$	1	$\dfrac{1}{e}$
1	$\dfrac{1}{e}$	0	0
2	$\dfrac{1}{e}\dfrac{1}{2!}$	1	$\dfrac{1}{e}\dfrac{1}{2!}$
3	$\dfrac{1}{e}\dfrac{1}{3!}$	2	$\dfrac{1}{e}\dfrac{2}{3!}$
4	$\dfrac{1}{e}\dfrac{1}{4!}$	3	$\dfrac{1}{e}\dfrac{3}{4!}$
r	$\dfrac{1}{e}\dfrac{1}{r!}$	$r - 1$	$\dfrac{1}{e}\dfrac{r-1}{r!}$

$$\Sigma P(r)\lvert r - 1 \rvert = \frac{1}{e} + 0 + \frac{1}{e}\frac{1}{2!} + \frac{1}{e}\frac{2}{3!} + \frac{1}{e}\frac{3}{4!} + \dots + \frac{1}{e}\frac{r-1}{r!} + \dots$$

$$= \frac{1}{e}\left[1 + 0 + \frac{1}{2!} + \frac{2}{3!} + \frac{3}{4!} + \dots + \frac{r-1}{r!} + \dots\right]$$

$$= \frac{1}{e}\left[1 + \left(\frac{1}{1!} - \frac{1}{1!}\right) + \left(\frac{2}{2!} - \frac{1}{2!}\right) + \left(\frac{3}{3!} - \frac{1}{3!}\right) + \left(\frac{4}{4!} - \frac{1}{4!}\right) + \dots + \left(\frac{r}{r!} - \frac{1}{r!}\right) + \dots\right]$$

$$= \frac{1}{e}\left[1 + \frac{1}{1!} + \frac{2}{2!} + \frac{3}{3!} + \frac{4}{4!} + \dots + \frac{r}{r!} + \dots - \frac{1}{1!} - \frac{1}{2!} - \frac{1}{3!} - \frac{1}{4!} \dots - \frac{1}{r!} - \dots\right]$$

$$= \frac{1}{e}\left[1 + \left\{1 + \frac{1}{1!} + \frac{1}{2!} + \frac{1}{3!} + \dots + \frac{1}{(r-1)!} + \dots\right\} - \left\{1 + \frac{1}{1!} + \frac{1}{2!} + \frac{1}{3!} + \frac{1}{4!} + \dots + \frac{1}{r!} \dots\right\} + 1\right]$$

$$= \frac{1}{e}[1 + e - e + 1] = \frac{2}{e} = \frac{2}{e}(1) = \frac{2}{e}\,\text{S.D.} \qquad\qquad \textbf{Proved.}$$

13.6 RECURRENCE FORMULA FOR POISSON DISTRIBUTION

Solution. By Poisson distribution

$$P(r) = \frac{e^{-m}.m^r}{r!} \qquad \qquad ...(1)$$

$$P(r+1) = \frac{e^{-m}m^{r+1}}{(r+1)!} \qquad \qquad ...(2)$$

By dividing (2) by (1), we get

$$\frac{P(r+1)}{P(r)} = \frac{e^{-m}m^{r+1}}{(r+1)!} \frac{r!}{e^{-m}.m^r} = \frac{m}{r+1}$$

$$P(r+1) = \frac{m}{r+1}P(r) \qquad \qquad \textbf{Ans.}$$

Example 2. *If the variance of the Poisson distribution is 2, find the probabilities for r = 1, 2, 3, 4 from the recurrence relation of the Poisson distribution. Also find P(r ≥ 4).*

Solution.   Variance = *m* = 2;
   Mean = 2

$$P(r+1) = \frac{m}{r+1}P(r) \qquad \qquad \text{[Recurrence relation]}$$

Now $$P(r+1) = \frac{2}{r+1}P(r) \qquad \qquad (m = 2)$$

If $r = 0$, $P(1) = \dfrac{2}{0+1}P(0) = \dfrac{2}{0+1}(0.1353) = 0.2706$ $\qquad P(0) = e^{-m} = e^{-2} = 0.1353$

If $r = 1$, $P(2) = \dfrac{2}{1+1}P(1) = \dfrac{2}{2}(0.2706) = 0.2706$

If $r = 2$, $P(3) = \dfrac{2}{2+1}P(2) = \dfrac{2}{3}(0.2706) = 0.1804$

If $r = 3$, $P(4) = \dfrac{2}{3+1}P(3) = \dfrac{1}{2}(0.1804) = 0.0902$

$$P(r \geq 4) = P(4) + P(5) + P(6) + ... = 1 - [P(0) + P(1) + P(2) + P(3)]$$
$$= 1 - [0.1353 + 0.2706 + 0.2706 + 0.1804] = 1 - 0.8569 = 0.1431 \textbf{ Ans.}$$

Example 3. *Assume that the probability of an individual coal miner being killed in a mine accident during a year is $\dfrac{1}{2400}$. Use appropriate statistical distribution to calculate the probability that in a mine employing 200 miners, there will be at least one fatal accident in a year.*

(A.M.I.E., Summer 2001)

Solution. $$p = \frac{1}{2400}, \quad n = 200$$

$$m = np = \frac{200}{2400} = \frac{1}{12}$$

$$P \text{ (At least one)} = P(1 \text{ or } 2 \text{ or } 3 \text{ or } ..... \text{ or } 200) = P(1) + P(2) + P(3) + ... + P(200)$$

$$= 1 - P(0) = 1 - \frac{e^{-m}.m^0}{0!} = 1 - e^{-\frac{1}{12}} = 1 - 0.92 = 0.08 \qquad \textbf{Ans.}$$

Example 4. *Suppose 3% of bolts made by a machine are defective, the defects occurring at random during production. If bolts are packaged 50 per box, find (a) exact probability and (b) Poisson approximation to it, that a given box will contain 5 defectives.*

Solution.
$$p = \frac{3}{100} = 0.03$$

(a)
$$q = 1 - p = 1 - 0.03 = 0.97$$

Hence the probability for 5 defective bolts in a lot of 50
$$= {}^{50}C_5 (0.03)^5 (0.97)^{45} = 0.013074 \quad \text{(Binomial Distribution)}$$

(b) To get Poisson approximation $m = n p = 50 \times \dfrac{3}{100} = \dfrac{3}{2} = 1.5$

Required Poisson approximation $= \dfrac{m^r e^{-m}}{r!} = \dfrac{(1.5)^5 e^{-1.5}}{5!} = 0.01412$ **Ans.**

Example 5. *The number of arrivals of customers during any day follows Poisson distribution with a mean of 5. What is the probability that the total number of customers on two days selected at random is less than 2?*

Solution.
$$m = 5$$
$$P(r) = \frac{e^{-m} m^r}{r!}, \quad P(r) = \frac{e^{-5} (5)^r}{r!}$$

If the number of customers on two days < 2 = 1 or 0

First day	Second Day	Total
0	0	0
0	1	1
1	0	1

Required probability $= P(0)\,P(0) + P(0)\,P(1) + P(1)\,P(0)$

$$= \frac{e^{-5}(5)^0}{0!} \cdot \frac{e^{-5}(5)^0}{0!} + \frac{e^{-5}(5)^0}{0!} \cdot \frac{e^{-5}(5)^1}{1!} + \frac{e^{-5}(5)^1}{1!} \cdot \frac{e^{-5}(5)^0}{0!} = e^{-5} \cdot e^{-5} + e^{-5} \cdot e^{-5} \cdot 5 + e^{-5} \cdot 5 \cdot e^{-5}$$

$$= e^{-10}[1 + 5 + 5] = 11 e^{-10} = 11 \times 4.54 \times 10^{-5} = 4.994 \times 10^{-4}$$ **Ans.**

Example 6. *Using Poisson distribution, find the probability that the ace of spades will be drawn from a pack of well-shuffled cards at least once in 104 consecutive trials.*

Solution. Probability of the ace of spades $= P = \dfrac{1}{52}, \quad n = 104$

$$m = np = 104 \times \frac{1}{52} = 2$$
$$P(r) = e^{-m} \cdot \frac{m^r}{r!} = e^{-2} \cdot \frac{2^r}{r!} = \frac{1}{e^2} \frac{2^r}{r!}$$

$P(\text{At least once}) = P(1) + P(2) + P(3) + \dots + P((104)) = 1 - P(0) = 1 - \dfrac{1}{e^2} \times \dfrac{2^0}{0!} = 1 - \dfrac{1}{e^2} = 1 - 0.135 = 0.865$ **Ans.**

Example 7. *In a certain factory producing cycle tyres, there is a small chance of 1 in 500 tyres to be defective. The tyres are supplied in lots of 10. Using Poisson distribution, calculate the approximate number of lots containing no defective, one defective and two defective tyres, respectively, in a consignment of 10,000 lots.*

Solution.
$$p = \frac{1}{500}, \quad n = 10$$

$$m = np = 10 \cdot \frac{1}{500} = \frac{1}{50} = 0.02, \quad P(r) = \frac{e^{-m} \cdot m^r}{r!}$$

S.No.	Probability of defective	Number of lots containing defective
1	$P(0) = \dfrac{e^{-0.02}(0.02)^0}{0!} = e^{-0.02} = 0.9802$	$10{,}000 \times 0.9802 = 9802$ lots
2	$P(1) = \dfrac{e^{-0.02}(0.02)^1}{1!}$ $= 0.9802 \times 0.02 = 0.019604$	$10{,}000 \times 0.019604 = 196$ lots
3.	$P(2) = \dfrac{e^{-0.02}(0.02)^2}{2!}$ $= 0.9802 \times 0.0002 = 0.00019604$	$10{,}000 \times 0.000196 = 2$ lots

Ans.

Example 8. *A car hire firm has two cars which it hires out day by day. The number of demands for a car on each day is distributed as a Poisson distribution with mean 1.5. Calculate the number of days in a year on which (i) neither car is on demand (ii) a car demand is refused.*

(MDU, Dec. 2010, A.M.I.E., Summer 2004, Winter 2001, June 2009) $(e^{-1.5} = 0.2231)$

Solution. $m = 1.5$

(*i*) If the car is not used, then demand $(r) = 0$

$$P(r) = \frac{e^{-m}.m^r}{r!}, \quad P(0) = \frac{e^{-1.5}(1.5)^0}{0!} = e^{-1.5} = 0.2231$$

Number of days in a year when the demand is zero $= 365 \times 0.2231 = 81.4315$ **Ans. 81 days**

(*ii*) Some demand is refused if the number of demands is more than two *i.e.* $r > 2$.

$$P(r > 2) = P(3) + P(4) + \ldots = 1 - [P(0) + P(1) + P(2)]$$

$$= 1 - \left[\frac{e^{-1.5}(1.5)^0}{0!} + \frac{e^{-1.5}(1.5)^1}{1!} + \frac{e^{-1.5}(1.5)^2}{2!}\right]$$

$$= 1 - [e^{-1.5} + e^{-1.5} \times 1.5 + e^{-1.5} \times 1.125] = 1 - e^{-1.5}[1 + 1.5 + 1.125] = 1 - e^{-1.5} \times 3.625$$

$$= 1 - 0.2231 \times 3.625 = 1 - 0.8087375 = 0.1912625$$ **Ans.**

Number of days in a year when some demand of car is refused

$$= 365 \times 0.1912625 = 69.81 = 70 \text{ days}$$ **Ans.**

Example 9. *If the probability that an individual suffers a bad reaction from a certain injection is 0.001, determine the probability that out of 2000 individuals*

(a) exactly 3 (b) more than 2 individuals (c) None (d) More than one individual will suffer a bad reaction.   *(A.M.I.E.T.E., Winter, 2002, 2000)*

Solution. $p = 0.001,$ $n = 2000$

$$m = np = 2000 \times 0.001 = 2$$

$$\therefore \qquad P(r) = \frac{e^{-m}m^r}{r!} = e^{-2}\frac{2^r}{r!} = \frac{1}{e^2} \times \frac{2^r}{r!}$$

$$P(3) = \frac{1}{e^2} \cdot \frac{2^3}{3!} = \frac{1}{(2.718)^2} \times \frac{8}{6} = (0.135) \times \frac{4}{3} = 0.18$$

(*b*) P (more than 2) $= P(3) + P(4) + P(5) + \ldots + P(2000)$

$$= 1 - [P(0) + P(1) + P(2)] = 1 - \left[\frac{e^{-2}(2)^0}{0!} + \frac{e^{-2}(2)^1}{1!} + \frac{e^{-2}(2)^2}{2!}\right]$$

$$= 1 - e^{-2}[1 + 2 + 2] = 1 - \frac{5}{e^2} \quad = 1 - 5 \times 0.135 = 1 - 0.675 = 0.325 \qquad \textbf{Ans.}$$

(c) $\qquad P \text{ (none)} = P(0) = \dfrac{e^{-2}(2)^0}{0!} = 0.135$

(d) $P \text{ (more than 1)} = P(2) + P(3) + P(4) + ... + P(2000) \ = 1 - [P(0) + P(1)]$

$$= 1 - \left[\frac{e^{-2}(2)^0}{0!} + \frac{e^{-2}(2)^1}{1!}\right] = 1 - 3e^{-2} = 1 - 3 \times 0.135 = 1 - 0.405 = 0.595 \qquad \textbf{Ans.}$$

Example 10. *A manufacturer knows that the razor blades he makes contain on an average 0.5% of defectives. He packs them in packets of 5. What is the probability that a packet picked at random will contain 3 or more faulty blades?*

Solution. $\qquad\qquad\qquad P = 0.5\% = 0.005, \qquad\qquad n = 5$

$$m = nP = 5 \times 0.005 = 0.025$$

$$P(r) = \frac{e^{-m} . m^r}{r!} = \frac{e^{-0.025}(.025)^r}{r!}$$

$$P \text{ (3 or more)} = P(3) + P(4) + P(5) = \frac{e^{-0.025}(0.025)^3}{3!} + \frac{e^{-0.025}(0.025)^4}{4!} + \frac{e^{-0.025}(0.025)^5}{5!}$$

$$= \frac{e^{-0.025}(0.025)^3}{5!}[20 + 5(0.025) + (0.025)^2] = \frac{0.975 \times 0.000015625 \times 20.125625}{120}$$

$$= 0.000002555. \qquad \textbf{Ans.}$$

Example 11. *In a certain factory turning out razor blades, there is a small chance of 0.002 for any blade to be defective. The blades are supplied in packets of 10. Use appropriate and suitable distribution to calculate the approximate number of packets containing no defective, one defective and two defective blades respectively in a consignment of 50000 packets.*

(K.U. 2009, R.G.P.V., Bhopal, II Semester, June 2006)

Solution. $\quad$ Here, $p = 0.002, \qquad\qquad n = 10$

$$m = np \qquad \Rightarrow \qquad m = 10 \times 0.002 = 0.020$$

	$p(r) = \dfrac{e^{-m}.(m)^r}{r!}$	
r	$p(r) = \dfrac{e^{-0.02}(0.02)^r}{r!}$	Number of packets $= 50000\, p$
0	$p(0) = \dfrac{e^{-0.02}(0.02)^0}{0!} = 0.980$	$50000 \times (0.980) = 49000$
1	$p(1) = \dfrac{e^{-0.02}(0.02)^1}{1!} = 0.0196$	$50000 \times (0.0196) = 980$
2	$p(2) = \dfrac{e^{-0.02}(0.02)^2}{2!} = 0.000196$	$50000 \times (0.000196) = 9.8$

Hence, number of packets containing no defective razor blades = 49000.

Number of packets containing one defective razor blade = 980

Number of packets containing two defective razor blades = 9.8 $\qquad \textbf{Ans.}$

Example 12. *Suppose that a book of 600 pages contains 40 printing mistakes. Assume that these errors are randomly distributed throughout the book and x, the number of errors per page has a Poisson distribution. What is the probability that 10 pages selected at random will be free of errors?*

Solution. $P = \dfrac{40}{600} = \dfrac{1}{15}, \quad n = 10,$

$$m = np = 10 \times \dfrac{1}{15} = \dfrac{2}{3} \qquad \Rightarrow \qquad P(r) = \dfrac{e^{-m}.m^r}{r!} = \dfrac{e^{\frac{-2}{3}}\left(\dfrac{2}{3}\right)^r}{r!}$$

$$P(0) = \dfrac{e^{\frac{-2}{3}}\left(\dfrac{2}{3}\right)^0}{0!} = e^{\frac{-2}{3}} = 0.51 \qquad\qquad\qquad \textbf{Ans.}$$

Example 13. *If there are 3 misprints in a book of 1000 pages find the probability that a given page will contain*

   (i) no misprint   (ii) more than 2 misprints.   *(U.P., III Semester, Dec. 2009)*

Solution.   Total number of pages = 1000

No. of misprints = 3

$$p = \dfrac{3}{1000} = 0.003, \qquad\qquad n = 1, \qquad m = np = 1 \times 0.003 = 0.003$$

Poisson distribution

$$P(r) = \dfrac{e^{-m}.m^r}{r!}, \qquad\qquad P(0) = \dfrac{e^{-0.003}(0.003)^0}{0!} = e^{-0.003} = 0.997$$

$$P(r > 2) = P(3) = \dfrac{e^{-0.003}(0.003)^3}{3!} = 0.0000000045$$

Hence (*i*) the probability that a page will contain no error = 0.997

   (*ii*) the probability that a page will contain more than two misprints = 0.0000000045   **Ans.**

Example 14. *A manufacturer knows that the condensers he makes contain on an average 1% of defectives. He packs them in boxes of 100. What is the probability that a box picked out at random will contain 4 or more faulty condensers?*   *(MDU, May 2007)*

Solution. $p = 1\% = 0.01, n = 100, \quad m = np = 100 \times 0.01 = 1$

$$P(r) = \dfrac{e^{-m}.(m)^r}{r!} = \dfrac{e^{-1}(1)^r}{r!} = \dfrac{e^{-1}}{r!}$$

$P(4 \text{ or more faulty condensers}) = P(4) + P(5) + \dots + P(100) = 1 - [P(0) + P(1) + P(2) + P(3)]$

$$= 1 - \left[\dfrac{e^{-1}}{0!} + \dfrac{e^{-1}}{1!} + \dfrac{e^{-1}}{2!} + \dfrac{e^{-1}}{3!}\right] = 1 - e^{-1}[1 + 1 + \dfrac{1}{2} + \dfrac{1}{6}] = 1 - \dfrac{8}{3e} = 1 - 0.981 = 0.019 \quad \textbf{Ans.}$$

Example 15. *An insurance company found that only 0.01% of the population is involved in a certain type of accident each year. If its 1000 policy holders were randomly selected from the population, what is the probability that not more than two of its clients are involved in such an accident next year? (given that e^{0.1} = 0.9048)*

Solution. $$p = 0.01\% = \dfrac{1}{100} \times \dfrac{1}{100} = \dfrac{1}{10000}, \qquad\qquad n = 1000$$

$$m = np = (1000) \times \dfrac{1}{10000} = \dfrac{1}{10} = 0.1$$

$$P(r) = \frac{e^{-m}m^r}{r!}$$

P (not more than 2) $= P(0, 1 \text{ and } 2) = P(0) + P(1) + P(2)$

$$= \frac{e^{-0.1}(0.1)^0}{0!} + \frac{e^{-0.1}(0.1)^1}{1!} + \frac{e^{-0.1}(0.1)^2}{2!} = e^{-0.1}\left(1 + 0.1 + \frac{0.01}{2}\right)$$

$$= 0.9048 \times 1.105 = 0.9998$$ **Ans.**

Example 16. *Fit a Poisson distribution to the set of observations :*

x	0	1	2	3	4
f	122	60	15	2	1

(MDU 2006, Dec. 2007, May 2008, R.G.P.V., Bhopal, II Semester, Dec. 2007, June 2007)

Solution. The mean number $= \dfrac{\Sigma f \cdot x}{\Sigma f}$.

x	f	fx
0	122	0
1	60	60
2	15	30
3	2	6
4	1	4
Total	200	100

$$\text{Mean} = \frac{\Sigma f x}{\Sigma f} = \frac{100}{200} = \frac{1}{2}$$

x	$P(x) = \dfrac{e^{-1/2}(1/2)^x}{x!}$	Theoretical frequency	Given frequency
0	$P(0) = \dfrac{e^{-\frac{1}{2}}\left(\dfrac{1}{2}\right)^0}{0!} = 0.6065$	$0.6065 \times 200 = 121.3$	121
1	$P(1) = \dfrac{e^{-\frac{1}{2}}\left(\dfrac{1}{2}\right)^1}{1!} = \dfrac{0.6065}{2} = 0.3033$	$0.3033 \times 200 = 60.7$	60
2	$P(2) = \dfrac{e^{-\frac{1}{2}}\left(\dfrac{1}{2}\right)^2}{2!} = \dfrac{0.6065}{8} = 0.0758$	$0.0758 \times 200 = 15.2$	15
3	$P(3) = \dfrac{e^{-\frac{1}{2}}\left(\dfrac{1}{2}\right)^3}{3!} = \dfrac{0.6065}{48} = 0.0126$	$0.0126 \times 200 = 2.5$	2
4	$P(4) = \dfrac{e^{-\frac{1}{2}}\left(\dfrac{1}{2}\right)^4}{4!} = \dfrac{0.6065}{384} = 0.0016$	$0.0016 \times 200 = 0.32$	1

Ans.

Example 17. *A skilled typist, on routine work , kept a record of mistakes made per day during 300 working days.*

Mistakes per day	0	1	2	3	4	5	6
No. of days	143	90	42	12	9	3	1

Fit a Poisson distribution to the above data and hence calculate the theoretical frequencies.

Solution.   The mean number of mistakes

$$= \frac{1}{300}(143 \times 0 + 90 \times 1 + 42 \times 2 + 12 \times 3 + 9 \times 4 + 3 \times 5 + 1 \times 6)$$

$$= \frac{1}{300}(90 + 84 + 36 + 36 + 15 + 6) = \frac{267}{300} = 0.89$$

Number of mistakes	Probability $P(r) = \dfrac{e^{-0.89} \times (0.89)^r}{r!}$	Theoretical frequency	Given frequency
0	$\dfrac{e^{-0.89} \times (0.89)^0}{0!} = 0.411$	$0.411 \times 300 = 123.3 = 123$ (say)	143
1	$\dfrac{e^{-0.89} \times (0.89)^1}{1!} = 0.365$	$0.365 \times 300 = 109.5 = 110$ (sa)	90
2	$\dfrac{e^{-0.89} \times (0.89)^2}{2!} = 0.163$	$0.163 \times 300 = 48.9 = 49$ (say)	42
3	$\dfrac{e^{-0.89} \times (0.89)^3}{3!} = 0.048$	$0.048 \times 300 = 14.4 = 14$ (say)	12
4	$\dfrac{e^{-0.89} \times (0.89)^4}{4!} = 0.011$	$0.011 \times 300 = 3.3 = 3$ (say)	9
5	$\dfrac{e^{-0.89} \times (0.89)^5}{5!} = 0.002$	$0.002 \times 300 = 0.6 = 1$ (say)	3
6	$\dfrac{e^{-0.89} \times (0.89)^6}{6!} = 0.0003$	$0.0003 \times 300 = 0.09 = 0$ (say)	1

Example 18. *Fit a Poisson distribution to the following data which gives the number of yeast cells per square for 400 squares.*

No. of cells per square (x)	0	1	2	3	4	5	6	7	8	9	10	Total
No. of squares (f)	103	143	98	42	8	4	2	0	0	0	0	400

It is given that $e^{-1.32} = 0.2674$ *(A.M.I.E., Summer 2000)*

Solution.

x	0	1	2	3	4	5	6	7	8	9	10	Total
f	103	143	98	42	8	4	2	0	0	0	0	400
f·x	0	143	196	126	32	20	12	0	0	0	0	529

$$m = \text{Mean} = \frac{\Sigma f.x}{\Sigma f} = \frac{529}{400} = 1.32$$

But Poisson distribution is $P(x) = \dfrac{e^{-m}.m^x}{x!} = \dfrac{e^{-1.32}(1.32)^x}{x!}$ or $P(x) = \dfrac{0.2674(1.32)^x}{x!}$

No. of cells	Probability $P(x) = \dfrac{0.2674(1.32)^x}{x!}$	Theoretical frequency		Given frequency
0	$\dfrac{0.2674(1.32)^0}{0!} = 0.2674$	$0.267 \times 400 = 107$		103
1	$\dfrac{0.2674(1.32)^1}{1!} = 0.353$	$0.353 \times 400 = 141$		143
2	$\dfrac{0.2674(1.32)^2}{2!} = 0.233$	$0.233 \times 400 = 93.2$		98
3	$\dfrac{0.2674(1.32)^3}{3!} = 0.1025$	$0.1025 \times 400 = 41$		42
4	$\dfrac{0.2674(1.32)^4}{4!} = 0.0338$	$0.0338 \times 400 = 13.52$	i.e., 14	8
5	$\dfrac{0.2674(1.32)^5}{5!} = 0.00893$	$0.00893 \times 400 = 3.57$	i.e., 4	4
6	$\dfrac{0.2674(1.32)^6}{6!} = 0.00196$	$0.00196 \times 400 = 0.784$	i.e., 1	2
7	$\dfrac{0.2674(1.32)^7}{7!} = 0.00037$	$0.00037 \times 400 = 0.148$	i.e., 0	0
8	$\dfrac{0.2674(1.32)^8}{8!} = 0.00006$	$0.00006 \times 400 = 0.024$	i.e., 0	0
9	$\dfrac{0.2674(1.32)^9}{9!} = 0.00000897$	$0.00000897 \times 400 = 0.003588$	i.e., 0	0
10	$\dfrac{0.2674(1.32)^{10}}{10!} = 0.00000118$	$0.00000118 \times 400 = 0.000472$	i.e., 0	0

Example 19. *Data was collected over a period of 10 years, showing number of deaths from horse kicks in each of the 200 army corps. The distribution of deaths was as follows:*

No. of deaths	0	1	2	3	4	Total
Frequency	109	65	22	3	1	200

Fit a Poisson distribution to the above data and hence calculate the theoretical frequencies.

(MDU, May 2009)

Solution. Mean of given distribution $= \dfrac{\Sigma f x}{\Sigma f} = \dfrac{65 + 44 + 9 + 4}{200} = \dfrac{122}{200} = 0.61$

This is the parameter (*m*) of the Poisson distribution.

$\therefore$ Required Poisson distribution is $N \cdot \dfrac{m^r \, e^{-m}}{r!}$, where $N = \Sigma f = 200$

$$= 200 e^{-0.61} \left(\frac{0.61}{r!} \right)^r = 200 \times 0.5433 \frac{(0.61)^r}{r!} = 108.7 \times \frac{(0.61)^r}{r!}$$

r	P (r)	Theoretical Frequency
0	108.7	109
1	$108.7 \times 0.61 = 66.3$	66
2	$108.7 \times \dfrac{(0.61)^2}{2!} = 20.2$	20
3	$108.7 \times \dfrac{(0.61)^3}{3!} = 4.1$	4
4	$108.7 \times \dfrac{(0.61)^4}{4!} = 0.7$	1
	Total = 200	Total = 200

Ans.

EXERCISE 13.1

1. Find the probability that at most 5 defective fuses will be found in a box of 200 fuses if experience shows that 2 per cent of such fuses are defective. **Ans. 0.785**

2. The number of accidents during a year in a factory has the Poisson distribution with mean 1.5. The accidents during different years are assumed independent. Find the probability that only 2 accidents take place during 2 years time. **Ans. 0.224**

3. A manufacturer of cotter pins knows that 5% of his product is defective. If he sells cotter pins in boxes of 100 and guarantee that not more than 10 pins will be defective, what is the approximate probability that a box will fail to meet the guaranteed quality. [$e^{-.05} = 0.006738$] **Ans. 0.0136875**

4. Suppose the number of telephone calls on an operator received from 9.00 to 9.05 follow a Poisson distribution with mean 3. Find the probability that

   (i) the operator will receive no calls in that time interval tomorrow,

   (ii) in the next three days the operator will receive a total of 1 call in that time interval.
   [$e^{-3} = 0.04978$]   **Ans.** (i) e^{-3} (ii) $3 \times (e^{-3})^2 \, (e^{-3}, 3)$

5. On the basis of past record it has been found that there is 70% chance of power cut in a city on any particular day. What is the probability that from the first to the 10th date of the month, there are 5 or more days without power cut.

   (*A.M.I.E.T.E., Summer 2001*) **Ans.** $\left[\dfrac{3^5}{5!} + \dfrac{3^6}{6!} + \dfrac{3^7}{7!} + \dfrac{3^8}{8!} + \dfrac{3^9}{9!} + \dfrac{3^{10}}{10!} \right] e^{-3}$

6. The distribution of typing mistakes committed by a typist is given below. Assuming a Poisson model, find out the expected frequencies:

Mistakes per pages	0	1	2	3	4	5
No. of pages	142	156	69	27	5	1

Ans. 147, 147, 74, 25, 6, 1 pages.

7. Let x be the number of cars per minute passing a certain crossing of roads between 5.00 P.M. and 7.00 P.M. on a holiday. Assume x has a Poisson distribution with mean 4. Find the probability of observing atmost 3 cars during any given minute between 5.00 P.M. and 7 P.M. (given $e^{-4} = 0.0183$) **Ans. 0.4331**

8. Number of customers arriving at a service counter during a day has a Poisson distribution with mean 100. Find the probability that at least one customer will arrive on each day during a period of five days. Also find the probability that exactly 3 customers will arrive during two days.   **Ans.** $(1-e^{-100})^5$, $e^{-200} \times \dfrac{4(100)^3}{3}$

9. In a normal summer, a truck driver gets on an average one puncture in 1000 km. Applying Poisson distribution, find the probability that he will have
   (i) no puncture,   (ii) two punctures in a journey of 3000 kms.   **Ans.** (i) e^{-3} (ii) $4.5\, e^{-3}$

10. Wireless sets are manufactured with 25 soldered joints each. On an average, 1 joint in 500 is defective. How many sets can be expected to be free from defective joints in a consignment of 10000 sets ?
                                                                                           Ans. 9512

11. In a certain factory turning out razor blades, there is small chance $\dfrac{1}{500}$ for any blade to be defective. The blades are supplied in packets of 10. Using Poisson's distribution, calculate the approximate number of packets containing (i) no defective (ii) one defective and (iii) two defective blades respectively in a consignment of 10,000 packets. ($e^{-0.02} = 0.9802$).   **Ans.** (i) 9802 (ii) 196 (iii) 2

12. If m and μ_r denote by the mean and central rth moment of a Poisson distribution, then prove that
   $$\mu_{r+1} = r\,m\,\mu_{r-1} + m\frac{d\mu_r}{dm}.$$
   $$\left[\textbf{Hint.}\ \mu_r = \sum_{r=0}^{\infty}(x-m)^r \frac{e^{-m}m^x}{x!},\ \text{find}\ \frac{d\mu_r}{dm} \right]$$

13. A certain screw-making machine produces an average 2 defective screws out of 100, and pack them in boxes of 500. Find the probability that a box contains 15 defective screws.
                                   (MDU., Dec. 2005, A.M.I.E., Winter 2005) **Ans.** 0.0347

14. The distribution of the number of road accidents per day in a city is Poisson with mean 4. Find the number of days out of 100 days when there will be:
   (i) no accident   (ii) at least 2 accidents
   (iii) at most 3 accidents   (iv) between 2 and 5 accidents   **Ans.** (i) 2 days (ii) 91 days (iii) 43 days (iv) 39 days.

15. **Fill in the blanks :**
   (a) If a random variable x follows Poisson distribution such that $P(x=1) = P(x=2)$ then the mean of the distribution is ....   **Ans.** 2
   (b) Mean and variance of a Poisson distribution are ....   **Ans.** equal
   (c) If the probability of a defective fuse is 0.05, the variance for the distribution of defective fuses in a total of 40 is ....   **Ans.** 2
   (d) The probability of the king of hearts drawn from a pack of cards once in 52 trials is ....   **Ans.** $\dfrac{1}{e}$
   (e) If the standard deviation of the Poisson distribution is $\sqrt{2}$, the probability for $r=2$ is .... **Ans.** $\dfrac{2}{e^2}$
   (f) If x has a modified Poisson distribution
       $$P_k = P_r(x=k) = \frac{(e^m-1)^{-1}m^k}{k!},\ (k=1,\,2,\,3,....),\ \text{then the expectation of } x \text{ is ....}\quad \textbf{Ans.}\ \frac{me^m}{e^m-1}$$
   (g) If x has a poisson distribution such that $P(x=k) = P(x=k+1)$ for some positive integer k then mean of x is ....
                                                    (A.M.I.E., Summer 2000)   **Ans.** $k+1$

Choose the correct answer:

16. In the Poisson distribution if $P(x=k) = P(x=k+1)$, then the mean is :
   (a) k   (b) 2k   (c) k+1   (d) k−1 *(R.G.P.V. Bhopal, II Semester, June 2007)* **Ans.** (c)

17. The value of measure of skewness of Poisson distribution is :

   (a) m (b) $\sqrt{m}$ (c) $\dfrac{1}{m}$ (d) $\dfrac{1}{\sqrt{m}}$ (R.G.P.V., Bhopal, II Semester, June 2006)   **Ans.** (d)

18. Poisson distribution with unit mean, mean-deviation about the mean is :

   (a) $\dfrac{1}{e}$ (b) $\dfrac{\sigma}{e}$ (c) $\dfrac{2\sigma}{e}$ (d) $\dfrac{2}{e}$ (R.G.P.V., Bhopal, II Semester, Feb 2006) **Ans.** (d)

19. In the Poisson distribution if $2P(x=1)=P(x=2)$, then the variance is :

   (a) 0   (b) −1   (c) 4   (d 2   (R.G.P.V., Bhopal, II Semester, June 2007) **Ans.** (c)

20. Let X be a Poisson random variable, such that $2\,P(X=0)=P(X=2)$. Then standard deviation of x is

   (a) 4.   (b) 2.   (c) $-\sqrt{2}$ (d) $\sqrt{2}$ **Ans.** (d)

21. A card is drawn from a well shuffled pack of cards. A sequence of 156 consecutive trials are made. Using Poisson distribution, the probability that the Queen of clubs will be drawn at least once is obtained as

   (a) e^{-3} (b) $1-e^{-3}$ (c) $e^{-\frac{1}{3}}$ (d) $1-e^{-\frac{1}{3}}$ **Ans.** (b)

22. The random variable x has a Poisson distribution. If $P(x=3)=\dfrac{1}{6}$, $P(x=2)=\dfrac{1}{3}$, then $P(x=0)$ is

   (a) exp (−3/2)   (b) exp (3/2)   (c) exp (−3)   (d) exp (−1/2)   **Ans.** (a)

23. For the Poisson's distribution if $p(x=2)=\dfrac{2}{3}p(x=1)$, then mean of Poisson's distribution will be

   (a) 0   (b) $\dfrac{4}{3}$ (c) $\dfrac{3}{4}$ (d) None of these   **Ans.** (b)

24. In the Poisson distribution if $2P(x=1)=P(x=2)$, then the variance is :

   (a) 0   (b) −1   (c) 4   (d) 2   **Ans.** (c)

   (R.G.P.V., Bhopal, II Semester, June 2007)

25. In poision distribution standard deviation is equal to..................

   (a) Mean   (b) Square of mean
   (c) Square root of mean   (d) Variance   (Bihar SBTE, 2012) **Ans.** (c)

26. Which result is true?

   In poisson distribution:

   (a) Mean = Variance   (b) Mean $=\sqrt{\text{Variance}}$

   (c) Mean = Standard deviation   (d) Mean $=\sqrt{\text{Standar deviation}}$ (Bihar SBTE, 2011) **Ans.** (c)

27. Standard deviation of Poisson Distribution is equal to ....

   (a) $\sqrt{np}$ (b) np

   (c) npq   (d) $\sqrt{npq}$ (Bihar SBTE, 2011) **Ans.** (d)

28. Mean and Variance of the Poisson Distribution is equal to ....

   (a) Unequal   (b) Equal
   (c) Both (a) & (b)   (d) None of these   (Bihar SBTE, 2010) **Ans.** (a)

GROUP B

Management Techniques

Introduction to Operations Research

14.1 OPERATIONS RESEARCH

(Bihar SBTE, 2010, 2014)

Work on Operations Research was initiated in England during world war II. The British Scientists developed a method to make scientifically based decision regarding the best utilisation of war material. After the war the ideas were introduced in civil sector i.e.; industry, business and civil government. In this chapter we will learn the basic terminology of operations research including linear models.

Operations research introduces logic into decision analysis. It applies analytical techniques to study the behaviour of the system in relation to its over all working consituents. So, OR techniques are used in management cadre. Large amount of computation are solved by computer facilities. Some techniques of OR are used for linear programming inventor control network management etc.

In India Operations Research came into existence with the opening of an OR unit in 1949 at the regional research laboratory in Hyderabad. At the same time another OR unit is set up at the then defence science laboratory to tackle the problems of stores, purchase and weapon evaluation.

14.2 DEFINITION OF OPERATIONS RESEARCH

(Bihar SBTE, 2008)

(1) Operations Research is defined as the research of operations. An operation is called a set of actions required to achieve a desired outcome.

(2) OR is a scientific method of providing executive departments with a quantitative basis for decisions regarding the operations under their control. — *Morse and Kimball*

14.3 SCOPE OF OPERATIONS RESEARCH

(Bihar SBTE, 2008)

OR is mainly concerned with the techniques of applying scientific knowledge to understand and explain the Phenomenon of operating system. The technique of OR can be applied to solve any engineering problem. Some typical applications from different engineering disciplines are as below:

(*i*) Allocation of resources of services among several activities to maximize the benefit.

(*ii*) Planning the best strategy to obtain maximum profit in the presence of competitor.

(*iii*) Inventory control.

(*iv*) Design of civil engineering structures like foundations, bridges, towers, dams etc.

(*v*) Design of water resources system for maximum benefit.

(*vi*) Optimum design of machine tools, gears, etc.

(*vii*) Optimum design of electrical machinery like motors, generators and transformers.

(*viii*) Optimal production planning controlling and scheduling.

Operations Research helps the manager and the executives in taking better decision. It provides an understanding which gives the decision maker new insight and capability to determine better solution in his decision making problem with competence and confidence.

The process of operations research begins with careful observations and formulating the problem and then constructing a mathematical model that contains the essence of the real problem.

14.4 MODEL IN OPERATIONS RESEARCH

A model, as used in operations research is defined as an ideal representation of the real problem. It represents few aspects of problem.

14.5 CLASSIFICATION SCHEMES OF MODELS

(1) By degree of abstractions   (2) By a function   (3) By structure

(4) By nature of environment   (5) By extent of generality   (6) By the time horizon

1. By Degree of Abstraction

Mathematical Model (Linear Programming formulation) of the blending problem or transportation problem are the most abstract type. Since it requires not only mathematical knowledge but also great concentration to get the idea of real life problem they represent.

Language models (Cricket or Hockey match commentary) are also abstract types.

14.6 LINEAR MODEL

The fundamental role of linear model in O.R. is to find the best possible solution to the given problem as long as the parameters are the linear. In linear model the assumption is that all its functions (objective functions and constraint functions) are linear. All though this assumption holds good for numerous practical problems. But it does not hold good in economic and financial problems where some degree of non-linearity is the rule and not the exception.

14.7 CHARACTERISTICS OF A GOOD MODEL

1.   The number of assumptions should be as few as possible.

2.   The number of variables should be as few as possible. It means the model should be simple yet close to reality.

3.   It should be easy and economical to construct.

14.8 ADVANTAGE OF A MODEL

1.   It provides a logical and systematic approach to the problem.

2.   It makes the over all structure of the problem more comprehensible and helps in dealing with the problem in its entirety.

14.9 LIMITATIONS OF A MODEL

1.   Models are only idealised representation of reality and should not be regarded as absolute in any case.

2.   The validity of a model for a particular situation can be ascertained only by conducting experiments on it.

14.10 CONSTRUCTING THE MODEL

The formulation of the problem requires analysis of the system under study. This analysis shows the various phases of the system. With the formulation of the problem the first stage in model construction is over. The next step is to construct a model in which effectiveness of the system is expressed as a function of the variables defining the system.

14.11 SIMPLIFICATION IN OR MODEL

While constructing a model one comes across two conflicting objectives

(i) The model should be as easy to solve as possible. The management must be able to understand the solution and use of the model.

(ii) It should be as accurate as possible and there is no significant loss of accuracy.

14.12 LIMITATIONS OF OPERATIONS RESEARCH

1. Qualitative factors and emotional factors which cannot be quantified find no place in mathematical models.

2. Mathematical models are applicable to only specific categories of the problems.

3. Some times all factors of the problem requires huge calculation for solution.

4. Being a new field there is a resistance from the employees and management to implement.

EXERCISE 14.1

1. Discuss the origin and development of the Operations Research.

2. What are the limitations of Operations Research.

3. Define Operations research.   (*Bihar SBTE, 2009, 2010*)

4. What is the role of decision making in OR.

5. Discuss the significance and scope of OR in modern management.

6. "Mathematics of OR is mathematics of optimization". Discuss.

7. State three advantages of OR models.

8. OR is the art of winning wars without actually fighting them. Comment.

9. Define a linear model. Discuss in detail.

10. "Model building is the essence of OR approach." Discuss.

Choose the correct alternative:

11. The operations research provides:

(a) Maximal solution   (b) Minimal solution   (c) Optimal solution   (d) None of these

(*Bihar SBTE, 2008*) **Ans.** (*c*)

CHAPTER 15

Linear Programming
(Graphical Method)

15.1. INTRODUCTION

It will be of interest to know that linear programming had its origin during the second world war (1939-45). To fight the war man and material (resources) have to be maintained. There has to be efficient and safe land, sea and airtransport etc.

The government in England studied the problems during war particularly problems of armed forces, civil defence and navel strategy etc. The study for the solutions of the above problems resulted the linear programming.

Linear programming is the most popular mathematical technique which involve the limited resources in an optimal manner.

The term *programming* means planning to maximize profit or minimize cost or minimize loss or minimum use of resources or minimizing the time etc. Such problems are called *optimization Problem*. The term linear means that all equations or inequations involved are linear.

Example 1. *A manufacturer produces two types of toys i.e., A and B. Each toy of type A requires 4 hours of moulding and two hours of polishing whereas each toy of type B requires 3 hours of moulding 5 hours of polishing. Moulder works for 80 hours in a week and polisher works for 180 hours in a week. Profit on a toy of type A is ₹ 3 and on a toy of type B is ₹ 4. In what way the manufacturer allocates his production capacity for the two types of toys so that he may make the maximum profit per week.*

Solution. Table. The above information can be written in tabular form as follows:

Toy \ operation	Moulding (in hours)	Polishing (in hours)	Profit (in ₹)
A	4	2	3
B	3	5	4
Time available (in hours)	80	180	

Let x be the number of toys of type A and y be the number of toys of type B produced per week.

Profit on one toy of type A = ₹ 3

Profit on x toys of type A = ₹ $3x$

Profit on one toy of type B = ₹ 4

Profit one y toys of type B = ₹ $4y$

Let Z be the weekly profit.

Then the weekly profit in ₹ is

$$Z = 3x + 4y$$

Here, Z is known as *objective function* which has to maximize/minimize.

One toy of type A on moulding requires = 4 hours.

x toys of type A on moulding requires = $4x$ hours

One toy of type B on moulding requires = 3 hours

y toys of type B on moulding requires = $3y$ hours

On moulding total time required = $4x + 3y$ hours
But moulder works for only 80 hours in a week.
So, $4x + 3y$ hours cannot exceed 80 hours.

$\Rightarrow$ $\qquad\qquad 4x + 3y \le 80$

This is known as *first constraint*:
One toy of type A on polishing requires = 2 hours
x toys of type A on polishing requires = $2x$ hours
One toy of type B on polishing requires = 5 hours
y toys of type B on polishing requires = $5y$ hours
On polishing total time required = $2x + 5y$ hours
But polisher works for only 180 hours in a week.
So, $2x + 5y$ hours cannot exceed 180 hours.

$\Rightarrow$ $\qquad\qquad 2x + 5y \le 180$

This is known as *second constraint*.
Since, the number of toys produced is non-negative.

$\Rightarrow$ $\qquad\qquad x \ge 0$ and $y \ge 0$

This is known as *third constraint*.
Under these three constraints (conditions) we have to plan the system to get the maximum profit.
Now, we summarize the above informations in mathematical form as follows :
To maximize $\qquad Z = 3x + 4y$ $\qquad\qquad\qquad\qquad\qquad\qquad\qquad\qquad\qquad\qquad$ ... (1)
Subject to the constraints :

$$4x + 3y \le 80 \qquad\qquad\qquad\qquad\qquad\qquad\qquad\qquad ...\,(2)$$

$$2x + 5y \le 180 \qquad\qquad\qquad\qquad\qquad\qquad\qquad\qquad ...\,(3)$$

$$\left.\begin{array}{l} x \ge 0 \\ y \ge 0 \end{array}\right\} \qquad\qquad\qquad\qquad\qquad\qquad\qquad\qquad ...\,(4)$$

The above mathematical expression is known as *mathematical formulation*.
From the above inequations we find out the values of x and y.
The values of x and y are substituted in the objective function $Z = 3x + 4y$.
The maximum/minimum value of the objective function is known as *optimal value*.

15.2. SOME DEFINITIONS

1. Linear Programming Problem
Here, we have to optimize the linear function Z subject to certain conditions. Such problems are called linear programming problems. As example 1 on page 210.

2. Objective functions
Objective function is a linear function of several variables, subject to the conditions that $Z = 3x + 4y$ in the previous example.

3. Optimal Value
Optimal value is a maximum or minimum value of a objective function to be calculated in a linear programming problem.

4. Non-negative Constants
Production of any item is always non-negative, so we write $x \ge 0, y \ge 0$.

5. Linear Relations
All the mathematical relations used in L.P.P. are linear relations.

6. Programming
Programming is the method of determining a particular programme.

7. Decision Variables

Decision variables are x and y which denote the required number of items/products.

8. Constraints

Constraints are linear inequalities or equations involved in linear programming problem. In the previous example (2), (3), (4) are constraints.

9. Optimization Problem

Optimization problem is a problem in which a objective function is to be maximize or minimize subject to the certain conditions. In the previous example we have to maximize the profit, so it is optimization problem.

15.3 CONVEX SET

Any set x is said to be convex, if for any two points x_1, x_2 in the set, the line segment joining these points is completely in the set x.

Convex Sets

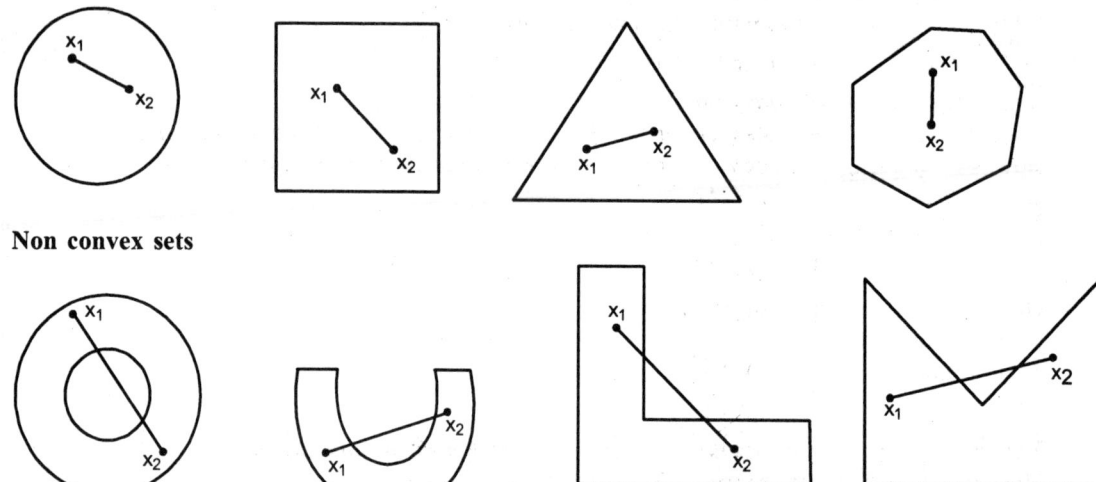

Non convex sets

15.4 MATHEMATICAL FORMULATION OF LINEAR PROGRAMMING PROBLEMS

In the previous section we have defined certain technical terms of L.P.P. Conversion of the verbal description of L.P.P. into algebraic equations/inequations is known as Mathematical Formultion.

Working Rule to formulate the L.P.P.

Step 1. Identify the decision variables to be determined and expressed them as x, y etc.

Step 2. Identify all the limitations or constraints in the given problem and then express them as linear inequalities or equations in terms of x, y etc.

Step 3. Identify the objective function (Z) which is to optimize (maximize or minimize) and express Z in terms of x, y.

Procedure:

The solution of the given L.P.P. should be divided under the following heads:

    1. Preparation of *table of information.*   **2.** Write down the *decision variables.*

    3. Form the *objective function.*   **4.** Write down the *constraints.*

    5. Write down Mathematical Formulation.

TYPE 1.   TO MAXIMIZE OBJECTIVE FUNCTION

Example 2. *A company produces two types of ornaments A and B that require gold and silver. Each unit of type A requires 1 gram of silver and 2 grams of gold. Type B requires (each unit) 2 grams of silver and 1 gram of gold. The company has only 100 grams of silver and 80 grams of gold. Each unit of type*

A brings a profit of ₹ 500 and each unit of type B brings a profit of ₹ 400. Formulate the problem as a linear programming problem to maximize the profit.

Solution. The above information is given in the following table:

1. Table:

Ornaments \ Metal	Silver (in grams)	Gold (in grams)	Profit (in ₹)
A	1	2	500
B	2	1	400
	100	80	

2. Decision variables. Decision variables are ornaments A and B.

Let the number of ornaments A be x and the number of ornaments B be y.

3. Objective function. To maximize profit.

Let Z be the objective function.

Profit of each unit of type A = ₹ 500

Profit of x units of type A = ₹ 500 x

Profit of each unit of type B = ₹ 400

Profit of y units of type B = ₹ 400 y

Total profit = 500 x + 400 y

$\Rightarrow$ $Z = 500 x + 400 y$

4. Constraint. (*i*) Company has only 100 grams of silver.

One unit of ornament A requires 1 gram of silver.

x units of ornament A require x grams of silver.

One unit of ornament B requires 2 grams of silver.

y units of ornament B require $2y$ grams of silver.

Total available quantity of silver for ornament A and B = 100 grams

$\Rightarrow$ $x + 2y \leq 100$

Constraint. (*ii*) Company has only 80 grams of gold.

One unit of ornament A requires 2 grams of gold.

x units of ornament A require $2x$ grams of gold.

One unit of ornament B requires 1 gram of gold.

y units of ornament B require y grams of gold.

Total available quantity of gold = 80 grams

$\Rightarrow$ $2x + y \leq 80$

Constraint. (*iii*) Production of ornament A and ornament B cannot be negative.

$\Rightarrow$ $x \geq 0$ and $y \geq 0$.

5. Mathematical Formulation. Hence, the linear programming problem for the given problem is as follows:

Maximize $Z = 500 x + 400 y$ ...(1)

Subject to the constraints:

$x + 2y \leq 100$ ...(2)

$2x + y \leq 80$ ...(3)

$x \geq 0$ ...(4)

$y \geq 0$ ...(5) **Ans.**

Example 3. *Two tailors A and B earn ₹ 150 and ₹ 200 per day respectively. A can stitch 6 shirts and 4 pants per day while B can stitch 10 shirts and 4 pants per day. Form a linear programming problem to minimise the labour cost to produce at least 60 shirts and 32 pants.*

Solution. The given data can be put in the tabular form as:

1. Table:

Earning per day Tailors Stitch	₹ 150 per day A	₹ 200 per day B	Minimum requirement
Shirts	6	10	60
Pants	4	4	32

2. Decision Variable. Let the tailors A and B work for x and y days respectively.

3. Objective function. The total labour cost for working x days of tailor A and y days of tailor B is ₹ $(150x + 200y)$.

Let Z denote the minimum labour cost, then

$Z = 150x + 200y$

4. Constraint (*i*). The minimum requirement of shirt is 60.

Tailor A stitches in one day = 6 shirts   Tailor A stitches in x days = $6x$ shirts

Tailor B stitches in one day = 10 shirts   Tailor B stitches in y days = $10y$ shirts

∴ $6x + 10y \geq 60$

Constraint (*ii*). The minimum requirement of pants is 32.

Tailor A stitches in one day = 4 pants   Tailor A stitches in x days = $4x$ pants

Tailor B stitches in one day = 4 pants   Tailor B stitches in y days = $4y$ pants

⇒ $4x + 4y \geq 32$.

Constraint (*iii*). The number of days worked by A or B is non negative.

∴ $x \geq 0$ and $y \geq 0$.

5. Mathematical formulation.

The mathematical formulation of given L.P.P. is as follows:

Minimize $Z = 150x + 200y$ ... (1)

subject to the constraints:

$6x + 10y \geq 60$ ... (2)

$4x + 4y \geq 32$ ... (3)

and $x, y \geq 0$.   ... (4)

Ans.

TYPE II. DIET PROBLEMS

The cost of food items containing different quantity of proteins, vitamins, fats required for a normal man should be minimum.

Example 4. *A dietician mixes together two kinds of food, say, X and Y in such a way that the mixture contains at least 6 units of vitamin A, 7 units of vitamin B, 12 units of vitamin C and 9 units of vitamin D. The vitamin contents of 1 kg of food X and 1 kg of food Y are given below:*

Cost	Vitamin A	Vitamin B	Vitamin C	Vitamin D
Food X	1	1	1	2
Food Y	2	1	3	1

One kg of food X costs ₹ 5, whereas one kg of food Y costs ₹ 8. Formulate the linear programming problem.

Solution.
1. **Decision Variables.** Decision Variables are units of food X and food Y. Let food X in the mixture be x kg. and food Y in the mixture be y kg.
2. **Objective Function.** To minimise the cost

1 kg of food X costs ₹ 5

x kg of food X costs ₹ $5x$

1 kg of food Y costs ₹ 8

y kg of food Y costs ₹ $8y$

Total cost of food X and $Y = 5x + 8y$

$\Rightarrow$ $Z = 5x + 8y$

3. **Constraint (i).** The mixture contains atleast 6 units of vitamin A.

1 kg of food X contains 1 unit of vitamin A.

x kg of food X contains x units of vitamin A.

1 kg of food Y contains 1 unit of vitamin A.

y kg of food Y contains $2y$ units of vitamin A.

$\Rightarrow$ $x + 2y \geq 6$

Constraint (ii). The mixture contains atleast 7 units of vitamin B.

1 kg of food X contains 1 unit of vitamin B.

x kg of food X contains x units of vitamin B.   1 kg of food Y contains 1 unit of vitamin B.

y kg of food Y contains y units of vitamin B.

$\Rightarrow$ $x + y \geq 7$

Constraint (iii). The mixture contains at least 12 units of vitamin C.

1 kg of food X contains 1 unit of vitamin C.

x kg of food X contains x units of vitamin C.   1 kg of food Y contains 3 units of vitamin C.

y kg of food Y contains $3y$ units of vitamin C.

$\Rightarrow$ $x + 3y \geq 12$

Constraint (iv). The mixture contains atleast 9 units of vitamin D.

1 kg of food X contains 2 units of vitamin D.

x kg of food X contains $2x$ units of vitamin D.

1 kg of food Y contains 1 unit of vitamin D.

y kg of food Y contains y units of vitamin D.

$\Rightarrow$ $2x + y \geq 9$

Constraint (v). The number of kg of food x and y is non-negative.

$x \geq 0$

$y \geq 0$

4. **Mathematical Formulation.** The linear programming problem of the given problem is as follows

To minimise $Z = 5x + 8y$ ...(1)

Subject to the constraints $x + 2y \geq 6$ ...(2)

$x + y \geq 7$ ...(3)

$x + 3y \geq 12$ ...(4)

$2x + y \geq 9$ ...(5)

$x \geq 0$ ...(6)

$y \geq 0$ ...(7)

Ans.

Example 5. *A firm is engaged in breeding goats. The goats are fed on various products grown on the farm. They need certain nutrients, named as X, Y, and Z. The goats are fed on two products A and B. One unit of product A contains 36 units of X, 3 units of Y and 20 units of Z, while one unit of product B contains 6 units of X, 12 units of Y and 10 units of Z. The minimum requirement of X, Y and Z is 108 units, 36 units and 100 units, respectively. Product A costs ₹ 20 per unit and product B costs ₹ 40 per unit. Formulate the L.P.P. to minimize the cost.*

Solution. The above information is give in the following table:

1. **Table:**

Food \ Nutrients	X	Y	Z	Cost (in ₹)
A	36	3	20	20
B	6	12	10	40
Minimum required units	108	36	100	

2. **Decision variables.** The decision variables are units of nutrients X, Y and Z. Let the units of food A be x the units of food B be y.

3. **Objective function.** To minimize the cost.

   cost of 1 unit of food $A = ₹\ 20$ cost of x units of food $A = ₹\ 20\ x$

   cost of 1 unit of food $B = ₹\ 40$ cost of y units of food $B = ₹\ 40\ y$

   Total cost $= 20\ x + 40\ y$ $\Rightarrow$ $Z = 20\ x + 40\ y$

4. **Constraint (i).** Minimum requirement of X nutrient $= 108$ units

   One unit of food A contains 36 units of X.

   x units of food A contains $36\ x$ units of X.

   One unit of food B contains 6 units of X.

   y units of food B contains $6y$ units of X.

   $\Rightarrow \quad 36x + 6y \geq 108$

   Constraint (ii). The minimum requirement of Y nutrients $= 36$ units.

   One unit of food A contains 3 units of nutrients Y.

   x units of food A contains $3x$ units of nutrients Y.

   One unit of food B contains 12 units of nutrients Y.

   y units of food B contains $12y$ units of nutrients Y.

   $\Rightarrow \qquad\qquad 3x + 12y \geq 36$

   Constraint (iii). The minimum requirement of nutrients $Z = 100$ units.

   One unit of food A contains 20 units of nutrients Z.

   x units of food A contains $20\ y$ units of nutrients Z.

   One unit of food B contains 10 units of nutrients Z.

   y units of food B contains $10\ y$ units of nutrients Z.

   $\Rightarrow \qquad\qquad 20x + 10y \geq 100$

   Constraint (iv). Units of food A and B are non-negative.

   $\therefore \qquad x \geq 0$ and $y \geq 0$.

4. **Mathematical Formulation.**

   To minimize, $Z = 20x + 40y$

   Subject to the constraints: $36x + 6y \geq 108$

   $\qquad\qquad\qquad 3x + 12y \geq 36$

   $\qquad\qquad\qquad 20x + 10y \geq 100$

   $\qquad\qquad\qquad x \geq 0,\ y \geq 0.$

EXERCISE 15.1

1. A furniture dealer deals in only two items viz., tables and chairs. He has ₹ 11,000 to invest and a space to store at most 40 pieces. A table costs him ₹ 500 and a chair ₹ 200. He can sell a table at a profit of ₹ 50 and a chair at a profit of ₹ 15. Assume that he can sell all the items that he buys. Formulate this problem as an L.P.P, so that he can maximize the profit.

Ans. Maximize, $Z = 50x + 15y$
Subject to the constraints :
$$x + y \le 40$$
$$500x + 200y \le 11000$$
$$x \ge 0$$
$$y \ge 0$$

2. A manufacturer produces nuts and bolts for industrial machinery. It takes 1 hour of work on machine A and 3 hours on machine B to produce a package of nuts; while it takes 3 hours on machine A and 1 hour on machine B to produce a package of bolts. He earns a profit of ₹ 2.50 per package on nuts and ₹ 1 per package on bolts. Form a linear programming problem to maximize his profit, if he operates each machine for at the most 12 hours a day.

Ans. Maximize, $Z = 2.5x + y$
Subject to the constraints :
$$x + 3y \le 12$$
$$3x + y \le 12$$
$$x \ge 0$$
$$y \ge 0$$

3. A person consumes two types of food, A and B, everyday to obtain 8 units of protein, 12 units of carbohydrates and 9 units of fat which is his daily minimum requirements. 1 kg of food A contains 2, 6, 1 units of protein, carbohydrates and fat, respectively. 1 kg of food B contains 1, 1 and 3 units of protein, carbohydrates and fat, respectively. Food A costs ₹ 8 per kg while food B costs ₹ 5 per kg. Form an LPP to find how many kg of each food should he buy daily to minimize his cost of food and still meet minimal nutritional requirements.

Ans. Maximize, $Z = 2.5x + y$
Subject to the constraints :
$$2x + 3y \ge 8$$
$$6x + y \ge 12$$
$$x + 3y \ge 9$$
$$y \ge 0$$

Choose the correct alternative:

4. Feasible region of a LPP is a ....
   (a) Convex set   (b) Concave set   (c) Universal set   (d) None   **Ans.** (a)

5. Circle is a convex set having .... corner points.
   (a) No   (b) One   (c) Two   (d) Three   **Ans.** (a)

6. A convex set bounded by lines or planes passes .... corner points.
   (a) One   (b) Two   (c) Finite   (d) Infinite   **Ans.** (d)

15.5 GRAPHICAL METHOD OF SOLVING LINEAR PROGRAMMING PROBLEMS

If a problem contains only two variables then we can solve the given problem by graphical method. There are two graphical method to solve a linear programming problem.
1. Corner point method   2.   Iso-profit or iso-cost method

15.6 CORNER POINT METHOD

This method is based on the fundamental extreme point theorem.

In previous class we have learnt how to formulate a system of linear inequalities involving two variables x and y mathematically.

Working Rule

Step 1. Formulate the given L.P.P. in mathematical form.

Step 2. The inequations are converted into equations.

In the equation on putting $y = 0$ we get x-coordinate on x-axis. Similarly, putting $x = 0$ we get y-coordinate on y-axis. Join these two points to get the graph of the equation.

Step 3. The inequation of a line divides the plane into two half planes, to choose the plane of the inequation we put $x = 0$ and $y = 0$ in the inequation. If origin satisfies the inequation then the region containing the origin is the region represented by the given inequation. Otherwise the half plane not containing the origin is the region represented by the given inequation.

Step 4. The region satisfying all the inequations is the feasible region.

Step 5. The vertices (corner points) of the required region are known as extreme points of the set of all feasible solutions of the L.P.P.

Step 6. By putting the values of x and y of each corner point in the objective function we get the values of the objective function at each of the vertices of the feasible region. Out of all the values of the objective function, we get a point at which the objective function is optimum (maximum or minimum).

Consider the following example:-

Example 6. *Solve the following LPP graphically*

$$Z_{max} = 5x_1 + 3x_2$$

Subject to $\quad 3x_1 + 5x_2 \leq 15.$

$$5x_1 + 2x_2 \leq 10$$

$$x_1, x_2, \geq 0$$

(Bihar SBTE, 2010)

Solution. We have to maximize

$$Z_{max} = 5x_1 + 3x_2 \qquad\qquad ... (1)$$

Subject to constraints

$$3x_1 + 5x_2 \leq 15 \qquad\qquad ... (?)$$

$$5x_1 + 2x_2 \leq 10 \qquad\qquad ... (3)$$

$$x_1, x_2 \geq 0$$

1. Conversion

$$3x_1 + 5x_2 = 15$$

$$5x_1 + 2x_2 = 10$$

2. Drawing of graph of lines.

<u>Region – represented by $3x_1 + 5x_2 \leq 15$</u>

$$3x_1 + 5x_2 = 15$$

x_1	0	5
x_2	3	0
	A	B

On joining A and B, we get the graph of equation $3x_1 + 5x_2 = 15$.

Put $x_1 = 0$, $x_2 = 0$ in $3x_1 + 5x_2 \leq 15$. $0 + 0 \leq 15$ which is true. So the half plane containing origin represents the solution set of the inequation $3x_1 + 5x_2 \leq 15$.

<u>Region – represented by $5x_1 + 2x_2 \leq 10$</u>

$$5x_1 + 2x_2 = 10$$

x_1	0	2
x_2	5	0
	C	D

On joining C and D, we get the graph of the equation $5x_1 + 2x_2 = 10$. Put $x_1 = 0$. $x_2 = 0$ in $5x_1 + 2x_2 \leq 10$ then $0 + 0 \leq 10$, which is true so the half-plane containing origin represents the solution set of the inequation $5x_1 + 2x_2 \leq 10$.

Region represented by $x_1 \geq 0$, $x_2 \geq 0$

First quadrant is the region of $x_1 \geq 0$, $x_2 \geq 0$

The shaded portion of the figure is feasible-region determined by the system of constraints (2) and (3) and

which is bounded by the co-ordinates of corner O (0, 0), D (2, 0), $E \left(\dfrac{20}{19}, \dfrac{45}{19}\right)$

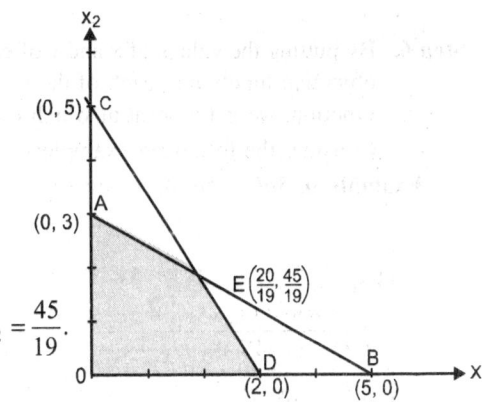

	$Z = 5x_1 + 3x_2$
$0\,(0, 0)$	$Z = 5 \times 0 + 3 \times 0 = 0$
$D\,(2, 0)$	$Z = 5 \times 2 + 3 \times 0 = 10$
$E\left(\dfrac{20}{19}, \dfrac{45}{19}\right)$	$Z = 5 \times \dfrac{20}{19} + 3 \times \dfrac{45}{19} = \dfrac{235}{19} = 12.36$
$A\,(0, 3)$	$Z = 5 \times 0 + 3 \times 3 = 9$

Hence the maximum value of $Z = 12.36$ at $x_1 = \dfrac{20}{19}$ and $x_2 = \dfrac{45}{19}$.

Example 7. *Solve the following LPP by graphically.*

$$Max\ Z = 3x_1 + 5x_2$$

Subject to $\qquad x_1 + 2x_2 \le 20;$

$\qquad\qquad\quad x_1 + x_2 \le 15$

$\qquad\qquad\quad x_2, \le 4$

(*Bihar SBTE, 2009*)

Solution.   We have,

$\qquad$ Maximize $\ Z = 3x_1 + 5x_2$ $\hfill$ ... (1)

$\qquad$ subject to constraints: $x_1 + 2x_2 \le 20$ $\hfill$ ... (2)

$\qquad\qquad\qquad\qquad\quad x_1 + x_2 \le 15$ $\hfill$ ... (3)

$\qquad\qquad\qquad\qquad\quad x_1 \ge 0,\ x_2 \ge 0$

1. Conversion $\qquad x_1 + 2x_2 = 20$

$\qquad\qquad\qquad\qquad x_1 + x_2 = 15$

2. Drawing of Graph of the lines

$\quad$ Region represented by $x_1 + 2x_2 \le 20$

$\qquad\quad x_1 + 2x_2 = 20$

x_1	0	20
x_2	10	0
	A	B

On giving A and B, we get the graph of the equation $x_1 + 2x_2 = 20$

Put $x_1 = 0$, $x_2 = 0$ in $x_1 + 2x_2 \le 20$, then $0 + 0 \le 20$, which is true. So the half plane containing origin represents the solution set of the inequation $x_1 + 2x_2 \le 20$.

Region represented by $x_1 + x_2 \le 15$

$\qquad\qquad\quad x + x_2 = 15$

x_1	0	15
x_2	15	0
	C	D

On joining C and D, we get the graph of the equation $x_1 + x_2 = 15$

Put $x_1 = 0$, $x_2 = 0$ in $x_1 + x_2 \le 15$ then $0 + 0 \le 15$, which is true, so the half plane containing origin represents the solution set of the inequation $x_1 + x_2 \le 15$

Region represented by $x_1 \geq x_2 \geq 0$

First quadrant is the region of $x_1 \geq 0$, $x_2 \geq 0$

The shaded portion of the figure is the feasible region ODEA determined by the system of constraints (2) and (3) which is bounded. The coordinates of corner are O (0, 0), D (15, 0), E (10, 5), A (0, 10).

3. Corner point Method

Now we have to maximize $Z = 3x_1 + 5x_2$ at these corner points.

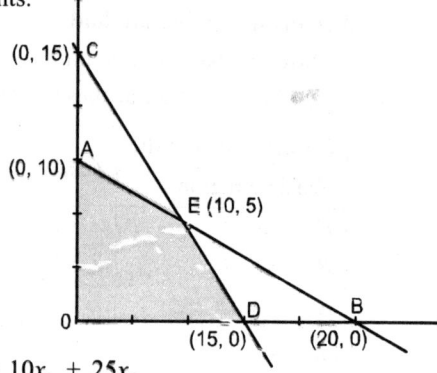

Corner point of the feasible region	$Z = 3x_1 + 5x_2$
D (15, 0)	$Z = 3(15) + 15(0) = 45$
E (10, 5)	$Z = 3(10) + 15(5) = 105$ = Maximum
A (0, 10)	$Z = 3(0) + 5(10) = 50$

Hence the maximum value of $z = 105$ **Ans.**

Example 8. *Solve the following LPP graphically Maximize $Z = 10x_1 + 25x_2$*

Subject to condition $x_1 + 3x_2 \leq 12$, $4x_1 + x_2 \geq 16$ and $x_1, x_2 \geq 0$ (*Bihar SBTE, 2012*)

Solution. We have,

Maximize $\qquad Z = 10x_1 + 25x_2$ ... (1)

Subject to constantraints.

$$x_1 + 3x_2 \leq 12 \qquad \text{... (2)}$$
$$4x_1 + x_2 \geq 16 \qquad \text{... (3)}$$
$$x_1, x_2 \geq 0$$

1. Conversion.

$x_1 + 3x_2 = 12$

$4x_1 + x_2 = 16$

2. Drawing of graph of the lines.

Region - represented by $x_1 + 3x_2 \leq 12$

$x_1 + 3x_2 = 12$

x_1	0	12
x_2	4	0
	A	B

On joining A and B, we get the graph of the equation $x_1 + 3x_2 = 12$

Put $x_1 = 0$, $x_2 = 0$ in $x_1 + 3x_2 \leq 12$, then $0 + 0 \leq 12$, which is true. So the the half plane containing origin reproscents the solution set of the inequation $x_1 + 3x_2 \leq 12$.

Region reproscented by $4x_1 + x_2 \geq 16$

$4x_1 + x_2 = 16$

x_1	0	4
x_2	16	0
	C	D

On joining C and D, we get the graph of the equation $4x_1 + x_2 = 16$

Put $x_1 = 0$, $x_2 = 0$ in $4x_1 + x_2 \geq 16$, $0 + 0 \geq 16$ which is not rue, so the half plane containing not origin represents the solution set of the inequation. $4x_1 + x_2 \geq 16$.

Region represented by $x_1 \geq 0$, $x_2 \geq 0$

First quadrant is the region of $x_1 \geq 0$, $x_2 \geq 0$ the shaded portion the figure is feasible-region determined by the system of constaints. (2) and (3) which is bounded by the co-ordinates of corner.

The co-ordinates of D (4, 0),   B (12.0),   E (3.27,2.9)

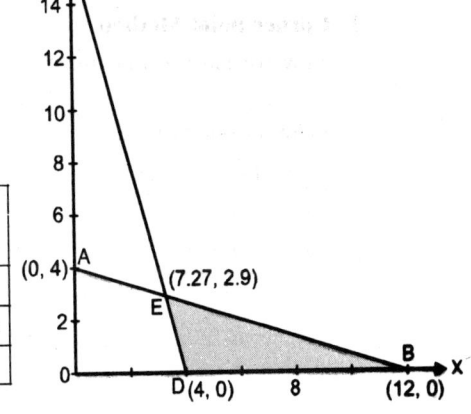

3. Cornder point method

Now we have to maximize.

$Z = 10x_1 + 25x_2$ at these corner points

Corner points of the feasible region	$Z = 10x_1 + 25x_2$
D (4, 0)	$Z = 10 \times 4 + 25 \times 0 = 40$
B (12, 0)	$Z = 10 \times 12 + 25 \times 0 = 120 = $ maximimum
E (3.27, 2.9)	$Z = 10 \times 3.27 + 25 \times 2.9 = 105.2$

Hence the maximum value of Z = 120 at $x_1 = 12$ and $x_2 = 0$

Example 9. *Solve the following LPP by graphically method maximize* $Z = 9x_1 + 6x_2$

Subject to $5x_1 + 10x_2 \leq 50$, $8x_1 + 2x_2 \geq 16$, $3x_1 - 2x_2 \geq 6$ *and* $x_1, x_2 \geq 0$

(Bihar SBTE, 2011)

Solution.   We have to maximize

$$Z = 9x_1 + 6x_2 \qquad \qquad \qquad ... (1)$$

Subject to constantraints.

$$5x_1 + 10x_2 \leq 50 \qquad \qquad ... (2)$$
$$8x_1 + 2x_2 \geq 16 \qquad \qquad ... (3)$$
$$3x_1 - 2x_2 \geq 6 \qquad \qquad ... (4)$$
$$x_1, x_2 \geq 0$$

1. Conversion

$$5x_1 + 10x_2 = 50$$
$$8x_1 + 2x_2 = 16$$
$$3x_1 - 2x_2 = 6$$

2. Drawing of the graph of the lines

Region represented by $5x_1 + 10x_2 \leq 50$

$$5x_1 + 10x_2 = 50$$

x_1	0	10
x_2	5	0
	A	B

On joining A and B, we get the graph of the equation. $5x_1 + 10x_2 = 50$. Put $x_1 = 0$, $x_2 = 0$ in $5x_1 + 10x_2 \leq 50$. Then, $0 + 0 \leq 50$. Which is true. So the half plane containing origin represents the solution set of the inequation $5x_1 + 10x_2 \leq 50$.

Region represented by $8x_1 + 2x_2 \geq 16$

$$8x_1 + 2x_2 = 16$$

x_1	0	2
x_2	8	0
	C	D

On joining C and D. We get the graph of the equation $8x_1 + 2x_2 = 16$. Put $x_1 = 0$ and $x_2 = 0$ in $8x_1 + 2x_2 \geq 16$ then $0 + 0 \geq 16$, which is not true. So, the half plane not containing origin represents the solution set of the inequation $8x_1 + 2x_2 \geq 16$.

Region – represented by $3x_1 - 2x_2 \geq 6$

$$3x_1 - 2x_2 = 6$$

x_1	2	0
x_2	0	-3
	E	F

On joining E and F. We get the graph of the equation $3x_1 - 2x_2 = 6$. Put $x_1 = 0$, $x_2 = 0$ in $3x_1 - 2x_2 \geq 0$ then $0 - 0 \geq 6$, which is not true. So, the half plane not containing origin represents the solution set of the inequation $3x_1 - 2x_2 \geq 6$.

Region represented by $x_1 \geq 0$, $x_2 \geq 0$.

First quadrant is the region of $x_1 \geq 0$, $x_2 \geq 0$. the shaded portion of the figure is the feasible region determined by the system of constants (2), (3) and (4) and which is bounded by the co-ordinates of corners D (2, 0), B (10, 0), G (4, 3).

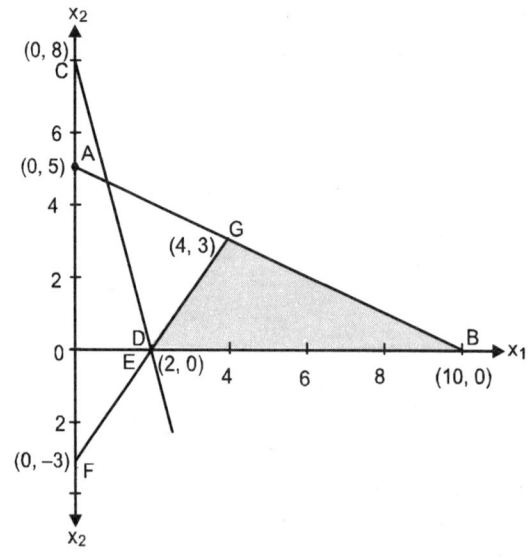

Corner points of the feasible region	$Z = 9x_1 + 6x_2$
D (2, 0)	$Z = 9(2) + 6(0) = 18$
B (10, 0)	$Z = 9(10) + 6(0) = 90$ Maximimum
G (4, 3)	$Z = 9(4) + 6(3) = 54$

Hence the maximum value of z = 90 at (10, 0).   **Ans.**

Example 10. *Solve the following LPP by Graphical method:*

$$\text{Maximize } Z = 10x_1 + 20x_2$$

with condition $2x_1 + 5x_2 \geq 10$, $7x_1 + 3x_2 \leq 49$, $3x_1 + 7x_2 \geq 12$ and $x_1, x_2 \geq 0$

(Bihar SBTE, 2014)

Solution.   We have to maximize

$$Z = 10x_1 + 20x_2 \qquad \qquad \dots (1)$$

Subject to constraints.

$$2x_1 + 5x_2 \geq 10 \qquad \qquad \dots (2)$$
$$7x_1 + 3x_2 \leq 49 \qquad \qquad \dots (3)$$
$$3x_1 + 7x_2 \geq 12 \qquad \qquad \dots (4)$$
$$x_1, x_2 \geq 0$$

1. Conversion

$$2x_1 + 5x_2 = 10$$
$$7x_1 + 3x_2 = 49$$
$$3x_1 + 7x_2 = 12$$

2. Drawing of the graph of the lines

Region – represented by $2x_1 + 5x_2 \geq 10$

$$2x_1 + 5x_2 = 10$$

x_1	0	5
x_2	2	0
	A	B

On joining A and B, we get the graph of the equation $2x_1 + 5x_2 = 10$. Put $x_1 = 0$, $x_2 = 0$ in $2x_1 + 5x_2 \geq 10$. Then, $0 + 0 \geq 10$. Which is not true. So the half plane not containing origin represents the solution set of the inequation $2x_1 + 5x_2 \geq 10$.

Region – represented by $7x_1 + 3x_2 \leq 49$

$$7x_1 + 3x_2 = 49$$

x_1	0	7
x_2	16.33	0
	C	D

On joining C and D, we get the graph of the equation $7x_1 + 3x_2 = 49$. Put $x_1 = 0$, $x_2 = 0$ in $7x_1 + 3x_2 \leq 49$, then, $0 + 0 \leq 49$, which is true. So the half plane containing origin represents the solution set of the inequation $7x_1 + 3x_2 \leq 49$.

Region – represented by $3x_1 + 7x_2 \geq 12$

$$3x_1 + 7x_2 = 12$$

x_1	0	4
x_2	1.71	0
	E	F

On joining E and F, we get the graph of the equation $3x_1 + 7x_2 = 12$. Put $x_1 = 0$, $x_2 = 0$ in equation $3x_1 + 7x_2 \geq 12$, then, $0 + 0 \geq 12$, which is not true. So the half plane not containing origin represents the solution set of the inequation $3x_1 + 7x_2 \geq 12$.

Region – represented by $x_1 \geq 0$, $x_2 \geq 0$.

Fist quardant is the region of $x_1 \geq 0$, $x_2 \geq 0$. The shaded portion of the figure is feasbile - region determined by the system of constaints (2), (3), (4) and which is bounded by the co-ordinates of corner points A (0, 1.7), B (4, 0), D (7, 0), C (0, 16.33).

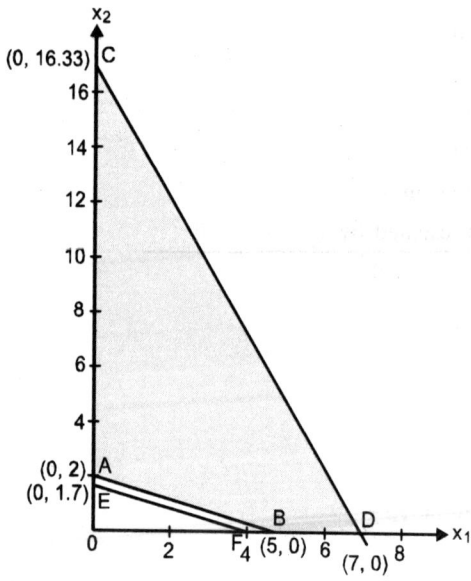

3. Corner point method

Now we have to maximize, $Z = 10x_1 + 20x_2$ at these corner points.

Corner points of feasible region	$Z = 10x_1 + 20x_2$
A (0, 2)	$Z = 10(0) + 20(2) = 40$
B (5, 0)	$Z = 10(5) + 20(0) = 50$
D (7, 0)	$Z = 10(7) + 20(0) = 70$
C (0, 16, 33)	$Z = 10(0) + 20(16.33) = 326.66$ maximum

Hence, the maximum value of $Z = 326.66$ at $x_1 = 0$, $x_2 = 16.33$.

Example 11. *Old hens can be bought for ₹ 2 each but young ones cost ₹ 5 each, the old hens lay 3 eggs per week and young ones 5 eggs per week. Each egg being worth 30 paise. A hen cost ₹ 1 per week to feed. If a person has only ₹ 80, how many of each kind should be bought to get a profit more than ₹ 6 per week so that he cannot house more than 20 hens?*

(Bihar SBTE, 2009)

Solution.   Number of old hens be bought $= x_1$

Number of young hens be bought $= x_2$

$$x_1 \geq 0, \quad x_2 \geq 0$$

Total profit $Z = 3x_1 (0.30) + 5x_2 (0.3) - (x_1 + x_2)$

$$= 0.9x_1 + 1.5x_2 - x_1 - x_2$$

$$= 0.5x_2 \, x_2 - 0.1 \, x_1 \text{ ₹}$$

To maximize $Z = 0.5x_2 - 0.1 \, x_1$

... (1)

Constraints:

$$x_1 + x_2 \leq 20 \qquad \text{... (2)}$$
$$2x_1 + 5x_2 \leq 80 \qquad \text{... (3)}$$
$$x_1 \geq 0, \ x_2 \geq 0$$

1. Conversion

$$x_1 + x_2 = 20$$
$$2x_1 + 5x_2 = 80$$

2. Drawing of graph o the lines.

Region – represented by $x_1 + x_2 \leq 20$

$$x_1 + x_2 = 20$$

x_1	0	20
x_2	20	0
	A	B

On joining A and B, we get the graph of the equation $x_1 + x_2 = 0$. Put $x_1 = 0$, $x_2 = 0$ in $x_1 + x_2 \leq 20$ then, $0 + 0 \leq 20$, which is true. So the half plane containing origin represents the solution set of the inequation $x_1 + x_2 \leq 20$.

Region – represented by $2x_1 + 5x_2 \leq 39$

$$2x_1 + 5x_2 = 80$$

x_1	0	40
x_2	16	0
	C	D

On joining C and D, we get the graph of the equation $2x_1 + 5x_2 = 80$. Put $x_1 = 0$, $x_2 = 0$ in $2x_1 + 5x_2 \leq 80$, the $0 + 0 \leq 80$, which is true. So the half plane containing origin represents the solution set of the inequation $2x_1 + 5x_2 \leq 80$.

Region represented by $x_1 \geq 0, \ x_2 \geq 0$

First quardant is the region of $x_1 \geq 0, x_2 \geq 0$.

The shaded portion of the figure is feasible region determined by the system of constraints **(2)** and **(3)** which is bounded by the co-ordinates of corner.

O $(0, 0)$, B $(20, 0)$, $E\left(\dfrac{20}{3}, \dfrac{40}{3}\right)$, C $(0, 16)$

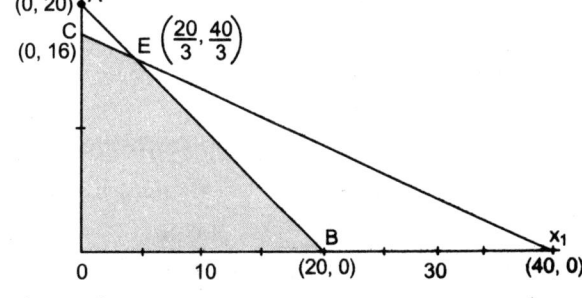

	$Z = 0.5x_2 - 0.1x_1$
A $(0, 0)$	$Z = 0.5 \times 0 - 0.1 \times 0 = 0$
B $(20, 0)$	$Z = 0.5 \times 20 - 0.1 \times 0 = 10$
$E\left(\dfrac{20}{3}, \dfrac{40}{3}\right)$	$Z = 0.5 \times \dfrac{20}{3} - 0.1 \times \dfrac{40}{3} = 2$
C $(0, 16)$	$Z = 0.5 \times 0 - 0.1 \times 16 = 1.6$

Hence maximum value of $Z = 10$ at $x_1 = 20$ and $x_2 = 0$. **Ans.**

EXERCISE 15.2

Solve graphically each of the following linear programming problems:

1. Maximize $Z = 10x + 6y$ subject to the constraints

$$3x + y \leq 12$$
$$2x + 5y \leq 34$$
$$x \geq 0, y \geq 0$$

> **Ans.** Maximum : 56; $x = 2, y = 6$

2. Maximize $Z = 60x + 15y$ subject to the constraints

$$x + y \leq 50$$
$$3x + y \leq 90$$
$$x \geq 0, y \geq 0$$

> **Ans.** Maximum : 1650, $x = 20, y = 30$

3. Minimize $Z = 18x + 10y$ subject to the constraints

$$4x + y \geq 20$$
$$x + y \geq 30$$
$$x \geq 0, y \geq 0$$

> **Ans.** Minimum : 134; $x = 3, y = 8$

4. Maximize $Z = 4x + 9y$ subject to the constraints

$$x + 5y \leq 200$$
$$2x + 3y \leq 134$$
$$x \geq 0, y \geq 0$$

> **Ans.** Maximum : 382; $x = 10, y = 38$

5. Maximize $Z = 5x + 7y$
subject to the constraints

$$x + y \leq 4$$
$$3x + 8y \leq 24$$
$$10x + 7y \leq 35$$
$$x, y \geq 0$$

> **Ans.** Maximum $Z = \dfrac{124}{5}, x = \dfrac{8}{5}, y = \dfrac{12}{5}$

6. Minimize $Z = 3x + 2y$
subject to the constraints :

$$x + y \geq 8$$
$$3x + 5y \leq 15$$
$$x \geq 0, y \geq 0$$ **Ans.** No feasible region.

7. Minimize $Z = 3x + 5y$
subject to the constraints

$$x + 3y \geq 3$$
$$x + y \geq 2$$
$$x, y \geq 0$$ **Ans.** Minimum $Z = 7, x = \dfrac{3}{2}, y = \dfrac{1}{2}$

8. Minimize $Z = 20x + 10y$
subject to the constraints

$$x + 2y \leq 40$$
$$3x + y \geq 30$$

> **Ans.** Minimum $Z = 200, x = 10, y = 0$

9. Minimize $Z = 30x + 20y$
subject to the constraints

$$x + y \leq 8$$
$$x + 4y \geq 12$$
$$x, y \geq 0$$

> **Ans.** Minimum $Z = 60, x = 0, y = 3$

10. Minimize $Z = x - 5y + 20$
subject to the constraints

$$x - y \geq 0$$
$$-x + 2y \geq 2$$
$$x \geq 3$$
$$y \leq 4$$
$$x, y \geq 0$$

> **Ans.** Minimum $Z = 4, x = 4, y = 4$

15.7 SOLUTION OF LINEAR PROGRAMMING PROBLEMS

Here we will solve the linear programming problems.

Working Rule

Step 1. Define the problem mathematically.

Step 2. Graph the constraint inequalities by converting them into equations. Find out their respective intercept on both the axes and connect them by straight lines.

Step 3. Find out the vertices of the feasible region.

Step 4. Find out the value of the objective function on the vertices.

Step 5. Find out the optimum value of the objective function.

Procedure . The solution of the given LPP should be divided under the following heads:

1. Prepare a *table* of the data given in the problem. **2.** Write down the *decision variables*.

3. Form the *objective function.* **4.** Write down the constraints.

5. Mathematical formulation. **6.** Region represented by inequations.

7. Apply Corner point method/Iso-cost, or ISO-profit method.

Type I. To maximize the objective Function (Z)

Example 12. *If a young man rides his motor cycle at 25 km/hour he had to spend ₹ 2 per km on petrol. If he rides at a faster speed of 40 km/hour, the petrol cost increases at ₹ 5 per km. He has ₹ 100 to spend on petrol and wishes to find what is the maximum distance he can travel within one hour, express this as an L.P.P. and solve it graphically.*

Solution:

The above information are given in the following table :

1. Table

S.N.	Speeds (km per hour)	Consumption of petrol per km.	Total amount Spent on petrol
1.	25	₹ 2	₹ 100
2.	40	₹ 5	

2. Decision Variables: Let the number of km riding motorcycle at the speed of 25k/h = x km

Let the number of km riding motor cycle at the speed of 40 km/hour = y km

3. Objective function

To maximize the distance of the journey.

$$Z = x + y$$

4. Constraint (*i*). The young man has ₹ 100 to spend on petrol.

When the speed is 25 km/hour cost of petrol for 1 km = ₹ 2

When the speed is 25 km/hour cost of petrol for x km = ₹ 2 x

When the speed is 40 km/hour cost of petrol for 1 km = ₹ 5

When the speed is 40 km/hour cost of petrol for y km = ₹ 5 y.

$\therefore$ $$2x + 5y \leq 100$$

Constraint (*ii*). $\qquad$ Time $= \dfrac{\text{distance}}{\text{speed}}$

Time taken in the first journey $= \dfrac{x}{25}$

Time taken in the second journey $= \dfrac{y}{40}$

Total time given $\qquad = 1$ hour

$\therefore$ $$\dfrac{x}{25} + \dfrac{y}{40} \leq 1$$

Constraint (*iii*) The distances in the journey are non-negative.

$\therefore$ $$x \geq 0 \text{ and } y \geq 0$$

5. Mathematical Formulation

To maximize, $\qquad Z = x + y$ $\qquad\qquad\qquad\qquad\qquad\qquad\qquad$ ... (1)

Subject to the constraints :

$$2x + 5y \leq 100 \qquad\qquad\qquad\qquad\qquad\qquad ... (2)$$

$$\dfrac{x}{25} + \dfrac{y}{40} \leq 1 \implies 8x + 5y \leq 200 \qquad\qquad\qquad ... (3)$$

$$x \geq 0, y \geq 0. \qquad\qquad\qquad\qquad\qquad\qquad\qquad ... (4)$$

6. Region represented by the contraints

x	50	0
y	0	20
Point	A	B

$2x + 5y = 100$

x	25	0
y	0	40
Point	C	D

$8x + 5y = 200$

We have drawn the graphs of the following lines :

$2x + 5y = 100$

$8x + 5y = 200$

$x = 0$

$y = 0$

Feasible region is represented by the shaded portion *OCEBO*.

7. Corner Point Method

The coordinates of the vertices of feasible region

OCEBO are O (0, 0), C (25, 0), $E\left(\dfrac{50}{3}, \dfrac{40}{3}\right)$ and

B (0, 20). The values of the objective function at these points are as follows :

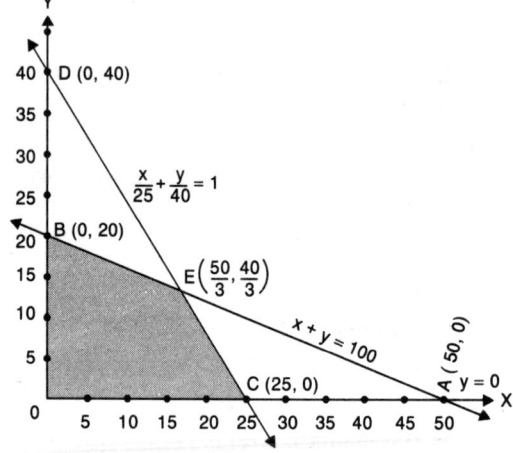

Corner Point (x, y) of the feasible region *OCEBO*	Value of the objective function $Z = x + y$
C (25, 0)	$25 + 0 = 25$
$E\left(\dfrac{50}{3}, \dfrac{40}{3}\right)$	$\dfrac{50}{3} + \dfrac{40}{3} = \dfrac{90}{3} = 30$ **Maximum**
B (0, 20)	$0 + 20 = 20$

Hence, $Z = 30$ is maximum when $x = \dfrac{50}{3}$ and $y = \dfrac{40}{3}$. **Ans.**

Example 13. *A factory owner purchased two types of machines, A and B for his factory. The requirements and the limitations for the machines are as follows:*

Machine	Area occupied	Labour force	Daily output (in units)
A	1000 m^2	12 men	60
B	1200 m^2	8 men	40

He has maximum area of 9000 m^2 available, and 72 skilled labourers who can operate both the machines. How many machines of each type should he buy to maximise the daily output ?

Solution.

1. Decision variables. Let x machines of type A and y machines of type B are bought to maximize the daily output.

2. Objective function. It is given that one machine of type A gives output 60 units so x machines of type A give output $60x$ units.

Similarly, y machines of type B give output $40y$ units.

Total output $= 60x + 40y$

$\Rightarrow \qquad Z = 60x + 40y$

3. Constraint (i).

∵ one machine of type A occupies 1000 m² area

∴ x machines of type A occupy 1000 x m² area

and ∵ one machine of type B occupies 1200 m² area

∴ y machines of type B occupy 1200 y m² area

Available area is 9000 m²

∴ $\qquad 1000\,x + 1200\,y \leq 9000.$

Constraint (ii).

∵ one machine of type A can be operated by 12 men

∴ x machines of type A can be operated by 12 x men

and ∵ one machine of type B can be operated by 8 men

∴ y machines of type B can be operated by $8y$ men

Total available labourers = 72.

∴ $\qquad 12\,x + 8\,y \leq 72.$

Constraint (iii). Number of machines cannot be negative.

∴ $\qquad x \geq 0 \text{ and } y \geq 0$

4. Mathematical Formulation

Thus the given L.P.P. is

Maximize $Z = 60x + 40y$

Subject to $\qquad 1000\,x + 1200\,y \leq 9000$

$\qquad\qquad 12\,x + 8\,y \leq 72$

$\qquad\qquad x \geq 0 \text{ and } y \geq 0$

5. Region Represented by the inequations

To solve this L.P.P. we draw the lines

$1000\,x + 1200\,y = 9000. \quad .... (1)$

$\qquad 12x + 8y = 72 \qquad ... (2)$

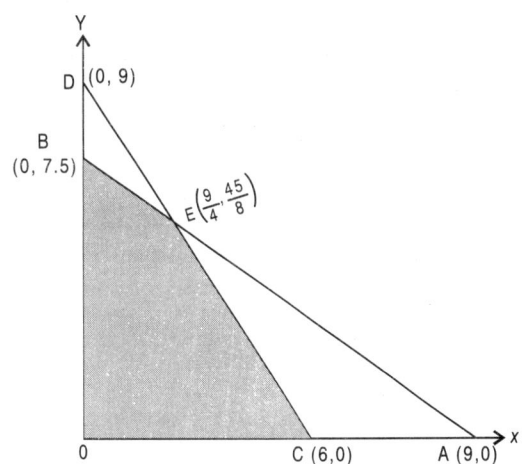

x	9	0
y	0	7.5
Points	A	B

x	6	0
y	0	9
Points	C	D

The feasible region of the L.P.P. is shaded in the adjoining figure.

6. Corner Point Method

On solving (1) and (2), we get the coordinates of $E\left(\dfrac{9}{4}, \dfrac{45}{8}\right)$.

The coordinates of corner points are $C\,(6, 0)$, $D\,(0, 7.5)$ and $E\left(\dfrac{9}{4}, \dfrac{45}{8}\right)$.

Now we evaluate $Z = 60\,x + 40\,y$ at the corner points.

Corner point of the feasible region	Corresponding value of $Z = 60x + 40y$
C (6, 0)	$Z = 60$ (6) $+ 0 = \mathbf{360}$
D (0, 7.5)	$Z = 60$ (0) $+ 40$ (7.5) $= 300$
$E\left(\dfrac{9}{4}, \dfrac{45}{8}\right)$	$Z = 60\left(\dfrac{9}{4}\right) + 40\left(\dfrac{45}{8}\right) = \mathbf{360}$

Hence, $Z = 360$ is maximum when $x = 6$, $y = 0$ or $x = \dfrac{9}{4}, y = \dfrac{45}{8}$.   **Ans.**

Example 14. *A dealer wishes to purchase a number of fans and sewing machines. He has only ₹ 5,760 to invest and has space for at most 20 items. A fan and sewing machine cost ₹ 360 and ₹ 240 respectively. He can sell a fan at a profit of ₹ 22 and sewing machine at a profit of ₹ 18. Assuming that he can sell whatever he buys, how should he invest his money in order to maximise his profit ? Translate the problem into LPP and solve it graphically.*

Solution.   The above information are given in the following table :

1. Table

Items	Cost (in ₹)	Profit (in ₹)	Space for total number of Items
Fan	360	22	
Sewing machine	240	18	
Total	5,760		20

2. Decision Variables. Let x and y be respectively the number of fans and sewing machines purchased.

3. Objective function. To maximize the profit
$$Z = 22x + 18y$$

4. Constraint (*i*). The available space is for at most 20 items.

∴ $x + y \le 20$

Constraint (*ii*). At most investment is ₹ 5,760.

∴ $360x + 240y \le 5,760 \quad \Rightarrow \quad 3x + 2y \le 48$

Constraint (*iii*). The number of fans and sewing machines are non-negative.

∴ $x \ge 0$ and $y \ge 0$.

5. Mathematical formulation.

Maximize $Z = 22x + 18y$ ... (1)

          Subject to the constraints

                   $x + y \le 20$ ... (2)

                   $3x + 2y \le 48$ ... (3)

                   $x \ge 0, y \ge 0$ ... (4)

6. Region represented by $x + y \le 20$ and $3x + 2y \le 48$

$x + y = 20$

x	20	0
y	0	20
Point	A	B

$3x + 2y \le 48$

x	16	0
y	0	24
Point	C	D

7. Corner Point Method

The feasible region, *OCEBO*, is shaded. Here C (16, 0), E (8, 12) and B (0, 20) are the corner points.

Now the value of $Z = 22x + 18y$.

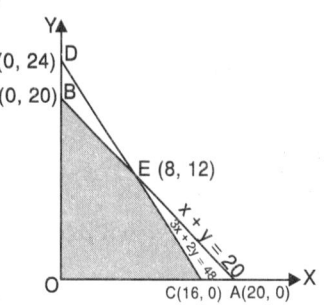

Corner point (x, y) of the feasible region *OCEBO*	Value of the objective function $Z = 22x + 18y$
C (16, 0)	$22\ (16) + 18\ (0) = 352$
E (8, 12)	$22\ (8) + 18\ (12) = 392$ **Maximum**
B (0, 20)	$22\ (0) + 18\ (20) = 360$

Thus the profit will be maximum, when 8 fans and 12 sewing machines are purchased and sold.

Also the maximum profit = ₹ 392.   **Ans.**

Example 15. *A diet is to contain at least 80 units of vitamin A and 100 units of minerals. Two foods F_1 and F_2 are available. Food F_1 costs ₹ 4 per unit and food F_2 costs ₹ 6 per unit. A unit of food F_1 contains at least 3 units of vitamin A and 4 units of minerals. A unit of food F_2 contains at least 6 units of vitamin A and 3 units of minerals. Formulate this as a linear programming problem. Find the minimum cost for diet that consists of mixture of these two foods and also meets the minimal nutritional requirements.*

Solution. The above information are given in the table below:

1. Table.

Food	Vitamin A	Mineral (*In units*)	Cost per unit (*in ₹*)
F_1	At least 3 units	4	4
F_2	At least 6 units	3	6
Diet	At least 80 units	100	

2. Decision Variables. Let x units of food F_1 and y units of food F_2 be mixed in the diet.

3. Objective function. To minimise the cost of the diet. $Z = 4x + 6y$

4. Constraint (*i*). Diet should contain at least 80 units of vitamin A. $3x + 6y \geq 80$

    Constraint (*ii*). Diet should contain at least 100 units of minerals $4x + 3y \geq 100$

    Constraint (*iii*). Number of units of food F_1 and F_2 are non-negative $x \geq 0$ and $y \geq 0$.

5. Region represented by $3x + 6y \geq 80$ **and** $4x + 3y \geq 100$.

$3x + 6y = 80$

x	$\dfrac{80}{3}$	0
y	0	$\dfrac{40}{3}$
Points	A	B

$4x + 3y = 100$

x	25	0
y	0	$\dfrac{100}{3}$
Points	C	D

7. Corner Point Method

The feasible region is the shaded portion.

The corner points are $A\left(\dfrac{80}{3}, 0\right)$

$E\left(24, \dfrac{4}{3}\right)$ and $D(0, 30)$.

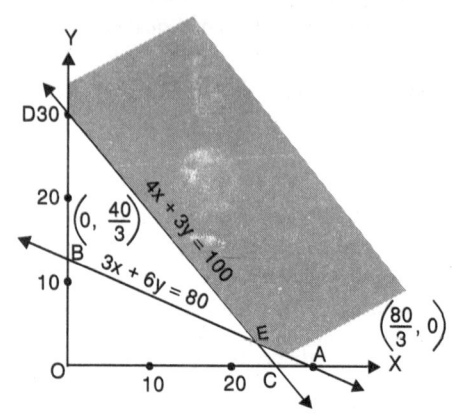

Corner points of the feasible region	Value of the objective function $Z = 4x + 6y$
$A\left(\dfrac{80}{3}, 0\right)$	$4 \times \dfrac{80}{3} + 6 \times 0 = \dfrac{320}{3}$
$E\left(24, \dfrac{4}{3}\right)$	$4 \times \dfrac{24}{3} + 6 \times \dfrac{4}{3} = 40$
$D(0, 30)$	$4 \times 0 + 6 \times 30 = 180$

Z is minimum for $x = 24$ and $y = \dfrac{4}{3}$.

The least cost of the mixture is ₹ 40 per unit.   **Ans.**

EXERCISE 15.3

1. A manufacturer produces two items X and Y. X needs two hours on machine A and two hours on machine B. Y needs 3 hours on machine A and 1 hour on machine B. If machine A can run for a maximum of 12 hours per day and machine B for 8 hours per day and profits from X and Y are ₹ 4 and ₹ 5 per items respectively. Formulate the problem as a linear programming problem and solve it graphically.

    Ans. Maximum profit = ₹ 22;   Number of items X = 3; Number of items Y = 2

2. An aeroplane can carry a maximum of 200 passengers. A profit of ₹ 400 is made on each first class ticket and a profit of ₹ 300 is made on each economy class ticket. The airline reserves at least 20 seats for first class. However, at least 4 times as many passengers prefer to travel by economy class to the first class. Determine how many each type of tickets must be sold in order to maximize the profit for the airline. What is the maximum profit?

    Ans. Maximum profit = ₹ 64000, First class tickets = 40. Economy class tickets = 160.

3. A manufacturer is trying to decide on the product quantities of two products-tables and chairs. There are 98 units of material and 80 labour-hours available. Each table requires 7 units of material and 10 labour-hours, while each chair requires 14 units of material and 8 labour-hours per chair. The profit on a table and a chair is ₹ 25 and ₹ 20, respectively. How many tables and chairs should be produced to have maximum profit?

    (**Hint**: Use Iso-profit method).

    Ans. Maximum profit = ₹ 200, Tables = 8, chairs = 0 or Tables = 4, chairs = 5.

4. A man owns a field of area 1000 sq. metre. He wants to plant trees in it. He has a sum of ₹ 1400 to purchase young trees. He has the choice of two types of trees. Type A requires 10 sq. metre of ground per tree and costs ₹ 20 per tree and type B requires 20 sq. metre of ground per tree and costs ₹ 25 per tree. When fully grown, type A produces an average of 20 kg. of fruits which can be sold at a profit of ₹ 2 per kg. and type B produces an average of 40 kg of fruits which can be sold at a profit of ₹ 1.50 per kg. How many trees of each type should be planted to achieve maximum profit when the trees are fully grown? What is the maximum profit ?

    Ans. Maximum profit = ₹ 32,00, Type A = 20. Type B = 40.

5. A farm is engaged in breeding goats. The goats are fed on various products grown on the farm. They need certain nutrients, named as X, Y and Z. The goats are fed on two products A and B. One unit of product A contains 36 units of X, 3 units of Y and 20 units of Z, while one unit of product B contains 6 units of X, 12 units of Y and 10 units of Z. The minimum requirement of X, Y and Z is 108 units, 36 units and 100 units respectively. Product A costs ₹ 20 per unit and product B costs ₹ 40 per unit. How many units of each product must be taken to minimize the cost?   **Ans.** Minimum cost = ₹ 160, Product A = 4 units, Product B = 2 units.

Choose the correct alternative:

6. Feasible region of LPP lies in the:

   (a) First quadrant   (b) Second quadrant   (c) Third Quadrant   (d) All quadrants

   (Bihar SBTE, 2003, 2009) **Ans.** (a)

7. A linear programming problem can be solved graphically if it involves .... variables.

   (a) One   (b) Two   (c) Three   (d) Any number of variables.

   (Bihar SBTE, 2004, 2010) **Ans.** (a)

8. In standard form of LPP which of the following is true?

   (a) All constraints are expressed in the form of equations of type ≤ 0

   (b) The right hand side of constraint equations are either negative or positive

   (c) all decision variables are non-negative.

   (d) The objective function must be of maximization type.   *(Bihar SBTE, 2008)* **Ans.** (a)

9. The optimal solution of a LPP may be ....

   (a) Unique   (b) Sometimes multiple(c) Sometime not   (d) All of these *(Bihar SBTE, 2005)* **Ans.** (d)

10. The optimum solution of a LPP occurs ....

    (a) Only at the boundary points   (b) Only at the corner points

    (c) At any point of the feasible region (d) All of them   **Ans.** (b)

11. The objective function of a LPP can be represented by .... which is known as isoprofit line.

    (a) One line   (b) Sets of parallel lines (c) Curve   (d) Family of curves   **Ans.** (b)

12. Feasible solution of LPP is represented by.................

    (a) Every point of the first quadrant   (b) Every point of the boundary lines

    (c) Every point of the feasible region   (d) None of them   **Ans.** (c)

13. If a line representing the objective function is parallel to any one of the boundary line of the feasible solution: There exists .... optimal solution.

    (a) Unique   (b) Multiple   (c) No   (d) Two   **Ans.** (a)

14. A graphical representation of nodes and connecting arows representing activities is called ....

    (a) Project   (b) Network   (c) Research   (d) None of these   **Ans.** (???)

15. Linear programming problem can be solved by

    (a) Newton's Rapson Method   (b) Simplex method

    (c) N.W. Corner method   (d) None of these   *(Bihar SBTE, 2014)* **Ans.** (???)

16. Linear Programming Problem can be solved by .... Method

    (a) Simpson 1/3 method   (b) N.W. corner method

    (c) NGauss-Jordan method   (d) None of these   *(Bihar SBTE, 2012)* **Ans.** (b)

16 Simplex Method

16.1 THEORY OF SIMPLEX METHOD

The basis of the complex method consists of two fundamental conditions :-

(*i*) The feasibility condition

It ensures that the starting solution is basic feasible, only basic feasible solution will be obtained during computation.

(*ii*) Optimality condition

It guarantees only better solution (as compared to the current solution)

16.2 SOME IMPORTANT DEFINITIONS

Consider the following problem

Maximize $Z = c_1x_1 + c_2x_2 + \ldots\ldots\ldots c_nx_n$,   Objective function

Subject to

$$a_{11}x_1 + a_{12}x_2 + \ldots\ldots a_{1n}x_n \leq b_1,$$
$$a_{21}x_1 + a_{22}x_2 + \ldots\ldots a_{2n}x_n \leq b_2,$$
$$a_{m1}x_1 + a_{m2}x_2 + \ldots\ldots, + a_{mn}x_n \leq b_m,$$

Constraints

where $x_1, x_2\ldots\ldots, x_n \geq 0$.

Introducing slack vairables $s, s_1, s_2, s_3\ldots\ldots\ldots s_m$ in m constraint equations. It can be put in the following standard form:

Maximize $Z = c_1x_1 + c_2x_2 + \ldots\ldots c_nx_n + s_1 + s_2 \ldots\ldots + s_m$.   .....(1)

Subject to

$$a_{11}x_1 + a_{12}x_2 + \ldots\ldots\ldots c_{1n}x_n + s_1 = b_1$$
$$a_{21}x_1 + a_{22}x_2 + \ldots\ldots\ldots a_{2n}x_n + s_2 = b_2 \qquad\qquad .....(2)$$
$$a_{m1}x_1 + a_{m2}x_2 + \ldots\ldots a_{mn}x_n + s_m = b_m$$

where $x_1, x_2, \ldots\ldots, x_n, s_1, s_2,\ldots\ldots s_m \geq 0$ .....(3)

1. **Solution :** To find out the values of $x_1, x_2 \ldots\ldots x_n, s_1, \ldots\ldots s_m$.

2. **Feasible solution :** The values of $x_1, x_2 \ldots\ldots x_n, s_1, \ldots\ldots s_m$ is known as feasible solution if these values satisfy the equations (2) and (3).

3. **Basic solution :** By making any n variables out of $n + m$ equal to zero, the values of the remaining m variable is called the basic solution, if the determinant of the coefficients of these slack variables is non negative.

4. **Basic variables :** The variables of the basic solution are called basic variables (Some of them may be zero). The other n variables are called non-basic variables.

5. **Basic feasible solution :** The basic solution of non-negative variable is called basic feasible solution.

6. **Non- degenerate basic feasible solution :** If all m basic feasible variables non-negative, then the set of these variables known as non degenerate basic feasible solution.

7. **Degenerate basic feasible solution :** If one or more basic feasible variable is zero, then the solution known as degenerate basic feasible solution.

8. **Optimal basic feasible solution :** It contains basic feasible solution that optimize the objective function.

9. **Sets of points :** Linear equation in two variables represents a line and a linear equation in three variables represents a plane. Both of them are considered as set of points.

Example 1. *A person wants to decide the constituents of a diet which will fulfill his daily requirements of proteins, fats and carbohydrates at a minimum cost. The choice are to be made from four different types of foods. The yields and cost per unit are given below:*

yields (per unit)			
	Proteins	*Fats*	*Carbohydrates*	*Cost (per unit)*
1	*3*	*2*	*6*	*45*
2	*4*	*2*	*4*	*40*
3	*8*	*7*	*7*	*85*
4	*6*	*5*	*4*	*65*
Minimum requirements	*800*	*200*	*700*	

Formulate linear programming problem.   (*Bihar SBTE, 2010*)

Solution. Let the units of food 1, 2, 3 and 4 be x_1, x_2, x_3 and x_4 respectively. Then the formulation of given LPP is as follows:

Maximize $Z = 45x_1 + 40x_2 + 85x_3 + 65x_4$

Subject to,

$$3x_1 + 4x_2 + 8x_3 + 6x_4 \leq 800$$
$$2x_1 + 2x_2 + 7x_3 + 5x_4 \leq 200$$
$$6x_1 + 4x_2 + 7x_3 + 4x_4 \leq 700 \qquad \textbf{Ans.}$$

Example 2. *Put the following linear programming problem in standard form:*

maximize $Z = 3x_1 + 2x_2 + 2x_3$

subject to $5x_1 + 7x_2 + 4x_3 \leq 7$; $4x_1 - 7x_2 - 5x_3 \geq -2$,

$$3x_1 + 4x_2 - 6x_3 \geq \frac{29}{7}; \qquad x_1, x_2, x_3 \geq 0 \qquad \text{(\textit{Bihar SBTE, 2010})}$$

Solution. Maximize $Z = (3x_1 + 2x_2 + 2x_3)$

Subject to $5x_1 + 7x_2 + 4x_3 + s_1 = 7$; $4x_1 - 7x_2 - 5x_3 - s_2 = -2$;

$$3x_1 + 4x_2 - 6x_3 - s_3 = \frac{29}{7}; \qquad x_1, x_2, x_3, s_1, s_2, s_3 \geq 0 \qquad \textbf{Ans.}$$

Working Rule: To construct table 1, 2, 3... with the help of Example 3.

Example 3. *Solve by using simplex method :*

Maximize $Z = 3x_1 + 4x_2$

Subject to $x_1 + x_2 \leq 450$

$$2x_1 + x_2 \leq 600$$

$$x_1, x_2 \geq 0$$

Solution.

Step 1. Express the the above objective function and inequations in the equation of standard form by adding slack variables.

$$\text{Maximize} \quad Z = 3x_1 + 4x_2 + 0s_1 + 0s_2$$
$$\text{Subject to} \quad x_1 + x_2 + s_1 = 450$$
$$2x_1 + x_2 + s_2 = 600$$

where $x_1, x_2, s_1, s_2 \geq 0$

Step 2. Find initial basic feasible solution.

Putting $x_1 = x_2 = 0$, in the constraints, we get

$s_1 = 450$ and $s_2 = 600$ and $Z = 0$

This is the initial basic feasible solution.

The above information can be expressed in the table 1 as follows.

Table 1

Coefficient	C_J	3	4	0	0	
C_B	Basis	body	matrix	identity	matrix	value
		x_1	x_2	s_1	s_2	b
0	s_1	1	1	1	0	450
0	s_2	2	1	0	1	600

Row, C_j represents the coefficients of the variables in the objective function.

Column, C_B represents the coefficients of the current basic variables. The basic variables are the slack variables s_1 and s_2.

Body matrix : The body matrix under non-basic variables x_1 and x_2 represents their coefficients in the constraints.

Indentity matrix : The identity matrix represents the coefficients of slack variables in the constraints.

Column b indicates the values of the basic variables, s_1 and s_2 in the initial basic feasible solution.

Table 2

C_B	C_j	3	4	0	0		
	Basis	x_1	x_2	s_1	s_2	b	θ
0	s_1	1	(1)	1	0	450	$\dfrac{450}{1}$
0	s_2	2	1	0	1	600	$\dfrac{600}{1}$
	$Z_j = \Sigma(C_B a_{ij})$	0	0	0	0	0	
	$C_j - Z_j$	3	4	0	0		

$\rightarrow$ (Key row) (Minimum)

$\uparrow$

K (Key column), (Maximum)

Row $Z_j : Z_j = \Sigma\ C_B a_{ij}$

where a_{ij} are the matrix elements in the i-th row and jth column. For example,

$Z_1 = 0 \times 1 + 0 \times 2 = 0$

Step 3. **Perform optimality test.** Here, since $C_J - Z_J$ is positive under x_1 and x_2 columns so initial basic problem is not optimal and can be improved.

If the elements in the $C_j - Z_j$ row are negative or zero, then the current solution is optimal. Since here, two elements 3 and 4 under x_1 and x_2 variable columns are positive, so the solution is not optimal and we proceed to the next step.

Step 4. **Iterate towards an optimal solution :**

Selection of incoming variable :

For this we choose the maximum positive values in $C_j - Z_j$ row. If more than one variable appears with the same **maximum value** then any one can be chosen arbitrarily. This variable is called **incoming variable.** This column is known as key column. The values of b are divided by elements of key column to get θ. Now we have to choose the minimum ratio of θ. The row containing the minimum non-negative value of θ is known as key row. The element lying at the intersection of key column and key row is called the key (or pivot) element and its is enclosed by

Selection of the outgoing variable

The outgoing variable from the basis in the row of key element is replaced by the incoming variable of the column of the key element.

Preparing table II:

We replace s_1 by x_2 from the basis. Corresponding to C_B coefficient is 0 and is replaced by 4 the coefficient of the incoming element. If Key element is not unity, made it unity and other elements in the key column are made zeroes by suitable row operations. Thus, we get the following table:

Table II

	C_j	3	4	0	0	
C_B	Basis	x_1	x_2	s_1	s_2	b
4	x_2	1	1	1	0	450
0	s_2	1	0	−1	1	150
	$Z_j = (\Sigma\, C_B a_{ij})$	4	4	4	0	1800
	$C_j - Z_j$	−1	0	−4	0	

Step 5. Since here all elements of $C_j - Z_j$ row are either zero or negative so the second feasible corresponding solution is optimal. The value of the variables in the column of basis is the corresponding under the column of b.

Hence, the optimal solution is

$x_1 = 0$

$x_2 = 450$

$Z = 3x_1 + 4x_2 = 3 \times (0) + 4\,(450) = 1800$

$Z_{max} = ₹\ 1800$

Otherwise prepare table III similar to table II. **Ans.**

Example 4. *Solve the following LPP by simplex method.*

Maximize $\qquad Z = 3x_1 + 2x_2$

Subject to $\qquad x_1 + x_2 \le 4;$

$\qquad\qquad\quad x_1 - x_2 \le 2;$

$\qquad\qquad\quad x_1, x_2 \ge 0$ *(Bihar SBTE, 2010)*

Solution.　We have to maximize $Z = x_1 + x_2 + 0S_1 + 0S_2$

Subject to
$$x_1 + x_2 + S_1 + 0S_2 = 4$$
$$x_1 - x_2 + 0S_1 + S_2 = 2$$
$$x_1, x_2, S_1, S_2 \geq 0$$

Let　$x_1 = x_2 = 0$ (Non basic)

　　$S_1 = 4$ and $S_2 = 2$ (Basic)

Tale 1

	C_j	3	2	0	0			
C_B	Basis	x_1	x_2	S_1	S_2	b		
o	S_1	1	1	1	0	4	$\dfrac{4}{1}$	
0	S_2	(1)	−1	0	1	2	$\dfrac{2}{1}$	← key row (min)
	$Z_g = \Sigma a i_j C_B$	0	0	0	0			
	$C_j - Z_j$	3	2	0	0			

↑
K (Key column)

Table II

	C_j	3	2	0	0		
C_B	Basis	x_1	x_2	s_1	s_2	b	
0	S_1	0	(2)	1	−1	2	$\dfrac{2}{2} = 1$ → Key row (Min)
3	x_1	1	−1	0	1	2	$\dfrac{2}{-1} = -2$
	$Z_j = \Sigma(C_B a_{ij})$	3	−3	0	3		
	$C_j - Z_j$	0	5	0	−3		

↑
K (Key column) (Max)

	C_j	3	2	0	0	
C_B	Basis	x_1	x_2	s_1	s_2	b
2	x_2	0	1	$\dfrac{1}{2}$	$\dfrac{1}{2}$	1
3	x_1	1	$\dfrac{1}{2}$	$\dfrac{1}{2}$	$\dfrac{1}{2}$	3
	$Z = \Sigma a_{ij} C_B$	3	$\dfrac{7}{2}$	$\dfrac{5}{2}$	$\dfrac{1}{2}$	
	$C_{ij} - Z_j$	0	$-\dfrac{3}{2}$	$-\dfrac{5}{2}$	$-\dfrac{1}{2}$	

Here, all the elements in $C_j - Z_j$ you are either negative or zero. Hence the optimal solution is for $x_1 = 3$, $x_2 = 1$, $s_1 = 0$, $s_2 = 0$

$Z = 3x_1 + 2x_2 + 0 + 0 = 3(3) + 2(1) = 11$　　　　　　　　　　　　**Ans.**

Example 5. *Use simplex method to solve the following problem.*

Maximize $\qquad$ $Z = 2x_1 + 5x_2$

Subject to $\qquad$ $x_1 + 4x_2 \le 24$

$\qquad\qquad\qquad 3x_1 + x_2 \le 21$

$\qquad\qquad\qquad x_1 + x_2 \le 9$

$\qquad\qquad\qquad x_1, x_2 \ge 0$

Solution. Step 1. First we express the objective funciton and inequations in equations of standard form by adding slack variables s_1, s_2 and s_3.

Maximize $\qquad$ $Z = 2x_1 + 5x_2 + 0s_1 + 0s_2 + 0s_3$.

Subject to $\qquad$ $x_1 + 4x_2 + s_1 = 24$

$\qquad\qquad\qquad 3x_1 + x_2 + s_2 = 21$

$\qquad\qquad\qquad x_1 + x_2 + s_3 = 9$

$\qquad\qquad\qquad x_1, x_2, s_1, s_2, s_3 \ge 0$

Step 2. Find initial basic feasible solution.

Putting $x_1 = x_2 = 0$, in the constraints, we get

$\qquad s_1 = 24, s_2 = 21$ and $s_3 = 9$ and $Z = 0$

This is initial basic feasible solution.

The above information can be expressed in matrix form as follow:

C_B	C_j Basis	2 x_1	5 x_2	0 s_1	0 s_2	0 s_3	b	θ
0	s_1	1	(4)	1	0	0	24	6 → (Key row) Min
0	s_2	3	1	0	1	0	21	21
0	s_3	1	1	0	0	1	9	9
	Z_j	0	0	0	0	0	0	
	$C_j - Z_j$	2	5 ↑	0	0	0	0	First feasible solution

key column (Max)

Step 3. Perform optimality test :

Here, since $C_j - Z_j$ is positive under x_1 and x_2 - columns so initial basic feasible solution is not optimal and can be improved.

Step 4. Iterate towards an optimal solution.

First we make pivot element unity for this we divide the key row by 4.

TABLE II A

C_B	C_j Basis	2 x_1	5 x_2	0 s_1	0 s_2	0 s_3	b	
5	x_2	$\dfrac{1}{4}$	1	$\dfrac{1}{4}$	0	0	6	$R_1 \to \dfrac{1}{4}R_1$
0	s_2	3	1	0	1	0	21	
0	s_3	1	1	0	0	1	9	

Now we make the other elements of key column 0 by suitable following row operations.

TABLE II B

C_B	C_j Basis	2 x_1	5 x_2	0 s_1	0 s_2	0 s_3	b	
5	x_2	$\dfrac{1}{4}$	1	$\dfrac{1}{4}$	0	0	6	
0	s_2	$\dfrac{11}{4}$	0	$-\dfrac{1}{4}$	1	0	15	$R_2 \rightarrow R_2 - R_1$
0	s_3	$\dfrac{3}{4}$	0	$-\dfrac{1}{4}$	0	1	3	$R_3 \rightarrow R_3 - R_1$

TABLE II

C_B	C_j Basis	2 x_1	5 x_2	0 s_1	0 s_2	0 s_3	b	θ	
5	x_2	$\dfrac{1}{4}$	1	$\dfrac{1}{4}$	0	0	6	24	
0	s_2	$\dfrac{11}{4}$	0	$-\dfrac{1}{4}$	1	0	15	$\dfrac{60}{4}$	
0	s_3	$\left(\dfrac{3}{4}\right)$	0	$-\dfrac{1}{4}$	0	1	3	$4 \rightarrow$	key row (Min)
	Z_j	$\dfrac{5}{4}$	5	$\dfrac{5}{4}$	0	0	30		
	$C_j - Z_j$	$\dfrac{3}{4}\uparrow$	0	$-\dfrac{5}{4}$	0	0			

key column (Max)

By the above table we observe that $\dfrac{3}{4}$ is the maximum positive value of $C_j - Z_j$ row so its column is the key column. Thus, here x_2 is entering variable.

Also 4 is the minimum ratio under θ, so row containing 4 is key row. Thus s_3 is leaving variable. At the intersection of key row and key column is $\dfrac{3}{4}$, so $\dfrac{3}{4}$ is pivot element.

C_B	Basis	x_1	x_2	s_1	s_2	s_3	b	
5	x_2	$\dfrac{1}{4}$	1	$\dfrac{1}{4}$	0	0	6	
0	s_2	$\dfrac{11}{4}$	0	$-\dfrac{1}{4}$	1	0	15	
0	x_1	1	0	$-\dfrac{1}{3}$	0	$\dfrac{4}{3}$	4	$R_3 \rightarrow \dfrac{4}{3} R_3$

C_B	Basis	x_1	x_2	s_1	s_2	s_3	b	
5	x_2	0	1	$\dfrac{1}{3}$	0	$-\dfrac{1}{3}$	5	$R_1 \to R_1 - \dfrac{1}{4}\,R_3$
0	s_2	0	0	$\dfrac{2}{3}$	1	$-\dfrac{11}{3}$	4	$R_2 \to R_2 - \dfrac{11}{4}\,R_3$
2	x_1	1	0	$-\dfrac{1}{3}$	0	$\dfrac{4}{3}$	4	

	C_j	2	5	0	0	0		
C_B	Basis	x_1	x_2	s_1	s_2	s_3	b	
5	x_2	0	1	$\dfrac{1}{3}$	0	$-\dfrac{1}{3}$	5	
0	s_2	0	0	$\dfrac{2}{3}$	1	$-\dfrac{11}{3}$	4	
2	x_1	1	0	$-\dfrac{1}{3}$	0	$\dfrac{4}{3}$	4	
	Z_j	2	5	1	0	1	33	Third feasible
	$C_j - Z_j$	0	0	-1	0	-1		solution

Since here all elements of $C_j - Z_j$ row are either zero or negative so the third feasible solution is optimal.

Hence, the optimal solution is

$$x_1 = 4$$
$$x_2 = 5$$
$$s_2 = 4$$
$$Z = 2.4 + 5 \times 2$$
$$Z = 2\,(4) + 5\,(5) = 33$$
$$Z_{max} = 33 \qquad \text{**Ans.**}$$

Example 6. *Use simplex method to solve the following linear programming problem:*

Max. $Z = 5x_1 + 7x_2$

Subject to $2x_1 + 3x_2 \le 13$, $3x_1 + 2x_2 \le 12$, $x_1 \ge 0, x_2 \ge 0$ (Bihar SBTE, 2005)

Solution.   **Step 1.** First we express the objective function and inequations in equations of standard form by adding slack variables s_1, s_2.

Max. $Z = 5x_1 + 7x_2 + 0\,s_1 + 0s_2$

$$2x_1 + 3x_2 + s_1 = 13$$
$$3x_1 + 2x_2 + s_2 = 12$$
$$x_1, x_2, s_1, s_2 \ge 0$$

Step 2. Find initial basic feasible solution by putting $x_1 = x_2 = 0$ in the constraints, we get

$$s_1 = 13,\ s_2 = 12 \text{ and } Z = 0$$

This is initial basic feasible solution. The above information can be expressed in matrix form as below:

Table – 1

C_B	C_j Basis	5 x_1	7 x_2	0 s_1	0 s_2	Soln.	θ
0	s_1	2	$\boxed{3}$	1	0	13	$\dfrac{13}{3}$ → key row
0	s_2	3	2	0	1	12	$\dfrac{12}{2}$
	$Z_j = \Sigma a_{ij}\, C_B$	0	0	0	0	0	0
	$C_j - Z_I$	5	7	0	0		

$\uparrow$
key column (Mix)

Step 3. Perform optimality test.

Here since $C_j - Z_j$ is positive under x_1 and x_2 columns so initial basic feasible solution is not pitimal and can be ignored.

Step 4. Iterate towards an optimal solution.

Here maximum positive entry in $C_j - Z_j$ row under x_2 column is 7, so x_2 is entring variable and minimum positive entry under θ column is $\dfrac{13}{3}$ along s_1 row, so s_1 is outgoing variable. At the intersection of key row and key column there is 3 which is pivot element. We make it unity by dividing s_1 row by 3. We make other elements of key column zero by corresponding elementary transformation.

Thus we obtain new simplex table as follows:

Table – 2

C_B	C_j Basis	5 x_1	7 x_2	0 s_1	0 s_2	Soln.	θ
7	x_2	$\dfrac{2}{3}$	1	$\dfrac{1}{3}$	0	$\dfrac{13}{3}$	$\dfrac{13}{2}$
0	s_2	$\boxed{\dfrac{5}{3}}$	0	$-\dfrac{2}{3}$	1	$\dfrac{10}{3}$	2 → key row (Min)
	Z_j	$\dfrac{14}{3}$	7	$\dfrac{7}{3}$	0		
	$C_j - Z_j$	$\dfrac{1}{3}$	0	$-\dfrac{7}{3}$	0		

$\uparrow$
key column (Max)

Step 5. Here since $C_j - Z_j$ row has a positive element $\dfrac{1}{3}$ under column x_1 so it is not optimal solution.

Maximum entry in $C_j - Z_j$ row is $\dfrac{1}{3}$ under x_1 column. So x_1 is incoming variable. Minimum entry in θ column is 2 along s_2 row. So, s_2 is outgoing variable. Here pivot element is $\dfrac{5}{3}$. We make pivot element unity by multiplying s_2 row by $\dfrac{3}{5}$ and make other elements of key column zero by corresponding elementary transformation. New simplex table is as under.

Table – 3

C_B	Basis	C_j	5	7	0	0	Soln.
			x_1	x_2	s_1	s_2	
7	x_2		0	1	$\dfrac{3}{5}$	$-\dfrac{2}{5}$	3
5	x_1		1	0	$-\dfrac{2}{5}$	$\dfrac{3}{5}$	2
	Z_j		5	7	$\dfrac{11}{5}$	$\dfrac{1}{5}$	2
	$C_j - Z_j$		0	0	$-\dfrac{11}{5}$	$-\dfrac{1}{5}$	

Here since all $C_j - Z_j \leq 0$. Therefore, it is optimal stage.

∴ The optimal solutions are $x_1 = 2$, $x_2 = 3$

∴ Max. $Z = 5x_1 + 7x_2 = 10 + 21 = 31$ **Ans.**

Example 7.   *Solve the following problem by simplex method:*

maximize $Z = 2x_1 + x_2$,

subject to $x_1 + 2x_2 \leq 10$

$x_1 + x_2 \leq 6$,

$x_1 - x_2 \leq 2$,

$x_1 - 2x_2 \leq 1$,

$x_1, x_2 \geq 0$ *(Bihar SBTE, 2010)*

Solution. Step 1. First we express the objective function and inequations in equations of standard form by adding slack variables s_1, s_2, s_3 and s_4.

Maximize $Z = 2x_1 + x_2$

Subject to $x_1 + 2x_2 + s_1 = 10$

$x_1 + x_2 + s_2 = 6$, $\quad x_1 - x_2 + s_3 = 2$, $\quad x_1 - 2x_2 + s_4 = 1$, $\quad x_1, x_2, s_1, s_2, s_3, s_4 \geq 0$

Step 2. Find initial basic feasible solution putting $x_1 = x_2 = 0$ in the constraints, we get $s_1 = 10$, $s_2 = 6$, $s_3 = 2$, $s_4 = 1$ and $Z = 0$. This is initial basic feasible solution. The above information can be expressed in matrix form as below:

Table – 1

C_B	Basis	C_j	1	0	0	0	0	b	θ	
		2								
		x_1	x_2	s_1	s_2	s_3	s_4			
0	s_1	1	2	1	0	0	0	10	10	
0	s_2	1	1	0	1	0	0	6	6	
0	s_3	1	– 1	0	0	1	0	2	2	
0	s_4	(1)	– 2	0	0	0	1	1	1	→ key row
$Z_j = \Sigma a_{ij} C_B$		0	0	0	0	0	0			
$C_j - Z_j$		2	1	0	0	0	0			

↑

key column

Step 3. Perform optimality test

Here since $C_j - Z_j$ is positive under x_1 and x_2 columns so initial basic feasible solution is not optimal and can be ignored.

Step 4. Iterate towards an optimal solution.

Here, maximum positive entry in $C_j - Z_j$ row under x_1 column is 2 so x_1 is entering variable and minimum positive entry under θ column is 1 along s_4 row so s_4 is outgoing variable.

Thus we obtain new simplex table as follows:

Table – 2

C_B	Basis	C_j	2	1	0	0	0	0		
			x_1	x_2	S_1	S_2	S_3	S_4	b	θ
0	s_1		0	4	1	0	0	−1	9	$\frac{9}{4}$
0	s_2		0	3	0	1	0	−1	5	$\frac{5}{3}$
0	s_3		0	(1)	0	0	1	−1	1	1 → key row
2	x_1		1	−2	0	0	0	1	1	$-\frac{1}{2}$
	Z_j		2	−4	0	0	0	2		
	$C_j - Z_j$		0	5	0	0	0	−2		

↑ key column

Step 5. Here since $C_j - Z_j$ row has a positive element so it is not optimal solution. Maximum entry in $C_j - Z_j$ row is 5 under x_2 column. So, x_2 is entering variable. Minimum entry in θ column is 1 in s_3 row. So, s_3 is outgoing variable. Here pivot element is 1.

We make other elements zero by corresponding elementary transformation. New simplex table is as under:

Table – 3

C_B	Basis	C_j	2	1	0	0	0	0			
			x_1	x_2	s_1	s_2	s_3	s_4	b	θ	
0	s_1		0	0	1	0	− 4	3	5	$\frac{5}{3}$	$R_2 \Rightarrow R_2 - 4R_4$
0	s_2		0	0	0	1	−3	(2)	2	1	→ key row
1	x_2		0	1	0	0	1	−1	1	−1	$(R_3 \rightarrow R_3 - 3R_1)$
2	x_1		1	0	0	0	2	−1	3	−3	
	Z_j		2	1	0	0	5	−3			
	$C_j - Z_j$		0	0	0	0	−5	3			

↑ key column

Step 6. Here $C_j - Z_j$ row has a positive element so it is not optimal solution.

Maximum entry in $C_j - Z_j$ row is 3 under s_4 column. So, s_4 is entering variable.

Minimum entry in θ column is 1 in s_2 row. So, s_2 is outgoing variable.

At the intersection of key row and key column we get 2 which is pivot element to make it unity we divide key row by 2 and make other elements zero of key column by corresponding elementary transformation. The new simplex table is as under:

Table – 4

C_B	C_j Basis	2 x_1	1 x_2	0 S_1	0 S_2	0 S_3	0 S_4	b
0	s_1	0	0	1	$-\dfrac{3}{2}$	$\dfrac{1}{2}$	0	2
0	s_4	0	0	0	$\dfrac{1}{2}$	$-\dfrac{3}{2}$	1	1
1	x_2	0	1	0	$\dfrac{1}{2}$	$-\dfrac{1}{2}$	0	2
2	x_1	1	0	0	$\dfrac{1}{2}$	$\dfrac{7}{2}$	0	4
	Z_j	2	1	0	$\dfrac{3}{2}$	$\dfrac{13}{2}$	0	
	$C_j - Z_j$	0	0	0	$-\dfrac{3}{2}$	$-\dfrac{13}{2}$	0	

Here since $C_j - Z_j$ row has no positive element. So we have reached to optimal solution:

$x_1 = 4$, $x_2 = 2$, $Z = 2x_1 + x_2 = 2 (4) + 2 = 10$ and $Z_{max} = 2 (4) + 2 = 10$ **Ans.**

Example 8. *Solve by simplex method the following L.P. problem.*

Minimize $Z = x_1 - 3x_2 + 3x_3$

Subject to $3x_1 - x_2 + 2x_3 \leq 7$

$2x_1 + 4x_2 \geq -12$

$-4x_1 + 3x_2 + 8x_3 \leq 10$

$x_1, x_2, x_3 \geq 0$

Solution.

Step 1. Right hand side of second constraint is made positive by multiplying -1.

Adding the slack variables s_1, s_2 and s_3 the above inequations can be written in standard form as follows:

Minimize $Z = x_1 - 3x_2 + 3x_3 + 0s_1 + 0s_2 + 0s_3$

Subject to $3x_1 - x_2 + 2x_3 + s_1 = 7$

$-2x_1 - 4x_2 + s_2 = 12$

$-4x_1 + 3x_2 + 8x_3 + s_3 = 10$

$x_1, x_2, x_3, s_1, s_2, s_3 \geq 0$

Step 2. To find initial basic feasible solution putting $x_1 = x_2 = x_3 = 0$ in the constraints, we get $s_1 = 7$, $s_2 = 12$, and $s_3 = 10$; $Z = 0$

The above information can be written in tabular form as follows:

TABLE I

C_B	C_j Basis	1 x_1	-3 x_2	3 x_3	0 s_1	0 s_2	0 s_3	b	$\dfrac{b}{\text{coefficient of } x_2}$ θ	
0	s_1	3	-1	2	1	0	0	7	$\dfrac{7}{2} \rightarrow$	Key Row (Min)
0	s_2	-2	-4	0	0	1	0	12		
0	s_3	8	3	(8)	0	0	1	10	$\dfrac{10}{8}$	
	Z_j	0	0	0	0	0	0			
	$C_j - Z_j$	1	-3	3	0	0	0			

$\uparrow$
key column (Max)

Step 3. Perform optimality test:

Here as we have to minimize Z, so if any $C_j - Z_j$ coefficient is negative the solution is not optimal The column having most negative $C_j - Z_j$ value will be the key column.

Since, here element 8 under x_3 variable column is most negative, so the solution is not optimal and we proceed to the next step.

Step 4. Iterate towards an optimal solution. Variable s_3 is replaced by x_3 by performing the row operations in the usual ways. Solution will become optimal when all $C_j - Z_j$ coefficients become zero or positive.

TABLE II

C_B	C_j Basis	1 x_1	-3 x_2	3 x_3	0 s_1	0 s_3	0 s_3	b	θ	
0	s_1	(4)	$-\dfrac{7}{4}$	0	1	0	$-\dfrac{1}{4}$	$\dfrac{9}{2}$	$\dfrac{9}{8}$	$\rightarrow$ key row (Min)
0	s_2	-2	-4	0	0	1	0	12	-6	
3	x_3	$-\dfrac{1}{2}$	$\dfrac{3}{8}$	1	0	0	$\dfrac{1}{8}$	$\dfrac{5}{4}$	$\dfrac{5}{32}$	
	Z_j	$-\dfrac{3}{2}$	$\dfrac{9}{8}$	3	0	0	$\dfrac{3}{8}$			
	$C_j - Z_j$	$\dfrac{5}{2}$	$\dfrac{-33}{8}$	0	0	0	$\dfrac{-3}{8}$			

$\uparrow$
key column (Max)

TABLE III

C_B	C_j Basis	1 x_1	-3 x_2	3 x_3	0 s_1	0 s_2	0 s_3	b	θ
1	x_1	1	$-\dfrac{7}{16}$	0	$\dfrac{1}{4}$	0	$-\dfrac{1}{16}$	$\dfrac{9}{8}$	$\dfrac{9}{8}$
0	s_2	0	$\dfrac{-39}{8}$	0	$\dfrac{1}{2}$	1	$-\dfrac{1}{8}$	$\dfrac{57}{4}$	
3	x_3	0	$\dfrac{5}{32}$	1	$\dfrac{1}{8}$	0	$\dfrac{3}{32}$	$\dfrac{29}{16}$	
	Z_j	1	$\dfrac{1}{32}$	3	$\dfrac{5}{18}$	0	$\dfrac{7}{32}$		
	$C_j - Z_j$	0	$-\dfrac{97}{32}$	0	$\dfrac{-5}{18}$	0	$\dfrac{-7}{32}$		

Here, all the elements in $C_j - Z_j$ row are either negative or zero, hence optimal solution is

for $x_1 = \dfrac{9}{8}$, $x_2 = 0$, $x_3 = \dfrac{29}{16}$

$z = x_1 - 3x_2 + 3x_3 = \dfrac{9}{8} - 3\,(0) + 3\left(\dfrac{29}{16}\right) = \dfrac{9}{8} + \dfrac{87}{16} = \dfrac{105}{16}$

$Z_{max} = \dfrac{105}{16}$ **Ans.**

Example 9. *Solve the following LPP using simplex method*

Maximize $Z = 10x_1 + 15x_2 + 20x_3$

with condition $2x_1 + 4x_2 + 6x_3 \le 24$

$3x_1 + 9x_2 + 6x_3 \le 30$

and x_1, x_2 or $x_3 \ge 0$ *(Bihar, SBTE, 2014)*

Solution. We have to maximize

$Z = 10x_1 + 15x_2 + 20x_3 + 0s_1 + 0s_2 + 0s_3$

subject to $2x_1 + 4x_2 + 6x_3 + s_1 = 24$

$3x_1 + 9x_2 + 6x_3 + s_2 = 30$

$x_1, x_2, x_3\ s_1, s_2 \ge 0$

TABLE I

C_B	Basis	C_j						b	θ
		10	15	20	0	0	0		
		x_1	x_2	x_3	s_1	s_2	s_3		
0	s_1	2	4	6	1	0	0	24	$\dfrac{24}{2}$
0	s_2	3	9	(6)	0	1	0	30	$\dfrac{30}{3}$ → key row (Min)
	Z_j	0	0	0	0	0	0	0	
	$C_j - Z_j$	10	15	20	0	0	0		

↑
key column (Max)

TABLE II

C_B		C_j						b	θ
		10	15	20	0	0	0		
		x_1	x_2	x_3	s_1	s_2	s_3		
0	s_1	-1	-5	0	1	-1	0	-6	
20	x_3	$\dfrac{1}{2}$	$\dfrac{3}{2}$	1	0	$\dfrac{1}{6}$	0	5	
	Z_i	10	30	20	0	$\dfrac{10}{3}$	0		
	$C_j - Z_j$	0	-15	0	0	$\dfrac{-10}{3}$	0		

Here all the elements in Cj – Zj row are either negative or zero, hernce optimal solution is

for $s_1 = -6$, $x_3 = 5$, $x_1 = 0$, $x_2 = 0$, $s_2 = 0$, $s_3 = 0$

$Z = 10x_1 + 15x_2 + 20x_3 + 0s_1 + 0s_2 + 0s_3$

$= 10\,(0) + 15\,(0) + 20\,(5) + 0 + 0 + 0 = 100$ **Ans.**

Example 10. *Three grades of coal A, B and C contain phosphorous and ash as impurities. In a particular industrial process, fuel upto 100 tons is required which should contain ash not more than 3% and phosphorous not more than 0.03%. It is desired to maximize the profit while satisfying these conditions. There is an unlimited supply of each grade. The percentage of impurities and percentage of profit are given below:*

Coal	Phosphorous (%)	Ash (%)	Profit (in ₹ /ton)
A	0.02	2.0	12.00
B	0.04	3.0	15.00
C	0.03	5.0	14.00

Find the proportion in which the three grades be used.   (Bihar SBTE, 2008)

Solution.   Let the three grades of coal A, B and C be the x_1, x_2 and x_3 tons respectively.

Max. $Z = 12 x_1 + 15x_2 + 14x_3$

$x_1 + x_2 + x_3 \leq 100$ $2x_1 + 3x_2 + 5x_3 \leq 3$

$0.02x_1 + 0.04x_2 + 0.03x_3 \leq 0.03$ $\Rightarrow$ $2x_1 + 4x_2 + 3x_3 \leq 3$

$x_1 \geq 0,$ $x_2 \geq 0,$ $x_3 \geq 0$

Step 1. First we express the objective function and inequations in equations of standard form by adding slack variables, s_1, s_2, s_3.

Max. $Z = 12x_1 + 15x_2 + 14x_3 + 0s_1 + 0s_2 + 0s_3$

Constraints $x_1 + x_2 + x_3 + s_1 = 100$

$2x_1 + 3x_2 + 5x_3 + s_2 = 3$

$2x_1 + 4x_2 + 3x_3 + s_3 = 3$

$x_1, x_2, x_3, s_1, s_2, s_3 \geq 0$

Step 2. Find initial basic feasible solution by putting $x_1 = x_2 = x_3 = 0$ in the constraints, we get

$s_1 = 100,$ $s_2 = 3,$ $s_3 = 3$ and $Z = 0$

This is initial basic feasible solution. The above information can be expressed in matrix form as below:

Table – 1

C_B	Basis	C_j 12	15	14	0	0	0		
		x_1	x_2	x_3	s_1	s_2	s_3	Soln.	θ
0	s_1	1	1	1	1	0	0	100	100
0	s_2	2	2	5	0	1	0	3	1
0	s_3	2	4	3	0	0	1	3	0.75 → key row
	Z_j	0	0	0	0	0	0	0	
	$C_j - Z_j$	12	15	14	0	0	0		

↑
key column

Step 3. **Perform optimality test:**

Here since $C_j - Z_j$ is positive under x_1, x_2 and x_3 columns so initial basic feasible solution is not optimal and can be ignored.

Step 4. **Iterate towards an optimal solution.**

Here maximum positive entry in $C_j - Z_j$ row under x_2 column is 15, so x_2 is incoming variable and minimum positive entry under θ column is 0.75 along s_3 row, so s_3 is outgoing variable. At the intersection of key row and key column there is 4 which is pivot element. We make it unity by dividing s_3 row by 4. We also make other elements of key column zero by corresponding elemeentary transformation.

Thus we obtain new simplex table as follows:

Table – 2

C_B	Basis	C_j 12 x_1	15 x_2	14 x_3	0 s_1	0 s_2	0 s_3	Soln.	θ	
0	s_1	$\dfrac{1}{2}$	0	$\dfrac{1}{4}$	1	0	$-\dfrac{1}{4}$	$\dfrac{397}{4}$	$\dfrac{397}{2}$	
0	s_2	$\left(\dfrac{1}{2}\right)$	0	$\dfrac{11}{4}$	0	1	$-\dfrac{3}{4}$	$\dfrac{3}{4}$	$\dfrac{3}{2}$	→ key row
15	x_2	$\dfrac{1}{2}$	1	$\dfrac{3}{4}$	0	0	$\dfrac{1}{4}$	$\dfrac{3}{4}$	$\dfrac{3}{2}$	
	Z_j	$\dfrac{15}{2}$	15	$\dfrac{45}{4}$	0	0	$\dfrac{15}{4}$	$\dfrac{45}{4}$		
	$C_j - Z_j$	$\dfrac{9}{2}$	0	$\dfrac{11}{4}$	0	0	$-\dfrac{15}{4}$			

↑ key column

Step 5. Here since $C_j - Z_j$ row has positive enteries under x_1 and x_3 columns. So, it is not a stage of optimality.

Maximum entry in $C_j - Z_j$ row is $\dfrac{9}{2}$ under x_1 column, so x_1 is incoming variable.

Minimum entry in θ column is $\dfrac{3}{2}$ along s_2 row, so s_2 is outgoing variable. Here pivot element is $\dfrac{1}{2}$. We make pivot element 1 by multiplying s_2 row by 2. We also make other elements of key column zero by corresponding elementary transformation.

New simplex table is as under:

Table – 3

C_B	C_j	12	15	14	0	0				
	Basis	x_1	x_2	x_3	s_1	s_2		Soln.	θ
0	s_1	0	0	$-\dfrac{5}{2}$	1	-1		$\dfrac{394}{4}$	197
12	x_1	1	0	$\dfrac{11}{2}$	0	2		$\dfrac{3}{2}$	—
										→ key row
	Z_j	12	15	36	0	9			
	$C_j - Z_j$	0	0	-22	0	-9			

↑
key column

Step 6. Here since $C_j - Z_j$ row has positive entry under s_3 column. So it is not stage of optimality. Maximum entry in $C_j - Z_j$ row is 3 under s_3 column so, s_3 is incoming variable.

Minimum entry in θ column is 0 along x_2 row so x_2 is outgoing variable. Here pivot element is 1. We make other elements of key column zero by corresponding elementary transformation.

New simplex table is as under:

Table – 4

C_B	C_j	12	15	14	0	0	0	
	Basis	x_1	x_2	x_3	s_1	s_2	s_3	Soln.
0	s_1	0	$-\dfrac{1}{2}$	$-\dfrac{3}{2}$	1	$-\dfrac{1}{2}$	0	$\dfrac{394}{4}$
12	x_1	1	$\dfrac{3}{2}$	$\dfrac{5}{2}$	0	$\dfrac{1}{2}$	0	$\dfrac{3}{2}$
0	s_3	0	1	-2	0	-1	1	0
	Z_j	12	18	30	0	6	0	
	$C_j - Z_j$	0	-3	-16	0	-6	0	

Here since all $C_j - Z_j \leq 0$. Therefore it is a optimal stage.

Thus optimal solutions are

$$x_1 = \frac{3}{2}, \qquad s_1 = \frac{394}{4}, \qquad s_3 = 0$$

and remaining other variables are

$$x_2 = 0, \qquad x_3 = 0, \qquad s_2 = 0$$

Maximum profit $Z = 12\, x_1 + 15 x_2 + 14 x_3 + 0 s_1 + 0 s_2 + 0 s_3$

$$= 12\left(\frac{3}{2}\right) + 15\,(0) + 14\,(0) + 0 + 0 + 0 = 18$$

$\Rightarrow \qquad Z = ₹\ 18$ per tons

Ans.

16.3 SPECIAL CASES IN THE SIMPLEX METHOD

(1) Tie in the choice of incoming variables.

A tie in the choice of incoming variable exists when more than one variable has the same largest positive (or negative) value. To break the tie any one of them is selected arbitrarily as the incoming variable, though a wrong choice may increase the number of iteration to reach the optimal solution. Unfortunately there is no method to predict this before hand. However if there is tie between a decision variable and slack/surplus variable, select the decision variable.

(2) Tie in the choice of outgoing variable (Degeneracy)

When ever one or more basic variable have zero value then degeneracy may arise:

(a) At the initial stage at least one basic variable is zero in the initial basic feasible solution. This will be so if the Right hand side of a constraint is zero.

(b) At any subsequent iteration when there is tie among the minimum non-negative replacement ratio arbitrary selection of the outgoing variable result in the other tied value variable becoming zero in the text table and the solution is said to be degenerate and the objective function may not improve. It may result cycling or circling the same result. Fortunatily cases of this type or less and there is no practical method in which cycling occurs. Here the objective function does not improve.

(Tie in Incoming Variables)

Example 11. Maximize $Z = x_1 + x_2$

Subject to $x_1 - 2x_2 \geq -8$

$3x_1 + x_2 \leq 11$

$x_1, x_2 \geq 0$

Solution. Before proceeding to the simplex method, one should keep in mind that the right hand side constant value of the constraint should be non-negative. So, converting the first constraint and introducing the black variables s_1, s_2 the above problem reduces into

Maximize $Z = x_1 + x_2 + 0s_1 + 0s_2$

Subject to the constraints

$-x_1 + 2x_2 + s_1 = 8$

$3x_1 + x_2 + s_2 = 11$

$x_1, x_2, s_1, s_2 \geq 0$

The tables in different iteration can be constructed as given on next page:

TABLE I

C_B	Basis	C_j	1	1	0	0		
			x_1	x_2	s_1	s_2	Sol.	θ
0	s_1		-1	2	1	0	8	$\dfrac{8}{-1}$
0	s_2		3	1	0	1	11	$\dfrac{11}{3}$ → key row (Min)
	Z_j		0	0	0	0	0	
	$C_j - Z_j$		1	1	0	0		

↑

key column (Max)

In this table, the coefficient value 1 occurs twice (tie) in $C_j - Z_j$ row. So either x_1 or x_2 may be selected as incoming variable. Let us select x_1 as an incoming variable.

New revised simplex table is as under

TABLE II

C_j		1	1	0	0		
C_B	Basis	x_1	x_2	s_1	s_2	Sol.	θ
0	s_1	0	$\dfrac{7}{3}$	1	$\dfrac{1}{3}$	$\dfrac{35}{3}$	7
1	x_1	1	$\dfrac{1}{3}$	0	$\dfrac{1}{3}$	$\dfrac{11}{3}$	11
	Z_j	1	$\dfrac{1}{3}$	$\dfrac{1}{3}$	0	$\dfrac{1}{3}$	
	$C_j - Z_j$	0	$\dfrac{2}{3}$	0	$-\dfrac{1}{3}$		

$\rightarrow$ key row (Min)

$\uparrow$
key column (Max)

Here x_2 is incoming variable and s_1 is outgoing variable.

New revised simplex table is as under:

TABLE III

C_j		1	1	0	0	
C_B	Basis	x_1	x_2	s_1	s_2	Sol.
1	x_2	0	1	$\dfrac{3}{7}$	$\dfrac{1}{7}$	5
1	x_1	1	0	$-\dfrac{1}{7}$	$\dfrac{2}{7}$	2
	Z_j	1	1	$\dfrac{2}{7}$	$\dfrac{3}{7}$	
	$G_j - Z_j$	0	0	$-\dfrac{2}{7}$	$-\dfrac{3}{7}$	

Since all $C_j - Z_j \leq 0$. So this is stage of optimal solution. Hence optimal solution is

$x_1 = 2$, $x_2 = 5$ and Max. $Z = 2 + 5 = 7$ **Ans.**

(Tie in out going variables)

Example 12. *Maximize $Z = 2x_1 + 3x_2 + 10x_3$*

Subject to $x_1 + 2x_3 = 0$, $x_2 + x_3 = 1$, $x_1, x_2, x_3 \geq 0$

Solution. Step 1. Set up the problem in the standard form:

The given problem can be expressed as

Maximize $Z = 2x_1 + 3x_2 + 10x_3$

Subject to $x_1 + 0x_2 + 2x_3 = 0$, $0x_1 + x_2 + x_3 = 1$, $x_1, x_2, x_3 \geq 0$

Here is no need to introduce artificial variables as x_1, x_2, are themeselves forming identity

matrix $\begin{bmatrix} 0 & 1 \\ 0 & 1 \end{bmatrix}$ and can be treated as basic variables.

Step 2. Find an initial Basic Feasible solution.

Putting $x_3 = 0$, the basic feasible (degenerate) solution to the problem is

$x_1 = 0$ (basic), $x_2 = 1$ (basic), $x_3 = 0$ (non-basic)

The above informations can be represented in the following table:

TABLE I

	C_j	2	3	10			
C_B	Basis	x_1	x_2	x_3	b	θ	
2	x_1	1	0	2	0	0	→ key row (Min)
3	x_2	0	1	1	1	1	
	Z_j	2	3	7	3		
	$C_j - Z_j$	0	0	3			

↑
key column (Max)

Step 3. Perform optimality Test

Since $C_j - Z_j$ is positive under x_3 column, so above table is not optimal.

Setp 4. Iterate towards an optimal solution

Performing iteration to get an optimal solution result in the following table.

TABLE II

	C_j	2	3	10	
C_B	Basis	x_1	x_2	x_3	b
10	x_3	$\dfrac{1}{2}$	0	1	0
3	x_2	$-\dfrac{1}{2}$	1	0	1
	Z_j	$\dfrac{7}{2}$	3	10	3
	$C_j - Z_j$	$-\dfrac{3}{2}$	0	0	

Table 2 is optimal and the optimal basic feasible solution is:

$x_1 = 0$ (non basic), $x_2 = 1$ (basic), $x_3 = 0$ (basic), $Z_{max} = 3$.

The above solution is, however, degenerate because basic variable x_2 has zero value. Note that the value of Z in table 1 as well as table 2 is same. Thus we have obtained the optimal degenerate solution from a degenerate solution without improving the value of Z. In other words, it is possible to move from one table to another without any improvement in the objective function.

EXERCISE 16.1

Solve the following problems by the Simplex method.

1. Max. $Z = 7x_1 + 5x_2$

    Subject to condition: $x_1 + 2x_2 \leq 6$; $x_1 + 3x_2 \leq 12$; $x_1, x_2 \geq 0$. *(Bihar SBTE, 2009)*

    Ans. $x_1 = 3$, $x_2 = 0$, $Z = 21$

2. Maximize $Z = 4x_1 + 10x_2$

Subject to $2x_1 + x_2 \leq 50$; $2x_1 + 5x_2 \leq 100$; $2x_1 + 3x_2 \leq 90$; $x_1, x_2 \geq 0$ **Ans.** $x_1 = 0$, $x_2 = 20$; $Z_{max.} = 200$

3. Minimize $Z = x_1 - 3x_2 + 2x_3$

Subject to $3x_1 - x_2 + 2x_3 \geq 7$; $-2x_1 + 4x_2 \leq 12$; $-4x_1 + 3x_2 + 8x_3 \leq 10$; $x_1, x_2, x_3 \geq 0$

Ans. $x_1 = 4$, $x_2 = 5$, $x_3 = 0$; $Z_{min} = -11$

4. Minimize $Z = 4x_1 + x_2$

Subject to $3x_1 + x_2 = 3$; $4x_1 + 3x_2 \geq 6$; $x_1 + 2x_2 \leq 3$; $x_1, x_2 \geq 0$ **Ans.** $x_1 = \dfrac{3}{5}$, $x_2 = \dfrac{6}{5}$; $Z_{min} = \dfrac{18}{5}$

5. A firm produces three products. These products are processed on three different machines. The time required to manufacture one unit of each of the three products and the daily capacity of the three machines are given in the table below:

Machine	Time per unit minutes			Machine capacity
	Product 1	Product 2	Product 3	(minutes/day)
M_1	2	3	2	440
M_2	4	—	3	470
M_3	2	5	—	430

It is required to determine the daily number of units to be manufactured for each product. The profit per unit for products 1, 2 and 3 is ₹ 4, ₹ 3 and ₹ 6 respectively. It is assumed that all the amounts produced are consumed in

the market.   **Ans.** $x_1 = 0$, $x_2, = 86$, $x_3 = \dfrac{470}{3}$, $Z_{max} = ₹\ 1198$

6. A firm manufacturing three products P_1, P_2 and P_3. The minimum number of units of P_1, P_2 and P_3 that must be produced are 100, 200 and 150 respectively. These products require two types of raw materials M_1 and M_2 which the firm can purchase upto a Maximum of 500 and 400 units respectively. Design a production plan so as to maximize the profit if the respective individual profits of P_1, P_2 and P_3 are ₹ 2, ₹ 5 and ₹ 4 respectively. consumption of raw materials is shown below:

Raw material	Consumption of raw materials per unit product		
	P_1	P_2	P_3
M_1	$\dfrac{1}{2}$	1	1
M_2	2	$\dfrac{1}{2}$	$\dfrac{1}{5}$

Ans. $x_1 = 100$, $x_2 = 300$, $x_3 = 150$

[**Hint:** Maximize $Z = 2x_1 + 5x_2 + 4x_3$

Subject to $\dfrac{x_1}{2} + x_2 + x_3 \leq 500$; $2x_1 + \dfrac{x_2}{2} + \dfrac{x_3}{5} \leq 400$; $x_1 \geq 100$; $x_2 \geq 200$; $x_3\ x_3 \geq 150$]

7. A company manufactures three types of leather belts, namely A, B and C. The unit profits from these three varieties are ₹ 10, ₹ 5 and ₹ 7 respectively. Leather is sufficient for only 800 belts per day (All types together). Belt A requires thrice the time of belt B and belt C requires twice the time of belt B. If all the belts of type B are produced, a maximum of 1000 belts per day can be produced. Belt A requires a fancy buckle and 150 such buckles per day are available. There are sufficient number of buckles for the other varieties. Determine how many belts of each type be produced to maximize the total profit.

Ans. Max. $Z = 10x_1 + 5x_2 + 7x_3$;   Subject to $x_1 + x_2 + x_3 \leq 800$; $3x_1 + x_2 + 2x_3 \leq 1000$; $x_1 \leq 150$;

$x_1, \ x_2, \ x_3 \geq 0$

8. A small scale industrialist produces four types of machine components A, B, C and D made of steel and brass. The amounts of steel and brass required for each component and the number of men weeks of labour required to manufacture and assemble one unit of each component are as follows:

	A	B	C	D	Availability
Steel	6	5	3	2	100 kg
Brass	3	4	9	2	75 kg
Man-weeks	1	2	1	2	20 kg

The labour is restricted to 20 man-weeks, steel is restricted to 100 kg per week and brass to 75 kg per week. The industrialist profit on each unit of A, B, C and D is ₹ 6, ₹ 4, ₹ 7 and ₹ 5 respectively. How many each type of machine component should be produce to maximize his profit and how much is his profit.

Ans. $A = 15$, $B = 0$, $C = \dfrac{10}{3}$, $D = 0$; $Z_{max} = ₹ \dfrac{340}{3}$ per week

Choose the correct alternative

9. Artificial variable technique is useful if .... of the constraints involve "≥" sign and the right hand side value be positive.

(a) All   (b) At least one   (c) None   (d) At most one

(Bihar SBTE, 2003, 2009) **Ans.** (b)

10. Simplex method yields always an .... solution.

(a) Exact   (b) Not correct   (c) Approximate   (d) None

(Bihar SBTE, 2004, 2011) **Ans.** (a)

11. Total number of basic cells in a basic feasible solution is:

(a) $m + n$ (b) $m \cdot n$ (c) $m + n - 1$ (d) $m - n - 1$ **Ans.** (b)

12. A linear programming problem can be solved by ....

(a) North-West corner method   (b) Simpson $\dfrac{1}{3}$ Rule

(c) Simplex method   (d) Newton-Rapson Method   *(Bihar SBTE, 2011)* **Ans.** (c)

CHAPTER 17
Transportation Problem
(N-W corner Method, Least cost Method, Vogel's Approximation method, u-v method)

17.1 TRANSPORTATION PROBLEM

We have already solved the transportation problems involving two variables in chapter 15 of linear programming on page............ In this chapter we will discuss the transportation problem of three variables or more.

Example 1. *Suppose that the state food corporation (SFC) of Bihar has stored sugar at warehouses situated at two places, Patna and Ranchi. The sugar is in demand at three places at Dhanbad Dumka and Purnea. The stock of sugar available at each of Patna and Ranchi are 3 metric tonnes and the demand of sugar is 2 metric tonnes, 3 metric tonnes and 1 metric tonne at Dhanbad, Dumka and Purnea respectively. The transportation cost (in thousand of Rupees) of one metric tonne between cities i and j (c) are shown in the figure below.*

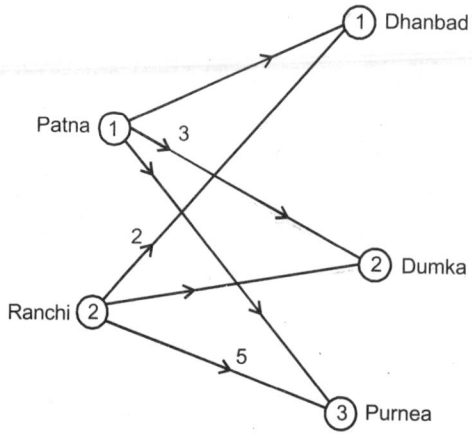

Formulate the problem mathematically and represent in tabular form.

Solution. Let x_{ij} $(i = 1, 2$ and $j = 1, 2, 3)$ be the number of metric tonnes being transported from ith source to jth destination. The above problem can be expressed mathematically to determine $x_{11}, x_{12}, x_{13}, x_{21}, x_{22},$ and x_{23}, so as to minimize

$$C_{11}\,x_{11} + C_{12}\,x_{12} + C_{13}\,x_{13} + C_{21}\,x_{21} + C_{22}\,x_{22} + C_{23}\,x_{23} \qquad \ldots (1)$$

Subject to the conditions that

$$\left.\begin{array}{l} x_{11} + x_{12} + x_{13} = 3 \\ x_{21} + x_{22} + x_{23} = 3 \end{array}\right\} \text{Availability} \qquad \ldots (2)$$

$$\left.\begin{array}{l} x_{11} + x_{21} = 2 \\ x_{12} + x_{22} = 3 \\ x_{13} + x_{23} = 1 \end{array}\right\} \text{Demand} \qquad \ldots (3)$$

and $x_{11}, x_{12}, x_{13}, x_{21}, x_{22}$ and $x_{23} \geq 0$

The above described problem can be put in the transportation table as below:

From \ To	Dhanbad	Dumka	Purnea	Availability
Patna	4	3	1	3
Ranchi	2	7	5	3
Requirements	2	3	1	

17.2 MATHEMATICAL FORMULATION

Let number of supply points (factories) = m the number of demand points (ware houses) = n

The cost of transportation of the items from ith factory to the jth ware house = C_{ij}

Number of units from ith factory to jth ware house = x_{ij}

Total cost of transportation under the distribution = Z

Factory   Truck   Ware house

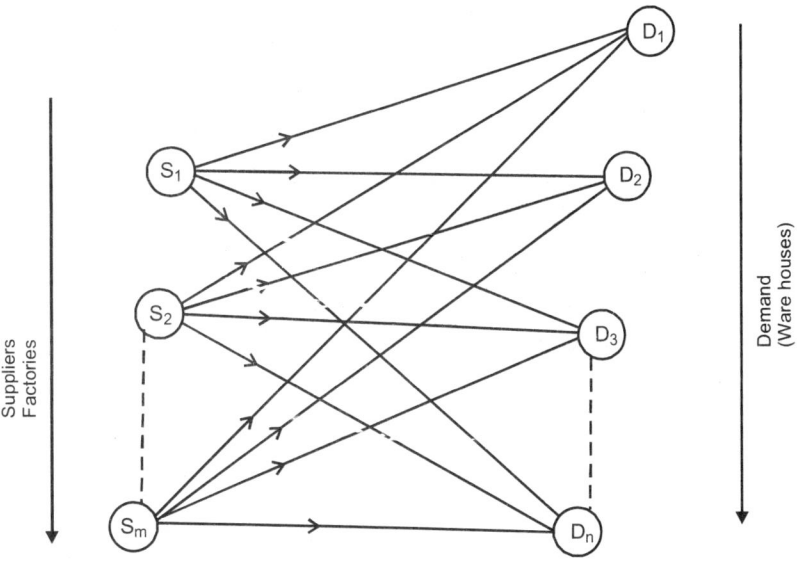

Mathematically,

Minimize $Z = \sum_{i=1}^{m} \sum_{j=1}^{n} x_{ij} \, C_{ij}$

Subject to $\sum_{j=1}^{n} x_{ij} \le S_i$ for $i = 1, 2, 3, \ldots\ldots\ldots m$

(supply)

[Total number of items transported to all the demand points (ware houses) $j \le$ Number of items from sources (factories) (i)]

$\sum_{i=1}^{m} x_{ij} \le D_j$ for $j = 1, 2, 3 \ldots\ldots\ldots n$

(Demand)

[Total number of items from different factories $\le$ Number of items required in Demand ware house j]

$x_{ij} \ge 0$, for all i and j

The above informations are shown in the following table.

Factories	Ware houses 1	2	3	.... n	Supply
1	C_{11} x_{11}	C_{12} x_{12}	C_{13} x_{13}	C_{1n} x_{1n}	S_1
2	C_{21} x_{21}	C_{22} x_{22}	C_{23} x_{23}	C_{2n} x_{2n}	S_2
3	C_{31} x_{31}	C_{32} x_{32}	C_{33} x_{33}	C_{3n} x_{3n}	S_3
$\vdots$	$\vdots$	$\vdots$	$\vdots$	$\vdots$	$\vdots$
m	C_{m1} x_{m1}	C_{m2} x_{m2}	C_{m3} x_{m3}	C_{mn} x_{mn}	S_m
Demand	D_1	D_2	D_3	.... D_n	

17.3 DEFINITIONS

1. Rim condition.

Total quantity of the item available at different sources = Total requirement at different destinations.

2. Feasible solution

A feasible solution to a transportation problem is a set of non negative allocations, x_{ij} that satisfy rim (row and column) restriction.

3. **Basic feasible solution.**

   If it contains not more than $m + n - 1$ non negative allocations where m is rows and n is the number of columns of T.P.

4. **Optimal solution.**

   A feasible solution that minimize the transportation cost is called an optimal solution.

5. **Non-degenerate basic feasible solution**

   A basic feasible solution to a $(m \times n)$ transportation problem is said to be non degenerate if:

   (*a*) the total number of non-negative allocation is exactly $m + n - 1$

   (*b*) these $m + n - 1$ allocations are in independent position.

6. **Degenerate basic feasible solution.**

   A basic feasible solution in which the total number of non-negative allocation is less than $m + n - 1$ is called degenerate basic feasible solution.

17.4 NORTH WEST CORNER RULE

North West Corner Method (NWCM)   (*Bihar SBTe, 2014*)

The method start at the north west corner cell of the table.

Step 1. Allocate as much as possible to the north west corner and adjust associated amount of (supply – demand)

(*a*) $D_1 < S_1$

   North west cell is to be filled by min $(D_1, S_1) = D_1$ then $x_{11} = D_1$.

   The balance supply $(S_1 - D_1)$ is to be filled in cell $(1, 2)$

D_1	S_1-D_1		S_1
			S_2
			S_3
D_1	D_2	D_3	

(*b*) If $D_1 = S_1$, set $x_{11} = D_1$

D_1			S_1
			S_2
			S_3
D_1	D_2	D_3	

(*c*) If $D_1 > S_1$, then $x_{11} = $ min $(D_1, S_1) = S_1$ the balance $(D_1 - S_1)$ is to be filled to cell $(2, 1)$. Proceed vertically.

S_1			S_1
D_1-S_1			S_2
			S_3
D_1	D_2	D_3	

Step 2. Continue in this manner, step by step away from North west corner until final value is reached in the south east corner and find the transportation.

Example 2. *Find feasible solution of the following transportation problem by North-West corner Rule*

	F_1	F_2	F_3	F_4	*Supply*
D_1	2	3	11	7	6
D_2	1	0	6	1	1
D_3	5	8	15	9	10
Demand	7	5	3	2	

                                                                         (*Bihar SBTE, 2009*)

Solution.   **North West corner rule**

(*i*) Allocate 6 to (1, 1) cell, this exhausted the demand at D_1

(*ii*) Move to cell (2.2) and assign 1. This exhausted the demand at D_2.

(*iii*) Move to cell (3.3) and allocate 2. Now move to cell (3, 2) and allocate 3. and move cell (3, 1) allocate (5 − 1) = 4. then move cell (3.1) allocate 7 − 6 = 1.

The given TP is

	F_1	F_2	F_3	F_4	Supply
D_1	2 ⑥	3 ×	11 ×	7 ×	6
D_2	1 ×	0 ①	6 ×	1 ×	1
D_3	5 ①	8 ④	15 ③	9 ②	10
Demand	7	5	3	2	$\Sigma a_i = 17 = \Sigma b_i$

Hence, the feasible solution is

$x_{11} = 6, x_{22} = 1, \quad x_{31} = 1, \quad x_{32} = 4, \quad x_{33} = 3, \quad x_{34} = 2$

Transportation cost = 2 (6) + 1 (0) + 5 (1) + 8 ((4) + 15 (3) + 9 (2)

$$= 12 + 0 + 5 + 32 + 45 + 18 = 112 \qquad \textbf{Ans.}$$

Example 3. *Find the solution corner of the following transportation problem by N-W corner method and determine the transportation cost.*

From \ To	D_1	D_2	D_3	D_4	Availability
S_1	21	16	25	13	11
S_2	17	18	14	23	13
S_3	32	27	18	41	19
Requirement	6	10	12	15	

(Bihar SBTE, 2005)

Solution.   The given TP is

From \ To	D_1	D_2	D_3	D_4	Availability
S_1	21 ⑥	16 ⑤	25 ×	13 ×	11
S_2	17 ×	18 ⑤	14 ⑧	23 ×	13
S_3	32 ×	27 ×	18 ④	41 ⑮	19
Requirement	6	10	12	15	

Here, we observe that,

$\Sigma a_i = 11 + 13 + 19 = 43$

$\Sigma b_j = 6 + 10 + 12 + 15 = 43$

Thus given transportation problem is balanced.

Start with N-W corner cell (1, 1).

Min (6, 11) = 6 = x_{11} $(D_1 = 6)$

Set $x_{21} = 0$, $x_{31} = 0$

Put a cross in the remaining cell of first column because demand of D_1 is complete.

Now availability at $S_1 = 11 - 6 = 5$

But $D_2 = 10$

Min (5, 10) = 5

Set $x_{12} = 5$ and set $x_{13} = x_{14} = 0$ and put cross in (1, 3) and (1, 4) cell.

D_2 still needs $10 - 5 = 5$ units which are to be supplied by next source S_2.

Min (5, 13) = 5

S_2 can supply 5 units to (2, 2). Set $x_{22} = 5$

Set $x_{32} = 0$ and put cross in (3, 2) because the demand of D_2 is exactly met.

In cell (2, 3) put $(13 - 5) = 8$ units

$x_{23} = 8$ and $x_{33} = 4$ and $x_{34} = 0$

Put cross in cell (3, 4) because S_2 is exhausted.

Next $D_4 = 15$, S_3 is left with $19 - 45 = 15$

Min (15, 15) = 15

S_3 can supply all 15 units to D_4.

Set (3, 4) = 15

So the solution of the given T.P. is

$x_{11} = 6$, $x_{12} = 5$, $x_{22} = 5$, $x_{23} = 8$, $x_{33} = 4$, $x_{34} = 15$

Transportation cost = 21 (6) + 16 (5) + 18 (5) + 41 (8) + 18 (4) + 41 (15)

= 126 + 80 + 90 + 112 + 72 + 615 = 1095   **Ans.**

Example 4. *Solve the following transportation problem by North-West corner Method (N-W Method) and find transportation cost.*

A	B	C	Availability	
11	21	16	14	S_1
7	17	13	26	S_2
11	23	21	31	S_3
18	28	25		
D_1	D_2	D_3		

(*Bihar SBTE, 2004*)

Solution.   By N-W method, we have

A	B	C	Availability	
11 ⑭	21 ✕	16 ✕	14	S_1
7 ④	17 ㉒	13 ✕	26	S_2
11 ✕	13 ⑥	21 ㉕	31	S_3
18	28	25		
D_1	D_2	D_3		

Min (18, 14) = 14 in cell (1, 1) $\Rightarrow$ x_{11} = 14, cell (1, 2) and cell (1, 3) should be crossed.

18 – 14 = 4 is required by D_1.

$\Rightarrow$ x_{22} = 26, cell (2,2) should be crossed.

26 – 4 = 22 is available at S_2.

Min (28, 22) = 22, so x_{22} = 22

Cell (2, 3), should be crossed 28 – 22 = 6

Min (31, 6) = 6, so so x_{32} = 6, 31 – 6 = 25

Min (25, 25) = 25, so x_{33} = 25

x_{11} = 0, x_{12} = 0, x_{13} = 0

x_{21} = 4, x_{22} = 22, x_{23} = 0

x_{31} = 0, x_{32} = 6, x_{33} = 25

Feasible solution x_{11} = 14, x_2 = 4, x_{22} = 22, x_{32} = 6, x_{33} = 25

Transportation cost = 11 (14) + 7 (4) + 17 (22) + 13 (6) + 21 (25)

$$= 154 + 28 + 374 + 108 + 525$$

$$= 1189$$ **Ans.**

Example 5. *Find an initial feasible solution by North-West corner method (N-W method) of the following transportation problem and find the transportation cost.*

2	11	10	3	7	4
1	4	7	2	1	8
3	9	4	4	12	9
3	3	4	5	6	
D_1	D_2	D_3	D_4	D_5	

(Bihar SBTE, 2003)

Solution.   We start N-W method

2	11	10	3	7	4	S_1
③	①	×	×	×		
1	4	7	2	1	8	S_2
×	②	④	②	×		
3	9	4	8	12	9	S_3
×	×	×	③	⑥		
3	3	4	5	6		
D_1	D_2	D_3	D_4	D_5		

Min (4, 3) = 3 so x_{11} = 3
4 – 3 = 1 so x_{12} = 1, $x_{13} = x_{14} = x_{15} = 0$
3 – 1 = 2 so x_{21} = 2
x_{32} = 0, so cell (3, 2) should be crossed 8 – 2 = 6
Min (4, 6) = 4, so x_{23} = 4, x_{33} = 0
8 – (2 + 4) = 2
Min (5, 2) = 2, x_{24} = 2
5 – 2 = 3
Min (9, 3) = 3
So, x_{34} = 3
9 – 3 = 6 left
Min (6, 6) = 6, so x_{35} = 6
x_{11} = 3, x_{12} = 1, x_{13} = 0, x_{14} = 0, x_{15} = 0
x_{21} = 0, x_{22} = 2, x_{23} = 4, x_{24} = 2, x_{25} = 0
x_{31} = 0, x_{32} = 0, x_{33} = 0, x_{34} = 3, x_{35} = 6
Transportation cost = 2 (3) + 11 (1) + 4 (2) + 7 (4) + 2 (2) + 8 (3) + 12 (6)
= 6 + 11 + 8 + 28 + 4 + 24 + 72 = 153 **Ans.**

EXERCISE 17.1

1. Determine an initial feasible solution to the following transportation problem using N-W corner rule.

To Available

	3	4	6	8	9	20	S_1
	2	10	1	5	8	30	S_2
From	7	11	20	40	3	15	S_3
	2	1	9	14	16	13	S_4
Demand	40	6	8	18	6		
	D_1	D_2	D_3	D_4	D_5		

Ans. x_{11} = 20, x_{21} = 20, x_{22} = 6, x_{23} = 4, x_{33} = 4, x_{34} = 11, x_{44} = 7 and x_{45} = 6

2. Find the feasible solution of the following transportation problem using N-W corner method:

W_1	W_2	W_3	W_4	Supply	
14	25	45	5	6	F_1
65	25	35	55	8	F_2
35	3	65	15	16	F_3
(Requirement) 4	7	6	13	30	(Total)

Ans. x_{11} = 4, x_{12} = 2, x_{22} = 5, x_{12} = 2, x_{22} = 5, x_{23} = 3, x_{33} = 3, x_{34} = 13

3. Determine an initial basic feasible solution to the following T.P. using N-W corner Rule.

					Supply	
2	11	10	3	7	4	S_1
1	4	7	2	1	8	S_2
3	9	4	8	12	9	S_3
Demand 3	3	4	5	6		
D_1	D_2	D_3	D_4	D_5		

Ans. $x_{11} = 3$, $x_{12} = 1$, $x_{22} = 2$, $x_{23} = 4$, $x_{24} = 2$, $x_{34} = 3$, $x_{35} = 6$

4. Determine an initial b.f.s. from the N-W corner Rule

			Supply	
4	3	1	3	S_1
2	7	5	3	S_2
Demand 2	3	1	6	(Total)
D_1	D_2	D_3		

Ans. $x_{11} = 2$, $x_{12} = 1$, $x_{22} = 2$, $x_{23} = 1$, Min. $Z = 30$

In a mxn transportation table, total number of cell is (m × n transportation)

(a) m + n   (b) m . n

(c) m + n – 1   (d) None of these   *(Bihar SBTE, 2000)* **Ans. (b)**

17.5 LEAST COST METHOD

The least cost method depends upon the concentration of cheapest routes. The method assigns as much as possible to the cell with the smallest cost unit. The satisfied row or column is cross out and the amount of supply and demand are adjusted accordingly.

Example 6. *Find the least cost of the transportation from the following table:*

				Supply
10	2	20	11	15
12	7	9	20	25
4	14	16	18	10
Demand 5	15	15	15	

Solution.

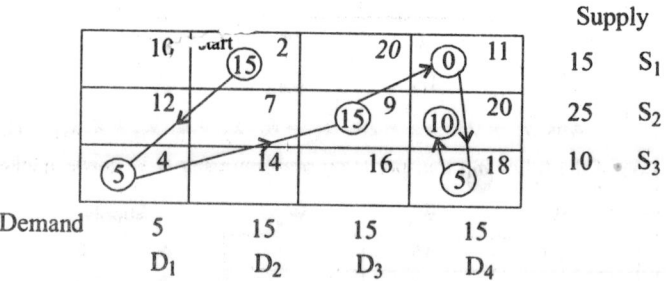

Demand   5   15   15   15
            D_1 D_2 D_3 D_4

The least cost method is applied in the following manner.

Step 1. Cell (1, 2) has the least unit cost (2).

x_{12} = min (15, 15) = 15

which happens to satisfy both row 1 and column 2 simultaneously. We arbitrarily cross out column 2 and adjust the supply in row 1 to (15 – 15) = 0.

Step 2. Cell (3, 1) has the smallest uncrossed out unit cost (4).

x_{31} = min (5, 10) = 5

and cross out column 1 because it is satisfied, and adjust the demand of row 3 to (10 – 5) = 5

Step 3. Continuing in the same manner:

x_{23} = 15, x_{14} = 0, x_{34} = 5, x_{24} = 10

Hence, the solution is x_{12} = 15, x_{14} = 0, x_{23} = 15, x_{24} = 10, x_{31} = 5 and x_{34} = 5

The associated objective value is

Z = 15 × 2 + 0 × 11 + 15 × 9 + 10 × 20 + 5 × 4 + 5 × 18 = 475   **Ans.**

Example 7. *Determine a solution to the following transportation problem using least cost method.*

↓ Source	D_1	D_2	D_3	D_4	Capacity
S_1	8	9	6	10	400
S_2	12	15	17	8	300
S_3	9	5	20	7	500
Demand	250	250	450	250	1200

Destination.   (*Bihar SBTE, 2012*)

Solution.

Source	D_1	D_2	D_3	D_4	Capacity
S_1	8	9	6	10	400
S_2	12	15	17	8	300
S_3	9	5	20	7	500
Demand	250	250	450	250	1200

Source	D_1	D_2	D_3	D_4	Capacity
S_1	8	9	400 6	10	400
S_2	250 12	15	50 17	8	300
S_3	9	250 5	20	7 250	500
Demand	250	250	450	250	1200

x_{13} = 400, x_{21} = 250, x_{23} = 50, x_{32} = 250, x_{24} = 250

Transportation cost = 400 × 6 + 250 × 12 + 50 × 17 + 250 × 5 + 250 × 7 = 9250   **Ans.**

Example 8. *Determine an initial basic feasible solution to the following transportation problem using North West corner rule & least cost method.*   (*Bihar SBTE, 2010*)

6	4	1	5	14	S_1
8	9	2	7	16	S_2
4	3	6	2	5	S_3
6	10	15	4		
D_1	D_2	D_3	D_4		

Solution.   (*i*) North West Corner Rule

⑥ 6	⑧ 4	1	× 5	14 S_1
× 8	② 9	⑭ 2	× 7	16 S_2
× 4	× 3	① 6	④ 2	5 S_3
6	10	15	4	
D_1	D_2	D_3	D_4	

$x_{11} = 6, \ x_{21} = 8, \ x_{22} = 2, \ x_{23} = 14, \ x_{33} = 1, \ x_{34} = 4$

Cost of transportation

$= 6 \times 6 + 8 \times 4 + 2 \times 9 + 14 \times 2 + 1 \times 6 + 4 \times 2$

$= 36 + 32 + 18 + 28 + 6 + 8 = 128$ **Ans.**

(*ii*) Least cost method

× 6	× 4	⑭ 1	5	14 S_1
⑥ 8	⑨ 9	① 2	7	16 S_2
4	① 3	6	④ 2	5 S_3
6	10	15	4	
D_1	D_2	D_3	D_4	

Here, $x_{13} = 14, \ x_{23} = 1, \ x_{34} = 4, \ x_{32} = 10, \ x_{21} = 6, \ x_{22} = 9$

Total cost $= 14 \times 1 + 1 \times 2 + 4 \times 2 + 1 \times 3 + 6 \times 8 + 9 \times 9$

$= 14 + 2 + 8 + 3 + 48 + 81 = 156$ **Ans.**

EXERCISE 17.2

1. Find the initial basic feasible solution to the following T.P. by least cost method:

			Supply	
2	7	4	5	S_1
3	3	1	8	S_2
5	4	7	7	S_3
1	6	2	14	S_4
Demand 7	9	18		
D_1	D_2	D_3		

Ans. $x_{12} = 2, \ x_{13} = 3, \ x_{23} = 8, \ x_{32} = 7, x_{41} = 7, \ x_{43} = 7$

2. Solve the following problem by least cost method:

			Available	
4	3	1	3	S_1
2	7	5	3	S_2
Demand 2	3	1		
D_1	D_2	D_3		

Ans. $x_{12} = 2, \ x_{13} = 1, \ x_{21} = 2, \ x_{22} = 1, \ \text{Min } Z = ₹ 18$

Choose the correct answer:

3. The sum of basic cell and no basic cell in a m × n transportation problem is always equal to:

   (a) m + n   (b) m . n   (c) m . n − 1   (d) m + n − 1 (*Bihar SBTE, 2009*) **Ans.** (d)

4. In a m × n transportation table, total number of cells is (m × n) transportation:

   (a) m + n   (b) m . n   (c) m + n − 1   (d) None of these

   (*Bihar SBTE, 2009*) **Ans.** (b)

5. Total number of squares, called cell in a n × n transportation table is

   (a) m + n   (b) m . n   (c) m + n − 1   (d) m . n − 1 (*Bihar SBTE, 2004*) **Ans.** (c)

Fill up the blanks:

6. In statistical quality control, it is difficult to identity and eliminate .... causes.

   (*Bihar SBTE, 2009*) **Ans.** (assignable)

Choose the correct alternative.

7. Transportation problem is solved by .... method

   (a) Simpson 1/3 method   (b) Least cost method

   (c) Bisection method   (d) None of these   (*Bihar SBTE, 2012*) **Ans.** (b)

8. Hugarian method is used to solve ....

   (a) Transportation problem   (b) Assignment problem

   (c) Linear programming problem   (d) None of these   (*Bihar SBTE, 2012*) **Ans.** (b)

9. Transportation problem in any company may be solved by ....

   (a) Newton Rapson Method   (b) Trepezoidal Method

   (c) Simpson 1/3 method   (d) Least cost method   (*Bihar SBTE, 2011*) **Ans.** (d)

10. A transportation problem becomes an assignment problem if

   (a) demand = 1, supply > 1   (b) demand > 1, supply = 1

   (c) demand = 1, suplly = 1   (d) demand > 1, supply > 1   **Ans.** (c)

17.6 VOGEL'S APPROXIMATION METHOD (VAM) OR PENALTY METHOD OR REGRET METHOD

Vogel's method is preferred to the methods described earlier. The solution found by VAM is very close if not equal to the optimal solution.

The working rule of this method is written below:

Working Rule

Step 1. Write the difference between the smallest and second smallest element in each column below the corresponding column and the difference between the smallest and second smallest element in each row on the right of the row.

Step 2. Identify the row or column with the greatest difference (penalty) and allocate min (D_1, S_1) in the lowest cost cell.

   If there is a tie in the minimum cost, then select that cell which will have maximum assignment.

Step 3. If whole D_1 is completed in lowest cost cell, then the remaining elements of the column are crossed. If the whole S_1 is written in lowest cost cell then the remaining elements of the row are crossed.

Step 4. In the new table replace the original supply/demand by the remaining supply/demand.

Step 5. Recompute the row and column difference in the reduced transportation table omitting crossed elements of rows or columns and identify the maximum difference (penalty).

Example 9. *Find initial basic feasible solution to the following transportation problem using Vogel's approximation method.*

	1	2	3	Availability
A	7	2	5	25
B	4	4	6	15
Requirement →	20	5	15	

(*Bihar SBTE, 2009*)

Solution.   Given the following T.P.

	1	2	3	Availability
A	7	2	5	25
B	4	4	6	15
Requirement	20	5	15	

Here availability = 25 + 15 = 40

    Requirement = 20 + 5 + 15 = 40

Thus, availability = Requirement

∴   The given TP is balanced.

Step 1.

	1	2	3	Availability	Penalty	
A	7	2 ⑤	5	25	3	←
B	4	4	6	15	2	
Requirement	20	5	15			
Penalty	3	2	1			

Here maximum **penalty** 3 is in first column as well as first row. But as minimum entry 2 of cell (1, 2) lies in first row, so **we select** this row. Allocate min (5, 25) = 5 in cell (1, 2). The requirement of second column is complete.

Step 2.

	1	2	3	Availability	Penalty
A	7	⑤	5	20	2
B	4 ⑮		6	15	2
Re-requirement	20	0	15		
Penalty	3		1		

      ↑

Here maximum penalty 3 is under first column and minimum entry in this column is 4 in cell (2, 1). Allocate min (20, 15) = 15 in this cell. The availability of second column is complete.

Step 3. Thus the reduced table is:

	1	2	3	Re-availability	Penalty
A	7	⑤	5	20	2
B	⑮			0	0
Re-requirement	5	0	15		
Penalty	7		5		

↑

Here maximum penalty is 7 under first column and remaining minimum entry in this column is 7 in cell (1, 1). Allocate min (5, 20) = 5 in this cell. The requirement of first column is complete.

Step 4. Again the reduced table is:

	1	2	3	Re-availability
A	⑤	⑤	⑮	0
B	⑮			0
Re-requirement	0	0	0	

Thus, the solution obtained through VAM is as below:

$x_{11} = 5$, $x_{12} = 5$, $x_{13} = 15$

$x_{21} = 15$ and all other variables i.e. $x_{22} = x_{23} = 0$ **Ans.**

Example 10. *Find initial basic feasible solution of the following transportation problem Vogel's approximation method.*

	W_1	W_2	W_3	W_4	Capacity
F_1	19	30	50	10	7
F_2	70	30	40	60	9
F_3	48	8	70	20	18
Requirement	5	8	7	14	

(Bihar SBTE, 2008)

Solution.

	W_1	W_2	W_3	W_4	Capacity a_i
F_1	19	30	50	10	7
F_2	70	30	40	60	9
F_3	48	8	70	20	18
Requirement b_j	5	8	7	14	

Here, $\Sigma a_i = \Sigma b_j = 34$

∴ Then given T.P. is balanced

By using VAM, we have

Step–I: To calculate penalty, which is the arithmetic difference between the smallest and next to the smallest unit cost, still remaining that row or column.

	W_1	W_2	W_3	W_4	Capacity	Penalty
F_1	19 ⑤	30	50	10	7	9
F_2	70	30	40	60	9	10
F_3	48	8	70	20	18	12
Requirement bj	5	8	7	14		
Penalty	29	22	10	10		

Step–II: Select the row or column for which the penalty is the largest and allocate the maximum feasible amount to that cell with least cost in that particular row or column.

	W_1	W_2	W_3	W_4	Capacity	Penalty
F_1	19 ⑤	30	50	10	7	9
F_2	70	30	40	60	9	10
F_3	48	8 ⑧	70	20	18	12
Re-Requirement bj	5	8	7	14		
Penalty	29	22	10	10		

Let us, construct reduced table as:

	W_1	W_2	W_3	W_4	Re-Capacity	Penalty
F_1	⑤	30	50	10	2	9
F_2	70 ×	30	40	60	9	10
F_3	48	⑧	70	20	18	12
Re-Requirement bj	0	0	7	14		
Penalty			10	10		

Again, let us construct reduced table as

	W_1	W_2	W_3	W_4	Re-Capacity	Penalty
F_1	⑤	30 ×	50	10	2	9
F_2	70 ×	30	40	60	9	10
F_3	48 ×	⑧	70	⑩	10	12
Re-Requirement bj	0	0	7	4		
Penalty		0	10	5		

	W_1	W_2	W_3	W_4	Re-Capacity	Penalty
F_1	⑤	30	50	②	0	
F_2	70 ×	30	7	60 ②	9	10
F_3	48 ×	⑧	70	⑩	10	12
Re-Requirement bj	0	0	17	2		
Penalty			10	5		

	W_1	W_2	W_3	W_4	Re-Capacity	Penalty
F_1	⑤	30 ×	50	②	0	
F_2	70 ×	30 ×	40 ⑦	②	9	
F_3	48 ×	⑧	70	⑩	0	
Re-Requirement bj	0	0	0	2		
Penalty	2					

	W_1	W_2	W_3	W_4	Re-Capacity
F_1	⑤	30 ×	50	②	0
F_2	70 ×	30 ×	⑦	②	0
F_3	48 ×	⑧	70	⑩	0
Re-Requirement bj	0	0	0	0	

Thus, the solution obtained through VAM is as

$x_{11} = 5$, $x_{14} = 2$, $x_{23} = 7$, $x_{24} = 2$, $x_{32} = 8$, $x_{34} = 10$ and another variable is zero.

Example 11. *Solve the following transportation problem by Vogel's approximation method and test its optimality*

From \ To	W_1	W_2	W_3	W_4	Available
F_1	1	2	1	4	30
F_2	3	3	2	1	50
F_3	4	2	5	9	20
Demand	20	40	30	10	

(Bihar SBTE, 2010)

Solution. Here, availability = 30 + 50 + 20 = 100

Demand = 20 + 40 + 30 + 10 = 100

Thus, availability = Demand

Hence the given T.P. is balanced.

Step 1.

From \ To	W_1	W_2	W_3	W_4	Available	Penalty
F_1	1	2	1	4	30	i
F_2	3	3	2	⑩ 1	50	1
F_3	4	2	5	9	20	2
Demand	20	40	30	10		
Penalty	2	1	1	3 ↑		

Here maximum penalty is 3 under fourth column. Minimum entry in the fourth column is 1 in cell (2, 4). Allocate min (10, 50) = 10 in cell (2, 4). Demand of fourth column is complete. We leave this column for further iterations.

Step 2.

From \ To	W_1	W_2	W_3	W_4	Available	Penalty
F_1	⑳ 1	2	1	× 4	30	1
F_2	3	3	2	⑩ 1	40	1
F_3	4	2	5	× 9	20	2
Demand	20	40	30	0		
Penalty	2	1	1			

↑

Here maximum penalty 2 lies in first column as well as third row. But we find minimum entry 1 in first column in (1, 1) cell. Allocate min (20, 30) = 20 in cell (1, 1). Demand of first column is complete.

Step 3.

From \ To	W_1	W_2	W_3	W_4	Available	Penalty
F_1	⑳ 1	2	1	× 4	10	1
F_2	× 3	3	2	⑩ 1	40	1
	×	⑳		×		
Demand	0	40	30	0		
Penalty		1	1			

Here, maximum penalty is 3 along third row. Minimum entry in third row is 2 in cell (3, 2). Allocate min (40, 20) = 20 in cell (3, 2).

Step 4.

From \ To	W_1		W_3	W_4	Available	Penalty
F_1	⑳ 1	⑩	1	× 4	10	1
F_2	× 3		2	⑩ 1	40	1
F_3	× 4	⑳	× 5	× 9	0	–
Demand	0		30	0		
Penalty	–		1	–		

↑

Here, all the penalties are same we can select any one.
Let we choose second column, then minimum entry in this column is 2 in cell (1, 2). Allocate min (20, 10) = 10 in cell (1, 2). Availability of first row is complete.

Step 5.

From \ To	W_1	W_2		W_4	Available	Penalty
F_1	⑳ 1	⑩ 2	×	× 4	0	–
F_2	× 3	3	㉚	⑩ 1	40	1
F_3	× 4	⑳ 2	×	× 9	0	–
Demand	0	10		0		
Penalty		1		–		

↑

Here, maximum penalty is 2 under third column.

Remaining minimum entry in this column is 2 in (2, 3) cell. Allocate min (30, 40) = 30 in this cell.

Step 6.

From \ To	W_1	W_2	W_3	W_4	Available	Penalty
F_1	⑳ 1	⑩ 2	× 1	× 4	0	–
F_2	× 3	⑩ 3	㉚ 2	⑩ 1	10	3
F_3	× 4	⑳ 2	× 5	× 9	0	–
Demand	0	10	0	0		
Penalty	–	1	–	–		

Here, maximum penalty 3 is under second column as well as second row. We can choose any one of them. Let us choose second column. Minimum remaining entry in second column is 3 in cell (2, 2).

Allocate min (10, 10) = 10 in this cell

The final table is as under

Step 7.

From \ To	W_1	W_2	W_3	W_4	Available
F_1	⑳ 1	⑩ 2	× 1	× 4	0
F_2	× 3	⑩ 3	㉚ 2	⑩ 1	0
F_3	× 4	⑳ 2	× 5	× 9	0
Demand	0	0	0	· 0	

Thus, the solution obtained through VAM is as below:

$x_{11} = 20$, $x_{12} = 10$, $x_{22} = 10$, $x_{23} = 30$, $x_{24} = 10$ and $x_{32} = 20$

Now $v_2 + v_j = c_{ij}$ for alloted cell

$v_i + v_1 = 1$, $u_2 + v_2 = 3$, $u_2 + v_4 = 1$

$v_i + v_3 = 1$, $u_2 + v_3 = 2$, $u_3 + v_2 = 2$

Let $u_2 = 0$

$v_2 = 3$, $v_3 = 2$, $v_4 = 1$, $v_1 = 2$

$v_3 = -1$, $u_1 = -1$

Now $w_{ij} = (v_i - v_j) = c_{ij}$ for unallocated cell.

$W_{12} = 0$, $W_{14} = -4$, $W_{21} = -1$, $W_{31} = -3$, $W_{33} = -4$

$W_{31} = -6$

All $W_{ij} \leq 0$

Current solution is optimal solution

Optimal transportation cost = 180.   **Ans.**

Example 12. *Find initial basic feasible solution of the following transportation problem by Vogel's approximation method and find transportation cost.*

	W_1	W_2	W_3	W_4	Capacity
F_1	19	30	50	10	7
F_2	70	30	40	60	9
F_3	48	8	70	20	18
Requirement b_j	5	8	7	14	

(*Bihar SBTE, 2008, 2010*)

Solution.   Here $\Sigma a_i = 7 + 9 + 18 = 34$

$\Sigma b_i = 5 + 8 + 7 + 14 = 34$

$\Rightarrow$ $\Sigma a_i = \Sigma b_i$

Thus rim condition is satisfied, so the given T.P. is balanced.

Step 1.

	W_1	W_2	W_3	W_4	Capacity	Penalty
F_1	19	30	50	10	7	[9]
F_2	70	30	40	60	9	[10]
F_3	48	8	70	20	18	[12]
Requirement b_j	5	8	7	14		
Penalty	[29]	[22]	[10]	[10]		
	↑					

Two least entries of first column are 19 and 48 whose difference (48 – 19) is 29. We put it below first column. Similarly other penalties of second, third and fourth columns are 22, 10 and 10 respectively.

Two least entries of first row are 19 and 10 whose difference (19 – 10) is 9. We put it along first row. Similarly other penalties of second and third rows are 10 and 12 respectively.

Step 2.   Select the row or column for which the penalty is the largest and allocate the maximum feasible amount to that cell with least cost in that particular row or column.

Here 29 is maximum penalty under first column and cell (1, 1) has minimum entry in this column. We allocate in this cell min (5, 7) = 5. The requirement of first column is complete, so we shade this column. We do not consider this column for further iterations.

	W_1	W_2	W_3	W_4	Capacity	Penalty
F_1	19 ⑤	30	50	10	7	9
F_2	70	30	40	60	9	10
F_3	48	8	70	20	18	12
Requirement b_j	5	8	7	14		
Penalty	29	22	10	10		

Step 3. We re-calculate penalties leaving first column.

	W$_1$	W$_2$	W$_3$	W$_4$	Re-Capacity	Penalty
F$_1$	⑤	30	50	10	2	20
F$_2$	70 ×	30	40	60	9	10
F$_3$	48 ×	8 ⑧	70	20	18	12
Re-requirement	0	8	7	14		
Penalty		22 ↑	10	10		

Here we observe that 22 is maximum penalty under second column and cell (3, 2) has minimum entry in this column. We allocate in this cell min (8, 18) = 8. The requirement of second column is complete, so we shade it. We do not consider this column for further iterations.

Step 4. We re-calculate penalties leaving first and second column.

	W$_1$	W$_2$	W$_3$	W$_4$	Re-Capacity	Penalty
F$_1$	⑤	30	50	10 ②	2	40
F$_2$	70 ×	30	40	60	9	20
F$_3$	48 ×	⑧	70	20	10	50
Re-requirement	0	0	7	14		
Penalty			10	10		

Here we observe that 40 is maximum penalty along first row and cell (1, 4) has minimum entry in this row. We allocate in this cell min (14, 2) = 2. The capacity of first row is complete, so we shade it. We do not consider this row for further iterations.

Step 5. We recalculate penalties leaving first second columns and first row:

	W$_1$	W$_2$	W$_3$	W$_4$	Re-Capacity	Penalty
F$_1$	⑤	30 ×	50	10 ②	0	0
F$_2$	70 ×	30	40	60	9	20
F$_3$	48 ×	⑧	70	20 ⑩	10	50
Re-requirement	0	0	7	12		
Penalty	0	0	30	40		

Here, we observe that 50 is maximum penalty along third row and cell (3, 4) has minimum entry in this row. We allocate in this cell min (12, 10) = 10. The capacity of third row is complete, so we shade it. We do not consider this row for further iterations.

Step 6. We re-calculate penalties of remaining rows and columns.

	W$_1$	W$_2$	W$_3$	W$_4$	Re-Capacity	Penalty
F$_1$	⑤	30 ×	50	②	0	
F$_2$	70 ×	30	40	60 ②	9	20
F$_3$	48 ×	⑧	70	⑩	0	
Re-requirement	0	0	7	2		
Penalty			10	60		

↑

We observe that maximum penalty is 60 and fourth column. We allocate min (2, 9) = 2 in (2, 4) cell. The requirement of fourth column is complete. We shade it.

Step 7. Let us construct reduced table as below:

	W$_1$	W$_2$	W$_3$	W$_4$	Re-Capacity	Penalty
F$_1$	⑤	30 ×	50	②	0	
F$_2$	70 ×	30 ×	40 ⑦	②	7	40
F$_3$	48 ×	⑧	70	⑩	0	
Re-requirement	0	0	7	0		
Penalty			30			

Step 8. Final reduced table is as under:

	W$_1$	W$_2$	W$_3$	W$_4$	Re-Capacity
F$_1$	⑤	30 ×	50 ×	②	0
F$_2$	70 ×	30	⑦	②	0
F$_3$	48 ×	⑧	70 ×	⑩	0
Re-requirement	0	0	0	0	

Thus, the solution obtained through VAM is as

$x_{12} = 5$, $x_{14} = 2$, $x_{23} = 7$, $x_{24} = 2$, $x_{32} = 8$, $x_{34} = 10$

Transportation cost = $5 \times 19 + 2 \times 10 + 7 \times 40 + 2 \times 60 + 8 \times 8 + 10 \times 2 \theta = ₹ 799$

Example 13. *Find the initial feasible solution of the following transportation problem by Vogel's approximation method and show whether the solution is optimal or not.*

From＼To	D$_1$	D$_2$	D$_3$	W$_4$	Availability
F$_1$	23	27	16	18	30
F$_2$	12	17	20	51	50
F$_3$	22	28	12	32	53
Requirement	22	35	25	41	

Solution. Consider the above table:

Enter the difference (penalty) between the smallest and the second smallest element in each column and each row.

Find the greatest penalty (14) in the fourth column. 18 is the lowest cost in this column. Allocate min (30, 41) = 30 in this cell. See table given below:

Iteration – I

To From	D_1	D_2	D_3	D_4	Availability	Penalty
F_1	23 ×	27 ×	16 ×	18 ㉚	30	2
F_2	12	17	20	51	40	5
F_3	22	28	12	32	53	10
Requirement	22	35	25	41		
Penalty	10	10	4	14 ↑		

Again find the greatest difference (19) in fourth column. In this column the least cost out of remaining is (32). Allocate min (53, 11) = 11 in this cell of (32). See table given below:

Iteration – II

To From	D_1	D_2	D_3	D_4	Availability	Penalty
F_1	23 ×	27 ×	16 ×	18 ㉚	0	2
F_2	12	17	20	51 ×	40	5
F_3	22	28	12	32 ⑪	53	10
Re-Requirement	22	35	25	11		
Penalty	10	10 -	4	19 ↑		

Find the greatest difference (11) in second column. In this column, the lowest cost is (17). Allocate min (40, 35) = 35 in the cell of (17).

To From	D_1	D_2	D_3	D_4	Availability	Penalty
F_1	23	27	16	18	0	0
F_2	12	17 ㉟	20	51	40	5
F_3	22	28	12	32 ⑪	42	10
Re-requirement	22	35	25	0		
Penalty	10	11 ↑	8	0		

Again find the greatest difference (10) in the third row. The lowest cost out of remaining is 12. Allocate min (42, 25) = 25 in the cell of (12).

Other greatest penalty is 10 in the first column. In this column the least cost is 12. Allocate min (5, 22) = 5 to the cell of 12.

From \ To	D_1	D_2	D_3	D_4	Re-availability	Penalty
F_1	23 ×	27 ×	16 ×	18 ㉚	0	0
F_2	12 ⑤	17 ㉟	20 ×	51	5	8
F_3	22	28 ×	12 ㉕	32 ⑪	42	10 ←
Re-requirement	22	0	25	0		
Penalty	10	0	8	0		

↑

The only cell of fourth row and first column is vacant. It is set by $53 - (25 + 11) = 17$

From \ To	D_1	D_2	D_3	D_4	Re-availability
F_1	23 ×	27 ×	16 ×	18 ㉚	0
F_2	12 ⑤	17 ㉟	20 ×	51 ×	0
F_3	22 ⑰	28 ×	12 ㉕	32 ⑪	0
Re-requirement	0	0	0	0	

Thus, $x_{11} = 0$, $x_{12} = 0$, $x_{13} = 0$, $x_{14} = 30$, $x_{21} = 5$, $x_{22} = 35$, $x_{23} = 0$, $x_{24} = 0$
$x_{31} = 17$, $x_{32} = 0$, $x_{33} = 25$, $x_{34} = 11$

Transport cost $= 18 \times 30 + 12 \times 5 + 17 \times 35 + 22 \times 17 + 12 \times 25 + 32 \times 11 = 2221$ **Ans.**

For Optimal Test

From \ To	D_1	D_2	D_3	D_4	Re-availability	u_i
F_1	23 / −15	27 / −14	16 / −18	18 ㉚	0	−14
F_2	12 ⑤	17 ㉟	20 / −18	51 / −24	0	−10
F_3	22 ⑰	28 / −1	12 ㉕	32 ⑪	0	0
Re-requirement	0	0	0	0		
vj	22	27	12	32		

Here, all $\Delta ij \le 0$.

∴ It is an optimal solution.   **Ans.**

Example 14. *Determine a solution to the following transportation problem using VAM method:*

Source	D_1	D_2	D_3	D_4	Capacity
S_1	20	25	30	40	500
S_2	22	40	35	50	500
S_3	25	20	38	20	500
Demand	300	450	350	400	*(Bihar SBTE, 2014)*

Solution.

Source	D_1	D_2	D_3	D_4	Capacity	Penalty
S_1	20	25	30	40	500	5
S_2	⟨200⟩ 22	40	35	50	500	13
S_3	25	20	38	⟨400⟩ 20	500	5
Demand	300	450	350	400		
Penalty	2	5	5	20		

TABLE–II

Source	D_1	D_2	D_3	D_4	Capacity	Penalty
S_1	20	25	30	40	500	5
S_2	⟨200⟩ 22	40	35	50	500	13
S_3	25	20	38	⟨400⟩	100	5
Demand	300	450	350	0		
Penalty	2	5	5			

TABLE–III

Source	D_1	D_2	D_3	D_4	Capacity	Penalty
S_1	20	25	⟨350⟩ 30	40	500	5
S_2	⟨300⟩	40	35	50	200	0
S_3	25	20	38	⟨400⟩	100	5
Demand	0	450	350	0		
Penalty		5	5			

TABLE–IV

Source	D_1	D_2	D_3	D_4	Capacity	Penalty
S_1	20	25	⟨350⟩	40	150	5
S_2	⟨300⟩ 22	40	35	50	200	0
S_3	25	⟨100⟩	38	⟨400⟩	100	5
Demand	0	450	0	0		
Penalty		5				

TABLE–V

Source	D_1	D_2	D_3	D_4	Capacity	Penalty
S_1	20	⟨150⟩ 25	⟨350⟩	40	150	5
S_2	⟨300⟩	⟨200⟩ 40	35	50	200	5
S_3	25	⟨100⟩	38	⟨400⟩	0	
Demand	0	350	0			
Penalty		5				

TABLE–VI

Source	D_1	D_2	D_3	D_4	Capacity	
S_1		⑮⓪ 25	㉟⓪ 30		0	
S_2	③⓪⓪ 22	②⓪⓪ 40			0	
S_3		①⓪⓪ 20		④⓪⓪ 20	0	
Demand	0	0		0		

$x_{12} = 150$, $x_{14} = 350$, $x_{21} = 300$, $x_{22} = 200$, $x_{32} = 100$, $x_{34} = 400$

Least transportation cost

$= 150 \times 25 + 350 \times 30 + 300 \times 22 + 200 \times 40 + 100 \times 20 + 400 \times 20$

$= 3750 + 10500 + 6600 + 8000 + 2000 + 8000$

$= 38850$ **Ans.**

Example 15. *Solve the following transportation problem by Vogel's approximation method and test its optimility.*

From \ To	D_1	D_2	D_3	Supply
01	2	7	4	5
02	3	3	1	8
03	5	4	7	7
04	1	6	2	14
Demand	7	9	18	

(*Bihar SBTE, 2009*)

Table 1

Solution.

From \ To	D_1	D_2	D_3	Supply	
01	⑤ 2	× 7	× 4	5	(2)
02	× 3	⑧ 3	⑥ 1	8	2
03	× 5	① 4	× 7	7	1
04	② 1	× 6	⑫ 2	14	1
Demand	7	① 9	18	34	
Penalty	1	1	1		
	(2)	1	1		
	–	1	1		
	–	1	6	Maximum penalty	

First row : $4 - 2 = 2$

Second row : $3 - 1 = 2$

Third row : $5 - 4 = 1$

Fourth row : $2 - 1 = 1$

First column : $2 - 1 = 1$

Second column $4 - 3 = 1$

Third column: $2 - 1 = 1$

Minimum transportation cost

$$= 5 \times 2 + 2 \times 3 + 6 \times 1 + .7 \times 4 + 2 \times 1 + 12 \times 2 = 10 + 6 + 6 + 28 + 2 + 24 = 76$$

No. of allocation $= m + n - 1 = 3 + 4 - 1 = 6$

For test of optimality:

If each $w_{ij} \leq 0$ then it proves optimality test.

Table 2

		v_1 D_1	v_2 D_2	v_3 D_3	Supply
U_1	01	⑤ 2	7	4	5
U_2	02	3	② 3	1	8
U_3	03	5	⑦ 4	7	7
U_4	04	② 1	6	2	14
	Demand	7	9	18	34

For filled cell:

Let $U_4 = 0$

$u_1 + v_1 = 2$ $v_1 = 1$

$u_2 + v_2 = 3$ $v_3 = 2$

$u_3 + v_3 = 1$ $u_1 = 1$

$u_3 + v_2 = 4$ $u_2 = -1$

$u_4 + v_1 = 1$ $v_2 = 4$

$u_4 + v_3 = 2$ $u_3 = 0$

For unfilled cell:

$W_{ij} = U_i + V_j - C_{ij}$

$W_{12} = u_1 + v_2 - C_{12} = 1 + 4 - 7 = -2$

$W_{13} = u_1 + v_3 - C_{13} = 1 + 2 - 4 = -1$

$W_{21} = u_2 + v_1 - C_{21} = -1 + 1 - 3 = -3$

$W_{31} = u_3 + v_1 - C_{31} = 0 + 1 - 5 = -4$

$W_{33} = u_3 + v_3 - C_{33} = 0 + 2 - 7 = -5$

$W_{42} = u_4 + v_2 - C_{42} = 0 + 4 - 6 = -2$

$\therefore$ Each $W_{ij} \leq 0$

So it gives an optimal solution.

Hence, it proves the optimality test.   **Proved.**

Example 16. *Find the total transportation cost for the following table by Vogel's Approximation Method.*

Source	D_1	D_2	D_3	D_4	Capacity
S_1	8	9	7	5	100
S_2	6	4	8	7	150
S_3	9	10	7	6	150
S_4	5	8	4	6	200
Demand	175	100	125	200	

(Bihar SBTE, 2011)

Solution.

	D_1	D_2	D_3	D_4	Capacity	Penalty
S_1	8	9	7	5	100	2
S_2	6	4 (100)	8	7	150	2
S_3	9	10	7	6	150	1
S_4	5	8	4	6	200	1
Demand	175	100	125	200		
Penalty	1	4	3	1		

↑

	D_1	D_3	D_4	Capacity	Penalty
S_1	8	7	5	100	2
S_2	6	8	7	50	1
S_3	9	7	6	150	1
S_4	5	4 (125)	6	200	1
Demand	175	125	200		
Penalty	1	3	1		

↑

	D_1	D_4	Capacity	Penalty
S_1	8	5 (100)	100	2
S_2	6	7	50	1
S_3	9	6	150	1
S_4	5	6	75	1
Demand	175	100		
Penalty	1	1		

	D_1	D_4	Capacity	Penalty
S_2	6	7	50	1
S_3	9	6 (100)	150	1
S_4	5	6	75	1
Demand	175	100		
Penalty	①	1		

	D$_1$		Capacity	Penalty
S$_2$	6	(50)	50	1
S$_3$	9	(50)	50	1
S$_4$	5	(75)	75	1
Demand	175			
Penalty	1			

Minimum Transportation cost

$$= 4 \times 100 + 4 \times 25 + 5 \times 100 + 6 \times 100 + 6 \times 500 + 9 \times 50 + 5 \times 75$$
$$= 400 + 500 + 500 + 600 + 300 + 450 + 375 = 3125$$

EXERCISE 17.3

1. Solve the following problem using Vagel's method.

						Supply	
5	10	15	8	9	7	30	S$_1$
14	13	10	9	20	21	40	S$_2$
15	11	13	25	8	12	10	S$_3$
9	19	12	8	6	13	100	S$_4$
Demand							
50	20	10	35	15	50		
D$_1$	D$_2$	D$_3$	D$_4$	D$_5$	D$_6$		

2. Solve the following problem by Vogel's approximation method.

4	3	1	3	S$_1$
2	7	5	3	S$_2$
Demand				
2	3	1		
D$_1$	D$_2$	D$_3$		

Ans. $x_{12} = 2$, $x_{13} = 1$, $x_{21} = 2$, $x_{22} = 1$, Z min = ₹ 18

3. Find the initial basic feasible solution by V.A.M.

				Supply	
19	30	50	10	7	S$_1$
70	30	40	60	9	S$_2$
40	8	70	20	18	S$_3$
Requirement					
5	8	7	14	(34)	Total
D$_1$	D$_2$	D$_3$	D$_4$		

Ans. $x_{11} = 5$, $x_{14} = 2$, $x_{23} = 7$, $x_{24} = 2$, $x_{32} = 8$, $x_{34} = 10$

4. Find an initial basic feasible solution to the following T.P. using VAM

					Supply	
3	5	8	9	11	20	S$_1$
5	4	10	7	10	40	S$_2$
2	3	8	7	7	30	S$_3$
Demand						
10	15	25	30	40		
D$_1$	D$_2$	D$_3$	D$_4$	D$_2$		

Ans. $x_{11} = 10$, $x_{12} = 5$, $x_{13} = 5$, $x_{22} = 10$, $x_{24} = 30$, $x_{35} = 30$, $x_{DM'3} = 20$, $x_{DM'5} = 10$, Min Z = 555

Choose the correct answer:

5.   The penalty of each row used in VAM is found by ....
      (*a*) sum of two unit cost   (*b*) difference of two unit cost
      (*c*) difference of least and next to least unit cost   (*d*) none of these (*Bihar SBTE, 2005, 2011*) **Ans.** (*c*)

17.7 TEST OF OPTIMALITY BY U-V METHOD (MODI METHOD)

$U - V$ method is an indirect method which is much simpler compared to simplex method. A matrix of order $m \times n$ is formed in all transportation problems.

The number of equation $= m + n - 1$

The number of unknowns $= m + n$

The number of equations $<$ The number of variables

Let ith source be U_i and jth destination be V_j and the cost of transportation between U_i source and V_j destination $= C_{ij}$

A table of non basic cells are filled by C_{ij} which is the cost between U_i source and V_j destination.

Working Rule

Step 1. Prepare a table by filling initial basic feasible solution.

Step 2. From the above matrix develope the value of $U - V$ satisfying the solution $U_i + V_j = C_{ij}$ for all basic cells. In the beginning we assume $U_1 = 0$.

Step 3. Compute the water values for all non basic cells using the relation $U_i + V_j = C_{ij}$. If all water values are negative, we get the optimal solution.

Step 4. Find out the most positive water value then term a closed loop with such a cell whose value is maximum positive as one of the corners; while all other corners are basic cells. Select the quantity and transfer this to the cell which had got maximum positive water value. If optimal solution is not obtained. Go to step 2 till the optimal solution is obtained.

Example 13. *A contractor Amarpali has three sets of heavy construction equipment available at Mumbai and Bengaluru. He has construction project in Chennai, Jaipur and Delhi that require 2, 3 and 1 sets of equipment respectively. The unit shipping cost C_{ij} per set between the source i and distribution j are given below.*

Mumbai and Chennai $C_{11} = 4$

Mumbai and Delhi $C_{12} = 3$

Mumbai and Jaipur $= C_{13} = 1$

Bengaluru and Chennai $C_{21} = 2$

Bengaluru and Delhi $C_{22} = 7$

Bengaluru and Jaipur $C_{23} = 5$

Find the total minimum shipping cost by using $U - V$ method.

Solution.

	Chennai	Delhi	Jaipur	Capacity
Mumbai	4	3	1	3
Bengaluru	2	7	5	3
Required	2	3	1	6

Step 1. The total capacity at Mumbai and Bengaluru = 6 units.

The requirement at Chennai, Delhi and Jaipur = 2 + 3 + 1 = 6 units.

Therefore, the requirement = total available capacity.

Hence, the problem is a balanced transportation problem.

By the North-West Corner Rule we form the following table:

	Chennai		Delhi		Jaipur		Capacity
Mumbai	4	②	3	①	1	×	3
Bengaluru	2	×	7	②	5	①	3
Required	2		3		1		6

Transportation cost = $4 \times 2 + 3 \times 1 + 7 \times 2 + 5 \times 1 = 8 + 3 + 14 + 5 = 30$

Step 2. To test whether the solution is optimal or not we calculate the water values for all non basic and basic cells by the formula $U_i + V_j = C_{ij}$

Here the number of basic cell = $m + n - 1 = 2 + 3 - 1 = 4$

Hence the rule for $m + n - 1$ is satisfied, therefore we have a non degenerate basic solution.

Now we apply the formula $U_i + V_j = C_{ij}$ for all basic cells.

$U_1 + V_1 = 4,$ $U_1 + V_2 = 3,$ $U_2 + V_2 = 7,$ $U_2 + V_3 = 5$

In this way we have four equations.

The number of unknowns are five i.e., U_1, U_2, V_1, V_2 and V_3.

Let $U_1 = 0$, then $V_1 = 2, V_2 = 3, U_2 = 4, V_3 = 1$ ... (1)

The values of all variables are found and a table is formed below:

	Chennai		Delhi		Jaipur		Capacity	U
Mumbai	4	②	3	①	1		3	0
Bengaluru	2		7	②	5	①	3	4
Required	2		3		1		6	
V	2		3		1			

Water value for Mumbai – Jaipur = $U_1 + V_3 - C_{13} = 0 + 1 - 1 = 0$ [Using (1)

Water value for Bengaluru – Chennai = $U_2 + V_1 - C_{21} = 4 + 2 - 2 = 4$

Step 3. Select the non basic cell which has maximum positive water value. Here non basic cell is Bengaluru to Chennai. This cell is connected to all basic cells.

(Bengaluru – Chennai) – (Mumbai – Delhi) + (Bengaluru – Delhi) – (Bengaluru – Chennai) = 0

We know that allotment in Mumbai – Chennai = 2 units and the allotment in Bengaluru – Delhi = 2

Minimum (2, 2) 2 which is for Mumbai – Chennai and Bengaluru – Delhi.

Hence, 2 units are alloted for Bengaluru to Chennai and the other changes are made accordingly.

	Chennai		Delhi		Jaipur		Capacity	U
Mumbai	4		3	③	1		3	0
Bengaluru ε	2	②	7	ε	5	①	3 + ε	4
Required ε		2		3 + ε		1	6 + ε	
V	– 2		3		1			

Again we test whether the solution is optimal or not by calculating water values for non basic cells by the formula $U_i + V_j = C_{ij}$

Here, the relation $U_i + V_j = C_{ij}$ is satisfied for all basic cells from the above table.

Number of basic cells = 3

But by the rule $m + n - 1 = 2 + 3 - 1 = 4$.

Hence the rule of m + n – 1 is not satisfied here such a situation is called degeneracy. Degeneracy is to be removed.

The following equations are for all basic cells

$$U_1 + V_2 = 3$$
$$U_2 + V_1 = 2$$
$$U_2 + V_3 = 5$$

There are three equations but the number of unknowns are 5 i.e.; U_1, U_2, V_1, V_2, V_3.

In order to solve these equations we put $U_1 = 0$ it gives $V_2 = 3$, with the help of this $V_2 = 3$ other variables could not be evaluated.

Here we convert one more non basic cell as a basic cell out of Mumbai – Chennai, Mumbai – Jaipur or Bengaluru – Delhi.

Arbitrarily we select Bengaluru – Delhi to be a basic cell and allot E units. From this we get $U_2 + V_2 = 7$.

Now, we have

$U_1 + V_2 = 3$, $U_2 + V_1 = 2$, $U_2 + V_3 = 5$ and $U_2 + V_2 = 7$ but we know $U_1 = 0$ it gives $V_2 = 3$

By putting $V_2 = 3$ in $U_2 + V_2 = 7$, $\Rightarrow$ $U_2 = 4$

By putting $U_2 = 4$ in $U_2 + V_3 = 5$, we get $V_3 = 1$

By putting $U_2 = 4$ in $U_2 + V_1 = 2$, we get $V_1 = -2$

Water values for Mumbai – Chennai = $U_1 + V_1 - C_{12} = 0 - 2 - 4 = - 6$

Water values for Mumbai – Jaipur = $U_1 + V_3 - C_{13} = 0 + 1 - 1 = 0$

Since all water values are negative or zero.

The optimal solution is obtained.

Now we put $\varepsilon = 0$.

Hence, the solution is from Mumbai – Delhi is equal to 3 sets

From Bengaluru to Chennai = 2 sets

From Bengaluru to Jaipur = 1 set

Total cost = $3 \times 3 + 2 \times 2 + 5 \times 1 = ₹ 18$ **Ans.**

In this way transportation cost is reduced from ₹ 30 to ₹ 18. So we save ₹ (30 – 18) = ₹ 12.

Objective Tpe Questions

1.   The penalty of each row, used in VAM is found by:

(a) Sum of two unit cost

(b) Difference of two unit cost

(c) Difference of least and next to least unit cost

(d) None of these   (*Bihar SBTE, 2010*) **Ans.** (*c*)

2.   The sum of basic cell and no basic cell in a m × n transportation problem is always equal to:

(a) m + n   (b) m.n.

(c) m.n – 1   (d) m + n – 1   (*Bihar SBTE, 2009*) **Ans.** (d)

18 Assignment Problem

18.1 ASSIGNMENT PROBLEM

In a publishing house, there are different jobs: for example;

Composing, printing, binding, selling, accounting. Different persons have different degree of proficiency to accomplish different works. Proficiency is not the same for all the persons performing a special job. The general manager of the organisation can wish to assign each of the work job to one agent in such a manner that the effective performance of the organisation at the whole be optimize. This is know as assignment problem.

18.2 ASSIGNMENT MODEL

An assignment problem concern to what happen to the effectiveness function when we associate each a number of origins. With each of the same number of destination each source or facility (origin) is to be associated with one and only one job (destination) and association are to be made in such a way so as to maximize or (minimize) the total effectiveness. The objective is to determine how all the assignment should be made in order to minimize total cost. The general assignment model with n workers and n-jobs.

Let C_i is the total cost of ith man performing the ith job. Now we define a variable x_{ij} such that

$$x_i = \begin{cases} 0, & \text{if the man is not assign the } i\text{th job} \\ 1, & \text{if the man is assigned the } j\text{th job} \end{cases}$$

Determine x_i

which minimize $\displaystyle\sum_{i=1}^{n} \sum_{j=1}^{n} c_{ij}\, x_{ij}$

$$\sum_{i=1}^{n} x_{ij} = 1 \quad \text{and} \quad \sum_{i=1}^{n} x_{ij} = 1$$

$$x_{ij} \geq 0$$

The assignment model is actually a special case of the transportation model. In which the workers represent the source and job represents the destination. The supply (demand) amount at each source (destination) exactly equals to 1. The cost of transporting from worker i to job j is c_{ij}. As all the supply and demand amounts equal to 1. This technique has led to development of a simple solution called a Hungarian method.

18.3 DIFFERENCE BETWEEN TRANSPORTATION PROBLEM AND ASSIGNMENT PROBLEM

(Bihar SBTE, 2008, 2009)

Transportation Problem	Assignment problem
1. No. of sources and number of destinations need not be equal. Hence the cost matrix is not necessarily a square matrix.	1. Since assignment is done on one to one basis, the number of sources and the number of destinations are equal. Hence, the cost matrix must be a square matrix.
2. x_{ij} the quantity to be transported from ith origin to jth destination can take any possible positive values, and satisfies the rim requirements.	2. x_{ij} the jth job is to be assigned to the ith person and can take either the value 1 or zero.
3. The capacity and the requirement value is equal to a_j and b_j for the ith source & jth destinations. $(i = 1, 2.... m; j = 1, 2, ...n)$	3. The capacity and the requirement value is exactly one i.e. for each source of each destination the capacity and the requirement value is exactly one.
4. The problem is unbalanced if the total supply and total demand are not equal.	4. The problem is unbalanced if the cost matrix is not a square matrix.

Example. *Write down difference between assignment model and transportation model.* (*Bihar SBTE, 2008*)

Solution. The assignment model is a special type of transportation mode where the resources are being allocated to the activities on a "one to one basis".

18.4 WORKING RULE

(Hungarian's Method)

Step 1. (a) Subtract the minimum cost of each row from all the elements of the respective row.

(b) From the resulting matrix, subtract the minimum cost of each column from all the elements of the respective column.

Step 2. (a) Row assignment

Choose the row containing single zero and draw square around it and draw a line in the column containing squared zero.

(b) Column Assignment

Choose the column containing single zero and make a square around it and draw a line in the row containing squared zero.

Step 3. Choose the smallest uncrossed element and subtract from all remaining uncrossed elements and add this smallest element to all the elements at the cross-section of two lines. Repeat the above process.

Step 4. If the number of lines drawn is not equal to number of rows repeat the above method. If the number of lines is equal to number of rows stop and get a optimal solution.

Example 1. *There are three machines and three jobs. The associated cost matrix is given below:*

	M_1	M_2	M_3
J_1	470	900	550
J_2	500	750	450
J_3	650	200	950

How should the jobs be allocated on, one to one basis, so as to minimize the total machine cost.

Solution. Step 1. Minimum element of the first row is 470. Subtract 470 from all elements of the first row.

Subtract minimum element 450 from each elements of second row.

Subtract minimum element 200 from each elements of third row.

Now matrix is as under

	M_1	M_2	M_3
J_1	$470 - 470 = 0$	$900 - 470 = 430$	$550 - 470 = 80$
J_2	$500 - 450 = 50$	$750 - 450 = 300$	$450 - 450 = 0$
J_3	$650 - 200 = 450$	$200 - 200 = 0$	$950 - 200 = 750$

Step 2. Since the smallest element in each column is zero, so the matrix remains unchanged.

Step 3. In the first row there is a single zero, make a square around it and draw a line through the column which contains the squared zero. Similarly make all the other assignments.

	M_1	M_2	M_3
J_1	[0]	430	80
J_2	50	300	[0]
J_3	450	[0]	750

Since there are squared zeroes in all the rows as well as columns. So, there is no need for going into column assignment.

We see that all the assignment are made at zero level. So, the solution is optimal. Thus the optimal solution is $J_1 \rightarrow M_1$, $J_2 \rightarrow M_3$, $J_3 \rightarrow M_2$ and the cost of assignment is $470 + 450 + 200 = 1120$ **Ans.**

Example 2. *Solve the following minimal assignment problem*

Jobs \ Men	I	II	III	IV
1	12	30	21	15
2	18	33	9	31
3	44	25	24	21
4	23	30	28	4

(Bihar SBTE, 2004)

Solution. Step 1. Subtract the smallest elements 12, 9, 21, 4 from first, second, third and fourth rows respectively, we get

Iteration – 1

Jobs \ Men	I	II	III	IV
1	0	18	9	3
2	9	24	0	22
3	23	4	3	0
4	19	26	24	0

Step 2. Subtract the smallest element 4 from second column. Since all other columns have zero so they need no subtraction.

Iteration – 2

Jobs \ Men	I	II	III	IV
1	0	14	9	3
2	9	20	0	22
3	23	0	3	0
4	19	22	24	0

Step 3. Row Assignment

In the first row there is a single zero in first column. Square around zero. Draw a line through first column.

In the second row there is a single zero in third column. Square around the zero. Draw a line through third column.

In the fourth row there is a single zero in fourth column. Square around zero. Draw a line through fourth column.

Column assignment. In the second column there is a single zero in third row. Square around zero. Draw a line through third row.

Iteration – 3

Jobs \ Men	I	II	III	IV
1	[0]	14	9	3
2	9	20	[0]	22
3	23	[0]	3	0
4	19	22	24	[0]

Step 4. Here we observe that,

Number of rows = Number of drawn lines.

Hence the solution is optimal.

Minimal assignment 1 – I,   2 – III,   3 – II,   4 – IV   **Ans.**

Example 3. *There are five jobs to be performed by five persons in an office. The cost matrix for these jobs is as under:*

	P_1	P_2	P_3	P_4	P_5
J_1	8	7	4	11	6
J_2	10	5	5	13	7
J_3	6	9	8	7	12
J_4	6	7	2	3	2
J_5	7	8	8	10	5

Carry out assignment for minimum cost.

Solution. Step 1. Subtract the minimum element 4 from all elements of J_1 row.

Subtract the minimum element 5 from all elements of J_2 row.

Subtract the minimum element 6 from all elements of J_3 row.

Subtract the minimum element 2 from all elements of J_4 row.

Subtract the minimum element 5 from all elements of J_5 row.

As shown in the following table.

Row Assignment.

	P_1	P_2	P_3	P_4	P_5
J_1	$8 - 4 = 4$	$7 - 4 = 3$	$4 - 4 = 0$	$11 - 4 = 7$	$6 - 4 = 2$
J_2	$10 - 5 = 5$	$5 - 5 = 0$	$5 - 5 = 0$	$13 - 5 = 8$	$7 - 5 = 2$
J_3	$6 - 6 = 0$	$9 - 6 = 3$	$8 - 6 = 2$	$7 - 6 = 1$	$12 - 6 = 6$
J_4	$6 - 2 = 4$	$7 - 2 = 5$	$2 - 2 = 0$	$3 - 2 = 1$	$2 - 2 = 0$
J_5	$7 - 5 = 2$	$8 - 5 = 3$	$8 - 5 = 3$	$10 - 5 = 5$	$5 - 5 = 0$

Column Assignment.

There is no zero in the fourth column so we subtract the smallest element 1 from the remaining elements of 4th column.

	P_1	P_2	P_3	P_4	P_5
J_1	4	3	0	6	2
J_2	5	0	0	7	2
J_3	0	3	2	0	6
J_4	4	5	0	0	0
J_5	2	3	3	4	0

Row Assignment.

In the first row there is a single zero in the third column. Square around it and draw a line in the third column.

In the second row there is a single zero in the second column. Square around it and draw a line in the second column.

In the fourth row there is a single zero in the fourth column. Square around it and draw a line in the fourth column.

Column Assignment.

In the first column there is a single zero in the third row square around it and draw a line in the third row.

	P_1	P_2	P_3	P_4	P_5
J_1	4	3	[0]	6	2
J_2	5	[0]	0	7	2
J_3	[0]	3	2	0	6
J_4	4	5	0	[0]	0
J_5	2	3	3	4	[0]

Minimal assignment

$$J_1 \to P_3, \; J_2 \to P_2, \; J_3 \to P_1, \; J_4 \to P_4, \; J_5 \to P_5,$$

Total cost $= 4 + 5 + 6 + 3 + 5 = 23$ **Ans.**

18.5 MODIFIED MATRIX

We have discussed three steps to solve an assignment problem it may happen that complete assignment are not made and unmarked zero are not left at all. In this case we apply the following steps.

Step 1. Choose the smallest uncrossed element and subtract from the remaining uncrossed elements.

Add the smallest element to the elements at the intersecting points of two lines.

The reduced matrix is treated as above.

If the lines drawn are less than the number of rows continue.

If the number of lines drawn is equal to the number of rows stop. Then find out the minimum cost.

Example 4. *Find optimal solution for the assignment problem with the following cost matrix:*

	1	2	3	4
A	5	6	1	8
B	7	9	2	6
C	6	4	5	7
D	5	7	7	6

(Bihar SBTE, 2009)

Solution. Step 1. In the given table the cost of worker with machine are as follows:

	1	2	3	4
A	5	6	1	8
B	7	9	2	6
C	6	4	5	7
D	5	7	7	6

(*i*) Subtracting the minimum element of each row from all elements of that row, we get

	1	2	3	4
A	$5 - 1 = 4$	$6 - 1 = 5$	$1 - 1 = 0$	$8 - 1 = 7$
B	$7 - 2 = 5$	$9 - 2 = 7$	$2 - 2 = 0$	$6 - 2 = 4$
C	$6 - 4 = 2$	$4 - 4 = 0$	$5 - 4 = 1$	$7 - 4 = 3$
D	$5 - 5 = 0$	$7 - 5 = 2$	$7 - 5 = 2$	$6 - 5 = 1$

(*ii*) Subtracting the minimum element of each column from all elements of that column, we get

	1	2	3	4
A	$4 - 0 = 4$	$5 - 0 = 5$	$0 - 0 = 0$	$7 - 1 = 6$
B	$5 - 0 = 5$	$7 - 0 = 7$	$0 - 0 = 0$	$4 - 1 = 3$
C	$2 - 0 = 2$	$0 - 0 = 0$	$1 - 0 = 1$	$3 - 1 = 2$
D	$0 - 0 = 0$	$2 - 0 = 2$	$2 - 0 = 2$	$1 - 1 = 0$

Step 2. Row assignment. In the first row there is an independent zero in the third column. Square around zero and cancel the third column by dotted line. In third row there is a single zero. We assign it making square around it and canceled the second column.

Column assignment.

There is a single zero in the first column. We assign it making square around it and canceled the row (fourth row) containing the assigned zero. The resulting matrix is given below:

	1	2	3	4
A	4	5	[0]	6
B	5	7	0	3
C	2	[0]	1	2
D	·0·	2	2	0

Step 3. (a) Choose the smallest element in the table not covered by straight lines and subtract this element from the remaining uncrossed elements not having straight line through them.

(b) Add this smallest element to all elements lying at the intersection of any two lines.

The resulting matrix is given below:

4 − 2 = 2	5	0	6 − 2 = 4
5 − 2 = 3	7	0	3 − 2 = 1
2 − 2 = 0	0	1	2 − 2 = 0
0	2 + 2 = 4	2 + 2 = 4	0

Step 4. Row assignment. In the first row there is a single zero in the third column. Square around zero and draw a line through third column.

Column assignment. In the second column there is a single zero in third row square around zero and third row is canceled by a line.

In the fourth column there is a single zero a single in the fourth row. Square around it and fourth row is canceled by a line.

The resulting matrix is given below:

2	5	[0]	4
3	7	0	1
0	[0]	1	0
0	4	4	[0]

Step 5. Choose the smallest element 1 not covered by straight lines and subtract this element from the remaining uncrossed elements.

Add this smallest element to all the elements at the intersection of two lines. See the table below:

2 − 1 = 1	5 − 1 = 4	0	4 − 1 = 3
3 − 1 = 2	7 − 1 = 6	0	1 − 1 = 0
0	0	1 + 1 = 2	0
0	4	4 + 1 = 5	0

Step 6. Row assignment. In the first row there is a single zero in the third column so cancel the third column by dotted line.

In the second row there is a single zero in fourth column. Square around the zero and draw a dotted line through fourth column.

	1	2	3	4
A	1	4	[0]	3
B	2	6	0	[0]
C	0	[0]	2	0
D	[0]	4	5	0

Minimal amount: $A \to 3, B \to 4, C \to 2, D \to 1$

Total cost = $1 + 6 + 4 + 5 = 16$ **Ans.**

Example 5. *Four jobs are to be done on four different machines. The set up and production times are prohibitively high for change over. Table given below indicates the cost of producing job i on machine j.*

Job	Machine			
	M_1	M_2	M_3	M_4
J_1	5	7	11	6
J_2	8	5	9	6
J_3	4	7	10	7
J_4	10	4	8	3

Assign jobs to different machines so that the total cost is minimized. *(Bihar SBTE, 2010)*

Solution.

Job	M_1	M_2	M_3	M_4
J_1	5	7	11	6
J_2	8	5	9	6
J_3	4	7	10	7
J_4	10	4	8	3

Step 1. Subtract the smallest element 5 from first row.

Subtract the smallest element 5 from second row

Subtract the smallest element 4 from third row.

Subtract the smallest element 3 from fourth row.

	M_1	M_2	M_3	M_4
J_1	0	2	6	1
J_2	3	0	4	1
J_3	0	3	6	3
J_4	7	1	5	0

Step 2. Subtract the smallest element 4 from third column as this column has no zero.

	M_1	M_2	M_3	M_4
J_1	0	2	2	1
J_2	3	0	0	1
J_3	0	3	2	3
J_4	7	1	1	0

Step 3. Row Assignment. In the first row there is a single zero square around it and draw a line in the first column.

In the fourth row there is a single zero in the fourth column. Square around it and draw a line in the fourth column.

Column Assignment

In the second column there is a single zero in the second row square around it and draw a line in second row.

0	2	2	1
3	0	0	1
0	3	2	3
7	1	1	0

Step 4. The uncrossed smallest element 1 is subtracted from the remaining uncrossed elements.

Add the smallest element 1 at the intersection points of the two lines.

Reduced matrix is as follows:

0	2	1	1
4	1	0	2
0	2	1	3
7	0	0	0

Step 5. In the first row there is a single zero in first column. Square around it and draw a line through first column. Similarly make other assignment

0	1	1	1
4	0	0	2
0	2	1	3
7	0	0	0

Step 6. The uncrossed smallest element 1 is subtracted from the remaining uncrossed elements. Add the smallest element 1 at the intersection point of the two lines.

Reduced matrix is as under:

0	0	0	0
5	0	0	2
0	1	0	2
8	0	0	0

Step 7. We make assignment arbitrarily in any zero and cross the remaining zeroes in the same column as well as the same row.

0	0	0	0
5	0	0	2
0	1	0	2
8	0	0	0

As there is assignment in each row and in each column, optimal assignment can be made in the current solution. Hence optimal assignment policy is

$$J_1 \rightarrow M_1; \qquad J_2 \rightarrow M_2; \quad J_3 \rightarrow M_3; \qquad J_4 \rightarrow M_4$$

and optimal cost = ₹ (5 + 5 + 10 + 3) = ₹ 23   **Ans.**

Example 6. *Find the complete assignment of the following cost matrix for minimum cost:*

	P	Q	R	S	T
A	2	6	4	5	8
B	1	7	1	3	5
C	1	5	4	2	8
D	6	3	8	7	7
E	5	5	6	1	2

(Bihar SBTE, 2006)

Solution. Step 1. Subtracting the smallest elements 2, 1, 1, 3, 1 from first, second, third, fourth and fifth rows respectively, we get

Iteration – I

	P	Q	R	S	T
A	0	4	2	3	6
B	0	6	0	2	4
C	0	4	3	1	7
D	3	0	5	4	4
E	4	4	5	0	1

Step 2.   Subtracting smallest element 1, from fifth column, we get

Iteration – II

	P	Q	R	S	T
A	0	4	2	3	5
B	0	6	0	2	3
C	0	4	3	1	6
D	3	0	5	4	3
E	4	4	5	0	0

Step 3. Row assignment.

In the first row there is a single zero in first column. Square around it and draw a line through first column.

Column assignment

In the second column there is a single zero in fourth row. Square around zero. Fourth row is canceled by a line.

In the third column there is a single zero in the second row. Square around zero. Second row is canceled by a line.

In the fourth column there is a single zero in the fifth row. Square around zero. Fifth row is canceled by a line.

Iteration – III

	P	Q	R	S	T
A	[0]	4	2	3	5
B	0	6	[0]	2	3
C	0	4	3	1	6
D	3	[0]	5	4	3
E	4	4	5	[0]	0

Step 4. Choose the smallest element 1 not covered by straight lines and subtract this element from the remaining uncrossed elements.

Add this smallest element to all the elements at the intersection of two lines. See the table below:

Iteration – IV

	P	Q	R	S	T
A	0	3	1	2	4
B	1	6	0	2	3
C	0	3	2	0	5
D	4	0	5	4	3
E	5	4	5	0	0

Step 5. Row assignment

In the first row there is a single zero in first column. Square around it and draw a line through first column.

In the second row there is a single zero in third column. Square around zero and draw a line through third column.

In the fourth row there is a single zero in the second column. Square around zero. Draw a line through second column.

Column assignment

In the fifth column there is a single zero in fifth row. Square around zero and draw a line through fifth row.

In the fourth column these is a single zero in third row. Square around zero and draw a line through third row.

Iteration – V

	P	Q	R	S	T
A	[0]	3	1	2	4
B	1	6	[0]	2	3
C	0	3	2	[0]	5
D	4	[0]	5	4	3
E	5	4	5	0	[0]

Step 6. We see that all the assignment are made at zero level. So the solution is optimal.

Thus, the optimal solution is

$$A \to P, \quad B \to R, \quad C \to S, \quad D \to Q, \quad E \to T$$

and the cost of assignment is

$$= C_{11} + C_{23} + C_{34} + C_{42} + C_{55} = 2 + 1 + 2 + 3 + 2 = 10 \qquad \textbf{Ans.}$$

Example 7. *Assign five operators to five machines so that total cost is minimized. The assignment cost are given below:*

Operators

	I	II	III	IV	V
A	10	5	13	15	16
B	3	9	18	3	6
C	10	7	2	2	2
D	5	11	9	7	12
E	7	9	10	4	12

(Bihar SBTE, 2008)

Solution. Step 1. Subtracting the smallest elements 5, 3, 2, 5 and 4 from first, second, third, fourth and fifth rows respectively, we get

	I	II	III	IV	V
A	5	0	8	10	11
B	0	6	15	0	3
C	8	5	0	0	0
D	0	6	4	2	7
E	3	5	6	0	8

Step 2. Since every column has zero in it, so there is no need for column operation. The table remains same.

Step 3. Row assignment

In the first row there is a single zero in second column. Square around zero. Draw a straight line through second column.

In the fourth row there is a single zero in first column. Square around zero. Draw a line through first column.

In the fifth row there is a single zero in fourth column. Square around zero. Draw a line through fourth column.

Column assignment. In the third column there is a single zero in third row. Square around zero. Draw a line through third row.

	I	II	III	IV	V
A	5	[0]	8	10	11
B	0	6	15	0	3
C	8	5	[0]	0	0
D	[0]	6	4	2	7
E	3	5	6	[0]	8

Here, we observe that, number of rows ≠ number of lines. So we go for next step.

Step 4. The uncrossed smallest element 3 is subtracted from the remaining uncrossed elements.

Add the smallest element 3 at the intersection points of the two lines. Leave other covered elements unchanged. Reduced table is as under.

	I	II	III	IV	V
A	5	0	5	10	8
B	0	6	12	0	0
C	11	8	0	3	0
D	·0	6	1	2	4
E	3	5	3	0	5

Step 5. **Row Assignment.** In the first row there is a single zero in second column. Square around zero. Draw a line through second column.

Column Assignment. In the third column there is a single zero in third row. Square around zero. Draw a line through third row.

In the fifth column there is a single zero in second row. Square around zero. Draw a line through second row.

Now, in the first column there is a single zero in fourth row. Square around zero. Draw a line through fourth row.

Now in the fourth column, there is a single zero in fifth row. Square around zero. Draw a line through fifth row.

	I	II	III	IV	V
A	5	[0]	5	10	8
B	0	6	12	0	[0]
C	11	8	[0]	3	0
D	[0]	6	1	2	4
E	3	5	3	[0]	5

Here we observe that, number of rows = number of lines

Optimal assignment can be made of this stage. So go to step 6.

Step 6. For making assignment proceed as follows:

	I	II	III	IV	V
A	5	[0]	5	10	8
B	⊠	6	12	⊠	[0]
C	11	8	[0]	3	⊠
D	[0]	6	1	2	4
E	3	5	3	[0]	5

The optimal assignment is given by

A – II,   B – V,   C – III,   D – I,   F – IV

Step 7. The minimum assignment cost is read from the original cost matrix as:

$$C_{12} + C_{25} + C_{33} + C_{41} + C_{54} = 5 + 6 + 2 + 5 + 12 = 30$$ **Ans.**

Example 8. *Find the optimal assignment of the following cost matrix:*

From \ To	I	II	III	IV	V
A	10	3	3	2	8
B	9	7	8	2	7
C	7	5	6	2	4
D	3	5	8	2	4
E	9	10	9	6	10

(Bihar SBTE, 2010)

Solution. Step 1. Subtracting the smallest elements 2, 2, 2, 2 and 6 from first, second, third, fourth and fifth rows respectively, we get

From \ To	I	II	III	IV	V
A	8	1	1	0	6
B	7	5	6	0	5
C	5	3	4	0	2
D	1	3	6	0	2
E	3	4	3	0	4

Step 2. Subtracting the smallest elements 1, 1, 1 , 0 and 2 from first, second, third, fourth and fifth columns, we get

From \ To	I	II	III	IV	V
A	7	0	0	0	4
B	6	4	5	0	3
C	4	2	3	0	0
D	0	2	5	0	0
E	2	3	2	0	2

Step 3. Row assignment

In the second row there is a single zero in fourth column. Square around zero. Draw a line through fourth column.

Now, in the third row there is a single zero in the fifth column. Square around zero. Draw a line through fifth column.

In the fourth row there is a single zero in first column. Square around zero. Draw a line through first column.

Column assignment

In the second column there is a single zero in first row. Square around zero and draw a line through first row.

To\From	I	II	III	IV	V
A	7	[0]	0	0	4
B	6	4	5	[0]	3
C	4	2	3	0	[0]
D	[0]	2	5	0	0
E	2	3	2	0	2

Step 4. The uncrossed smallest element 2 is subtracted from the remaining uncrossed elements. Add the smallest element 2 at the intersection of the two lines. Leave other covered elements unchanged. Reduced table is as under.

To\From	I	II	III	IV	V
A	9	0	0	2	6
B	6	2	3	0	3
C	4	0	1	0	0
D	0	0	3	0	0
E	2	1	0	0	2

Step 5. Row assignment

In the second row there is a single zero in fourth column. Square around zero and draw a line through fourth column.

Now, in the fifth row there is a single zero in third column. Square around zero and draw a line through third column.

Column Operation

In the first column there is a single zero in fourth row. Square around zero and draw a line through fourth row.

Now, in the fifth column there is a single zero in third row. Square around zero and draw a line through third row.

Now, in the second column there is a single zero in first row. Square around zero and draw a line through first row.

To\From	I	II	III	IV	V
A	9	[0]	0	2	6
B	6	2	3	[0]	3
C	4	0	1	0	[0]
D	[0]	0	3	0	0
E	2	1	[0]	0	2

There we observe that, number of rows = number of lines.

Hence this is case of optimality. The optimal assignment is given by

A → II,   B → IV,   C → V,   D → I,   E → III

The minimum assignment cost is read from the original cost matrix as:

$$C_{12} + C_{24} + C_{35} + C_{41} + C_{53} = 3 + 2 + 4 + 3 + 9 = 21$$

Ans.

Example 9. *Five jobs are to be processed and life machines are available. Any machine can process any job and the time taken (in hours) by each machine to do the job beings as follows:*

(Bihar SBTE, 2012)

Jobs	A	B	C	D	E
1	32	38	40	28	40
2	40	24	28	21	36
3	41	27	33	30	37
4	22	38	41	36	36
5	29	33	40	35	39

Find the assignment of jobs to the machines that will minimize the total time taken.

Solution.

Iteration – I

	A	B	C	D	E
1	4	10	12	0	12
2	19	3	7	0	15
3	14	0	6	3	10
4	0	16	19	14	14
5	0	4	11	6	10

Iteration – II

	A	B	C	D	E
1	40	10	6	0	2
2	19	3	1	0	5
3	14	0	0	3	0
4	0	16	13	14	4
5	0	4	5	6	0

Iteration – III

	A	B	C	D	E
1	4	10	6	[0]	2
2	19	3	1	0	5
3	14	[0]	0	3	0
4	[0]	16	13	14	4
5	0	4	5	16	[0]

Iteration – IV

	A	B	C	D
1	4	0	5	[0]
2	19	2	[0]	[0]
3	15	[0]	0	4
4	[0]	15	12	14

Iteration – V

	A	B	C	D	E
1	4	9	5	[0]	2
2	19	2	[0]	0	5
3	15	[0]	0	4	1
4	[0]	15	12	4	4
5	0	3	4	0	[0]

Assignment of jobs

$1 \to D, \quad 2 \to C \qquad 3 \to B, \quad 4 \to A \qquad 5 \to E$

and optimal solution is = 28 + 28 + 27 + 22 + 39 = 144.   **Ans.**

Example 10. *Using Hungarian Method, find assignment for the following problem.* (*Bihar SBTE, 2012*)

Mechanies →

↓ Jobs	A	B	C	D	E
1	32	38	40	28	40
2	40	24	28	21	36
3	41	27	33	30	37
4	22	38	41	36	36
5	29	33	40	35	39

Jobs	A	B	C	D	E
1	32	38	40	28	40
2	40	24	28	21	36
3	41	27	33	30	37
4	22	38	41	36	36
5	29	33	40	35	39

Find the assignment of jobs to the machines that will minimize the total time taken.

Solution.

Iteration – I

	A	B	C	D	E
1	4	10	12	0	12
2	19	3	7	0	15
3	14	0	6	3	10
4	0	16	19	14	14
5	0	4	11	6	10

Iteration – II

	A	B	C	D	E
1	40	10	6	0	2
2	19	3	1	0	5
3	14	0	0	3	0
4	0	16	13	14	4
5	0	4	5	6	0

Iteration – III

	A	B	C	D	E
1	4	10	6	[0]	2
2	19	3	1	0	5
3	14	[0]	0	3	0
4	[0]	16	13	14	4
5	0	4	5	16	[0]

Iteration – IV

	A	B	C	D
1	4	0	5	[0]
2	19	2	[0]	0
3	15	[0]	0	4
4	[0]	15	12	14

Iteration – V

	A	B	C	D	E
1	4	9	5	[0]	2
2	19	2	[0]	0	5
3	15	[0]	0	4	1
4	[0]	15	12	4	4
5	0	3	4	0	[0]

Assignment of jobs

$1 \to D, 2 \to C$ $3 \to B, 4 \to A$ $5 \to E$

and optimal solution is = 28 + 28 + 27 + 22 + 39 = 144.   **Ans.**

Example 11. *Find the optimal assignment for the following cost matrix.*

Jobs \ Men	M_1	M_2	M_3	M_4	M_5
J_1	11	17	8	16	20
J_2	9	7	12	6	15
J_3	13	16	15	12	16
J_4	21	24	17	28	26
J_5	14	10	12	11	15

(Bihar SBTE, 2003, 2009)

Solution. Step 1. Subtract the smallest elements 8 from first row, 6 from 2nd row, 12 from 3rd row, 17 from the fourth row and 10 from the 5th row.

Iteration – I

Jobs \ Men	M_1	M_2	M_3	M_4	M_5
J_1	3	9	0	8	12
J_2	3	1	6	0	9
J_3	1	4	3	0	4
J_4	4	7	0	11	9
J_5	4	0	2	1	5

Step 2. Subtract the smallest element '1' of first column from the remaining elements of this column to get zero. Subtract the smallest element '4' from fourth column from the remaining elements of this column.

Iteration – II

Jobs \ Men	M_1	M_2	M_3	M_4	M_5
J_1	2	9	0	8	8
J_2	2	1	6	0	5
J_3	0	4	3	0	0
J_4	3	7	0	11	5
J_5	3	0	2	1	1

Step 3. Row assignment

There is zero in the first row and third column. Square around it and draw a line in the third column containing zero.

There is zero in the second row and fourth column. Square around zero and draw a line in the fourth column containing zero.

Column assignment

There is zero in the first column and third row. Square around zero and draw a line in the third row. There is a single zero in second column and fourth row. Square around zero and draw a line in the fourth row containing zero.

Iteration – III

Men / Jobs	M_1	M_2	M_3	M_4	M_5
J_1	2	9	[0]	8	8
J_2	2	1	6	[0]	5
J_3	[0]	4	3	0	0
J_4	3	7	0	11	5
J_5	3	[0]	2	1	1

Step 4. Choose the uncrossed smallest element '1' from the remaining uncrossed elements. Subtract 1 from the remaining elements and add '1' to the elements at intersection of two lines.

Iteration – IV

	M_1	M_2	M_3	M_4	M_5
J_1	1	8	0	8	7
J_2	1	0	6	0	4
J_3	0	4	4	1	0
J_4	2	6	0	11	4
J_5	3	0	3	2	1

Step 5. First row contains zero in third column, draw a line in this column. Fourth row contains zero in the third column, draw a line in this column. Second column contains zero in the second row, draw a line in this row. Fifth column contains zero in the third row, draw a line in this row.

Iteration – V

Men / Jobs	M_1	M_2	M_3	M_4	M_5
J_1	1	8	0	8	7
J_2	1	[0]	6	[0]	4
J_3	[0]	4	4	1	0
J_4	2	6	[0]	11	4
J_5	3	[0]	3	2	1

Here, the number of rows ≠ number of lines, so we go for next step.

Step 6. Subtract the smallest uncovered element 1 from all remaining uncovered elements. Add the smallest uncovered element 1 to the elements at the intersection point of two lines. Other covered elements will remains same.

Iteration – VI

Men / Jobs	M_1	M_2	M_3	M_4	M_5
J_1	0	8	0	7	6
J_2	1	1	7	0	4
J_3	0	5	5	1	0
J_4	1	6	0	10	3
J_5	2	0	3	1	0

Step 7. Row assignment

In the second row there is a single zero in fourth column. Square around zero and draw a line through fourth column.

In the fourth row there is a single zero in third column, square around zero and draw a line through third column.

Column assignment

In the second column there is a single zero in fifth row. Square around zero and draw a line through fifth row.

In the fifth column there is a single zero in third row. Square around zero and draw a line through third row.

Now, in the first column there is a single zero in first row. Square around zero and draw a line through first row.

Iteration – VII

Jobs \ Men	M_1	M_2	M_3	M_4	M_5
J_1	[0]	8	0	2	6
J_2	1	1	7	[0]	4
J_3	0	5	5	1	[0]
J_4	1	6	[0]	10	3
J_5	2	[0]	3	1	0

Here, we observe that the number of rows = Number of lines

Hence, it is the case of optimality.

The optimal assignment is given

$$J_1 - M_1, \qquad J_2 - M_4, \qquad J_3 - M_5, \qquad J_4 - M_3, \qquad J_5 - M_2$$

Therefore, the minimum assignment cost.

$$= C_{11} + C_{24} + C_{35} + C_{43} + C_{52} = 11 + 6 + 16 + 17 + 10 = 60 \qquad \textbf{Ans.}$$

EXERCISE 18.1

1. The personal manager of ABC Company desires to assign Mr. X, Mr. Y and Mr. Z to regional offices. If they are posted in Delhi, Mumbai and Kolkata their salaries are given as per Table below At a later date the company has decided to open its office in Chennai and would send one of the three to that branch if it were more economical than to move to Delhi, Mumbai or Kolkata. The salaries of Mr. X, Mr. Y and Mr. Z if posted in Chennai would be 2000, 1600 and 3000 respectively. What is the optimal assignment to the offices?

Persons	Delhi	Mumbai	Kolkata
X	1600	2200	2400
Y	1000	3200	2600
Z	1000	2000	4600

Ans. X → Mumbai, Y → Chennai, Z → Delhi, None → Kolkata

Cost = 2000 + 1600 + 1000 = 4600

2. Find the optimal assignment for the assignment problem with the following cost matrix.

	I	II	III	IV
A	5	3	1	8
B	7	9	2	6
C	6	4	5	7
D	5	7	7	6

Ans. A – III, B – IV, C – II, D – 1; Z_{min} = 16

3. In an emporium four salesmen A, B, C and D are available in four counters W, X, Y and Z. Each salesman can handle any counter. The service time (in hour) of each counter when manned by each salesman is given below:

Salesman

Counters	A	B	C	D
W	41	72	39	52
X	22	29	49	65
Y	27	39	60	51
Z	45	50	48	52

How should the salesmen be allocated in the sales counters so as to minimize the service time. Each salesman is given only one counter.   **Ans.** W ⇒ C, X ⇒ B, Y ⇒ A, Z ⇒ D

4. A team of 5 horses and 5 riders has entered a jumping show contest. The number of penalty points to be expected when each rider rides any horse is shown below:

Rider

Horse	R_1	R_2	R_3	R_4	R_5
H_1	5	3	4	7	1
H_2	2	3	7	6	5
H_3	4	1	5	2	4
H_4	6	8	1	2	3
H_5	4	2	5	7	1

How should the sorses be allotted to the riders so as to minimize the expected loss of the team?

Ans. $H_1 - R_5$, $H_2 - R_1$, $H_3 - R_4$, $H_4 - R_3$, $H_5 - R_2$; Z_{min} = 8

5. There are four jobs a, b, c and d to be performed on machines 1, 2, 3 and 4. The processing for each job-man combination is as given in Table below. Find the optimal assignment that will result in minimum man-hours required in completing all the four jobs.

	a	b	c	d
1	5	3	2	8
2	7	9	2	6
3	6	4	5	8
4	5	7	7	8

Ans. 1 → C, 2 → D, 3 → B, 4 → A
Time = 2 + 6 + 4 + 5 = 17 hrs.

6. Five employees of a company are to be assigned to five jobs, which can be done by any of them. The workers get different wages per hour. These are ₹ 5 per hour for A, B and C each and ₹ 3 per hour for D and E each. The amount of time in hours taken by each employee to do a given job is given in the table below. Determine the assignment pattern that minimizes the total cost of getting the given jobs done.

Employee

	A	B	C	D	E
1	7	9	3	3	2
2	6	1	6	6	5
Job 3	3	4	9	10	7
4	1	5	2	2	4
5	6	6	9	4	2

(*IGNOU, MBA, 2002*) **Ans.** 1 – D, 2 – B, 3 – A, 4 – C, 5– E; ₹ 45*)*

7. Five new machines are to be located in a matching shop. There are five possible locations in which the machines can be located. C_{ij}, the cost of placing machine i in place j is given in the table below

Place

	1	2	3	4	5
1	15	10	25	25	10
2	1	8	10	20	2
Machine 3	8	9	17	20	10.
4	14	10	25	27	15
5	10	8	25	27	12

It is required to place the machines at suitable places so as to minimize the total cost.

(*i*) Formulate an L.P. model to find an optimal assignment. (*ii*) Solve the problem by assignment technique of L.P.

Ans. 1 – 5, 2 – 3, 3 – 4, 4 – 2, 5 – 1; Z_{min} = 60

8. A company is faced with the problem of assigning six different machines to five different jobs. The costs estimated in hundreds of rupees are given in the table below.

Jobs

	1	2	3	4	5
1	2.5	5	1	6	2
2	2	5	1.5	7	3
3	3	6.5	2	8	3
Machines 4	3.5	7	2	9	4.5
5	4	7	3	9	6
6	6	9	5	10	6

Solve the problem assuming that the objective is to minimize the total cost.

Ans. 1 – 4, 2 – 1, 3 – 5, 4 – 3, 5 – 2, 6 – Dummy; Z_{min} = ₹ 2,000

9. Six machines M_1, M_2, M_3, M_4, M_5 and M_6 are to be located in six places P_1, P_2, P_3, P_4, P_5, P_6. C_{ij}, the cost of locating machine M_i at place P_j is given in the matrix below:

	P_1	P_2	P_3	P_4	P_5	P_6
M_1	20	23	18	10	16	20
M_2	50	20	17	16	15	11
M_3	60	30	40	55	8	7
M_4	6	7	10	20	25	9
M_5	18	19	28	17	60	70
M_6	9	10	20	30	40	55

Formulate an L.P. model to determine an optimal assignment. Write the objective function and the constraints in detail. Define any symbol used. Find an optimal layout by assignment technique of linear programming.

(P.U.B.E (Mech.) Nov. 2002)

Ans. $M_1 - P_4$, $M_2 - P_6$, $M_3 - P_5$, $M_4 - P_3$, $M_5 - P_1$, $M_6 - P_2$; $Z_{min} = 67$

10. An automobile dealer wishes to put five mechanics on five different jobs. All five mechanics have different kinds of skills and they exhibit different levels of efficiency from one job to another. The dealer has established the number of man hours that would be required for each job-man combination which is given in the following matrix. Find the optimal assignment that will result in minimum man hours required in completing all five jobs.

	1	2	3	4	5
a	16	13	17	19	20
b	14	12	13	16	17
c	14	11	12	17	18
d	5	5	8	8	11
e	3	3	8	8	10

Ans. $a \to 2$, $b \to 5$, $c \to 3$, $d \to 4$, $E \to 1$, Time = 13 + 17 + 12 + 8 + 3 = 53 hours

Choose the correct answer:

11. There will be .... different solutions in a $n \times n$ assignments problem.

   (a) n (b) n !   (c) $(n-1)$!   (d) $(n+1)$!   *(Bihar SBTE, 2004, 2005)* **Ans.** (b)

12. An assignment is said to be complete if .... assignment are made at zero level.

   (a) equal to n (b) more than n (c) less than n (d) less than or equal to n. *(Bihar SBTE, 2011)* **Ans.** (a)

13. A transportation problem becomes an assignment problem if

   (a) demand = 1, supply > 1   (b) demand > 1, supply = 1

   (c) demand = 1, supply = 1   (d) demand > 1, supply > 1   *(Bihar SBTE, 2009)* **Ans.** (c)

14. In an optimal solution of a $n \times n$ assignment problem, total number of allocation be

   (a) less than n (b) equal to n   (c) more than n   (d) all of them   *(Bihar SBTE, 2003, 2009)* **Ans.** (c)

15. There will be .... different solutions in a n × n assignment problem.

   (a) $\lfloor n$ (b) n (c) $\lfloor n-1$ (d) $\lfloor n+1$ *(Bihar SBTE, 2003, 2009)* **Ans.** (a)

16. An assignment is said to be complete if .... assignment are made at zero level.

   (a) equal to n (b) more than n (c) less than n   (d) less than or equal to n

   (Bihar SBTE, 2003, 2009) **Ans.** (c)

17. Hungarian method is used to solve ....

   (a) Linear .... problem   (b) Transportation problem

   (c) Assignment problem   (d) None of these   *(Bihar SBTE, 2014, 2009)* **Ans.** (c)

18. An optimal solution of a $n \times n$ assignment problem, total number of allocation be

   (a) less than n (b) equal to n

   (c) more than n (d) all of them   *(Bihar SBTE, 2009)* **Ans.** (c)

19

Network Analysis in Project Planning

(PERT and CPM)

19.1 PROJECT

Project is defined as a combination of inter related activities which must be executed in certain order to complete the task.

19.2 REQUIREMENT OF PROJECT

1. Time. It should be completed within a fix time.

2. Man power and Material Resources. It should involve minimum man power and other resources.

3. Financial resources. It should require minimum investment.

Project management is responsible for the above requirements.

19.3 STEPS OF MANAGEMENT

1. Project planning

2. Project scheduling

3. Project controlling

The network Techniques of CPM and PERT are important in project management.

19.4 PROJECT PLANNING

The identification of activities (jobs) required and required resources are the part of project planning. Different types of resources are:

1. Man power

2. Machinery

3. Financial

4. Material

5. Time and space resources.

19.5 PROJECT SCHEDULING

(*Bihar, SBTE, 2010*)

1. Starting and finishing time of each activity should be fixed.

2. Special attention to critical activities.

3. Allocation of man, machine and material to each activity.

4. Limitations and constraints should kept in mind.

19.6 PROJECT CONTROLLING

Quality in terms of time and cost. Reviewing the progress from time to time. Completion of the project in time. Methods to rectify the deviation from the plan.

19.7 NETWORK

It is the graphical representation of nodes (events) and connecting arrows representing activities of a project.

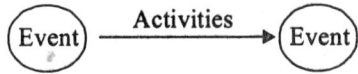

NETWORK OR ARROW DIAGRAM

19.8 TERMS USED IN A NETWORK

1.   **Activity.** An activity is represented by an arrow the tail of which represents the start and the head finish of the activity. An activity is an important part of the project.

2.   **Event (Nodes).** The arrow joins two nodes. Nodes are at the beginning point and at the end point of an activity. It is represented by a circle. Tail event is the ith event and head event is the jth event. An event indicates the completion of source activity and the starting of new one (EON network).

3.   **Path.** Path is the chain of activities and arrows connecting the initial events to other events.

19.9 CONSTRUCTION OF NETWORK

A project is divided into different activities. This process is called *Work Break-Down Structure (WBS)*. Start and finish events of the project are decided.

The activities should be in logical sequence in a network.

The order of activity in its chain should firmly be decided. Activity before a particular activity is called predecessor activity and the activity that follow a activity is called successor activity.

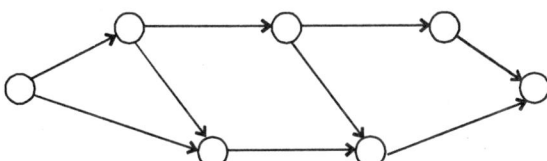

19.10 DRAWING OF A NETWORK

1. Each activity is represented by one and only one arrow.

2. Time flows from left to right.

3. Arrows should be in straight line not in a curve.

4. Angles between arrows should be obtuse as large as possible.

5. Arrow should not cross each other.

6. There must be no loop.

7. Each activity must have a tail and a head event.

8. Two or more activity should not have the same tail and head event.

9. In a network there must be only one initial event and only one end event.

10. If necessary dummy activity can be introduce.

19.11 DUMMY ACTIVITY

An activity which only determines the depending of one activity over the other, but does not consume any time is called a dummy activity. It is represented by dotted line arrow.

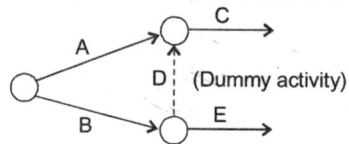

Kalpana is travelling by a train and she is reading a novel. These two activities, travelling and reading are going simultaneously. This is wrong represented in fig.1 and correct in fig.2 by introducing a dummy activity. The arrow of a dummy can be upward or downward.

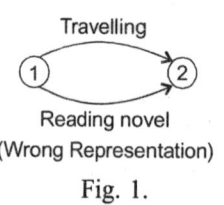

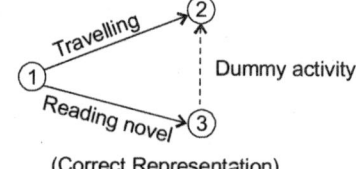

Fig. 1. Fig. 2.

19.12 LOOPING (CYCLING)

Three activities D, E and F form a loop (cycle). Activity D can not start until F is completed. And F can not start until each completed. This is a faulty network.

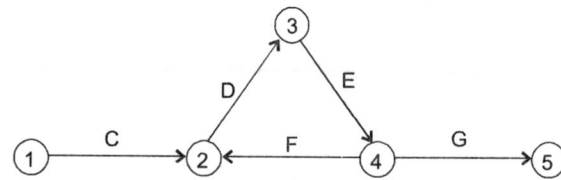

19.13 NUMBERING THE EVENT

When a network is drawn in a logical sequence and the events are written as 1,2,3 inside the node circle . The number sequence of the events should be such as to reflect the flow of the network. The rule is framed by Dr. Fulkerson is written by the following steps.

Step 1. The initial event which has all outgoing arrows with no incoming arrow is numbered as (1).

Step 2. Delete all the arrows coming out from node (1). In this way some nodes will become initial events. Number these events 2, 3.......

Step 3. Delete all the arrows coming out from these numbered arrows (2) and (3) by deleting the arrows coming out from (2) and (3) some more events will be the initial events. These events numbered as 4, 5.........

Step 4. Continue until the terminal node which has all arrows incoming but no arrow going out is numbered.

Example 1. *Using Fulkerson rule number the events*

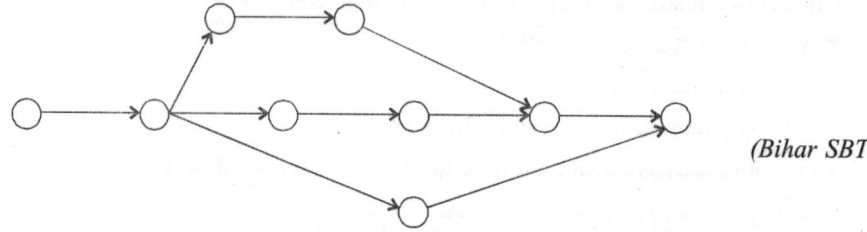

(Bihar SBTE, 2008)

Solution. We will use Fulkerson rule to number the events.

Step 1. Initial event is numbered as 1. Delete the arrow coming out of event 1.

Step 2. Now the initial event is numbered as 2. Delete all the arrows coming out from event 2.

Step 3. Now the initial events are numbered as 3,4,5. Delete the arrow coming out from 3. Then the next initial event is numbered as 6 remove the arrow coming out from event (4). So the initial event is numbered as 7. Remove the arrow coming out from 5th event (5).

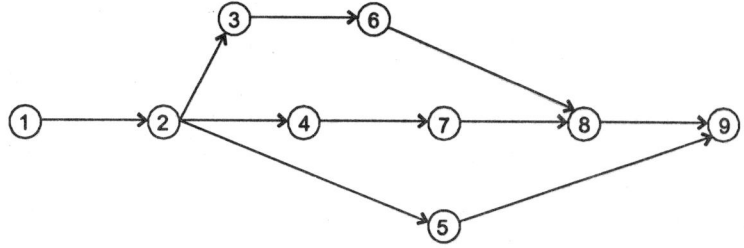

Step 4. Delete the arrows coming out from 6th event and 7th event. Then the next event will be the initial event and numbered as 8.

Step 5. Delete the arrow coming out from event 8. The final event will be the terminal event 9 as there is no arrow going out from event number 9.

Example 2. *Using Fulkerson rule number the events.*

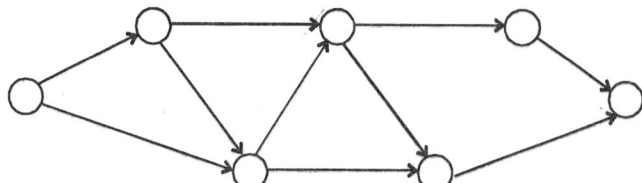

Solution. We will use Fulkerson rule to number the events in the given network.

Step 1. Initial event is numbered as 1. Delete the arrows coming out from the event 1.

Step 2. Now the initial events are numbered as 2,3. Remove the arrows coming out from the events 2 and 3. Now the initial events are numbered as 4,5. Now remove the arrows coming out from 4 and 5. The initial event is numbered as 6.

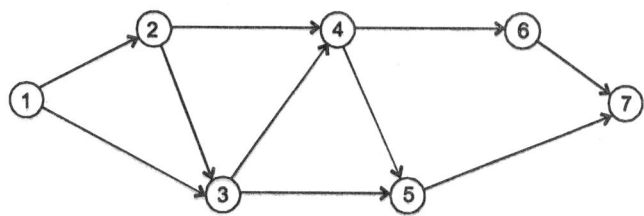

Delete the arrow coming out from the arrow 6, then the final event is numbered as 7 from which there is no going out arrow.

Example 3. *The various activities of a student are given below in random order*

        A: Admission in the college.   *B: Applying for the admission in the college*

        C: Preparation leave   *D: Attending the lectures*

        E: Appearing in the examination   *F: Getting scheme of examination.*

        G: Getting marksheet.   *H: Declaration of the result.*

    Construct a network diagram of the above project.

Solution.   The first step is to arrange the activities in a logical schedule. The person drawing the network must be enable to visualize that what activity precede and succeed a particular activity.

B: Applying for admission in the college   A: Admission in the college

D: Attending the lectures   C: Preparation leave

F: Getting scheme of examination   E: Appearing in the examination

H: Declaration of the result   G: Getting marksheet.

It is obvious that A can't be started **before** the completion of activity B. This is denoted as $B < A$.

With this notation, we can say that

   $A < D,$
   $D < C$
   $C < F$
   $F < E$
   $E < H$
   $H < G$

The logical order of activity of the given project can be shown in the following network diagram.

Fig. AON (Activity on Node)

The above fig. is an activity on node (AON) of network diagram. The fig. given below event on node (EON) is equivalent to (AON) of network diagram.

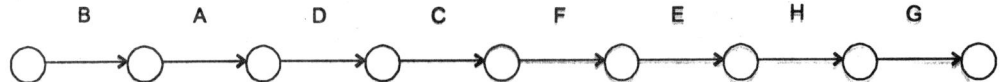

Fig. EON (Event on Node)

Example 4.   *Draw a network for the following project and number the events according to the Fulkerson's rule.*

A is the start event and K is the end event.

A precedes event B

J is the successor event to F

C and D are successor event to B

D is the preceding event to G

E and F occur after the event C

E precedes F

C restraints the occurrence of G and G precedes H

H precedes J and K succeeds J

F restraints the occurrence of H

Solution.   The logical order of events of the given project is shown in the following network diagram. Along the nodes are given their numbers as obtained by Fulkerson's Rule:

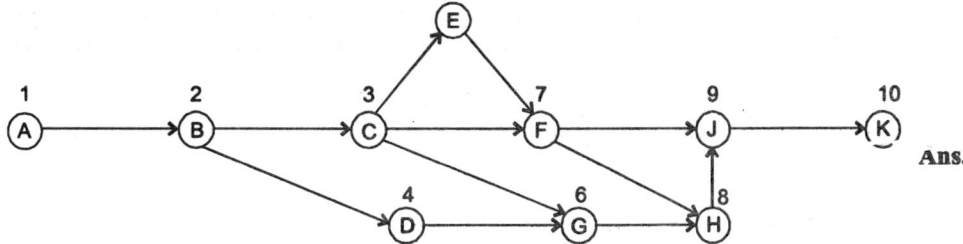

Ans.

Example 5. *Construct a network diagram for a project comprising of activities B, C, E, F, G, H, I, J, L, M, N, P and Q such that the following precedence relationships are satisfied: B < E, F; C, F < G; C < L; E, G < H; H, L < I; H < J; L < M; H, M < N; I, J < P; N < Q.*

Solution.   The required network is shown in the following figure. Four dummy activities have been used. Since activity L depends upon C and G depends upon both F and C, dummy D_1 is required. Dummies D_2 and D_3 establish that I depends on both H and L, while J depends only on H. Activity N depends on H and M both, while J depends on H. These precedence orders are established by employing D_4. The nodes of the network have been numbered using the Fulkerson's rule.

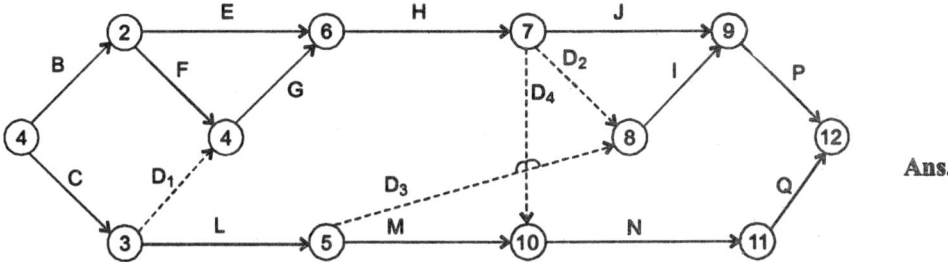

Ans.

Example 6. *The activities along with their dependency are given below. Draw an arrow diagram:*

Activities	A	B	C	D	E	F	G	H	I
Immediate Predecessor	–	–	–	C	A, B	C	E, F	D	E, H

Solution.   There is no predecessors of A, B and C so they can start from the same initial node. D depends upon C. Activity E depends upon A and B. Since E is to start from one node. So a dummy D_1 is required. Activity F depends on C and H depends upon D. Now activity G depends upon E and F. Both E and F could be made to mearge in one node. But E also control I. I is control by H but not F. Thus two more dummies D_2 and D_3 are required to complete the project as shown in the figure. Activities G and I are made to meet at the last node.

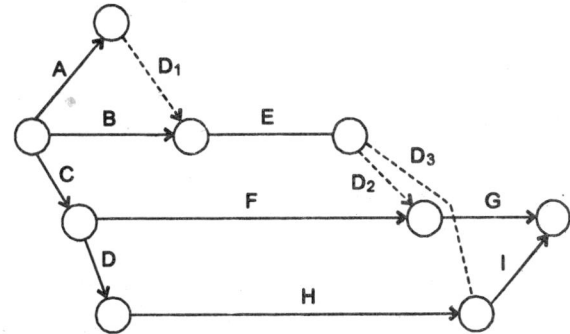

19.14 MEASURE OF ACTIVITIES

Each activity takes some time for its completion. This time duration depends upon the nature of activity.

There are two types of activities on the basis of time consumption.

1. Variable type activity

2. Deterministic type of activity.

1. Variable Type Activity

The time consumption for some activity involves a considerable degree of uncertainty. Such activity are known as variable activity. For example, research.

2. Deterministic Type Activity

The time consumption for a certain activities are accurately estimated. Such activities are called deterministic in nature. These activities are performed by skilled or experienced persons. The project which comprise of variable type activities are handled by project evaluation and review technique (PERT) version of the network and the projects comprising of deterministic type of activities are handled by critical path method (CPM) version of network.

19.15 CRITICAL PATH METHOD (CPM)

The critical path of a network gives the shortest time in which the whole project can be completed. Any activity is said to be critical if a delay in its start will cause a further delay in the completion of the project. And thus the target time will not be maintained in a network analysis.

Critical path method (CPM) can be defined as the identification of the sequence of the activities which will acquire greatest normal time to accomplish the period.

19.16 WORKING STEPS FOR CRITICAL PATH

Step 1. Write down the activity time just above the arrow representing that activity.

Step 2. Calculate earliest starting time (ES), earliest finishing time (EF), and latest finishing time (LF) for each activity.

Step 3. A table is prepared in which

(1) normal activity time

(2) earliest finish time

(3) latest finish time for each activity are written.

Step 4. The slack time is calculated by the formula LF_j-EF_i for each activity.

Step 5. The activities with zero slack are the critical activities. All critical activities are connected from beginning to end node by double arrows. This is the critical path.

19.17 COMPUTATION OF TIME IN CPM

We know the time completion of each activity and actual time of completion of each activity. This is known as activity time.

19.18 EARLIEST STARTING TIME

Earliest start time for an activity represents the time at which an activity can begin at the earliest.

19.19 RESOURCES SCHEDULING   (*Bihar, SBTE, 2009*)

Generally when a network of activities is developed to complete a project on the basis of sufficient resources and whatever the demands may be. During fluctuation of demand normally the cost of the project remains the same but when the demand goes down the hired labour and machinery can not be utilized, when the demand goes up the skilled labour and material will not be available easily. More money is to be spent to get resources. So the cost of project becomes more. Under these circumstances the scheduling of resources becomes necessary.

On scheduling the demand there are two benefits (1) cost of the project can be brought down, (2) the fluctuation of demand can be control.

19.20 PROJECT COST

The cost of the project depends upon the cost of resources required for each activity. Secondly the cost of the project also depends upon the project duration time. By increasing the cost of the project the duration time can be cut down to some extent. The cost can be cut down by increasing the project duration time. But we want a balance in cost and time. Scheduling the activities of the project implies lowest possible cost and optimum time for the project. There are two type of the cost of project (1) direct and (2) indirect cost.

19.21 DIRECT COST

The cost of the material and labour require to perform the activity of network of the project.

19.22 INDIRECT COST

The indirect cost of the project is divided into two parts

(1) fixed indirect cost i.e. cost of the machinery equipment.

(2) Variable indirect cost i.e; cost of the raw material.

19.23 PERT (PROJECT EVALUATION AND REVIEW TECHNIQUE)

Time estimation in CPM are taken as deterministic. But in practical problem project time completion of a project can not be definite with certainty. But there are certain project in which research work is involve and there is no experience to handle such projects in such cases an uncertainty remains regarding the duration of various activity. So in such cases we need probabilistic approach to handle such problems. In PERT network time estimate are based on probabilistic approach. The time duration in PERT analysis is random variable characterize by some probability distribution usually a β-distribution. To estimate the parameters of the β-distribution (mean and variance). The PERT system is based on three time estimate of the performance time of an activity as follows:

1. The optimistic time estimate:

The shortest possible time required for the completion of an activity in normal circumstances.

    2. The pessimistic time estimate. The maximum possible time of the activity will take, if every thing goes bad. However earth quake, flood, storm and labour trouble are not taken into account while estimating this time.

    3. The most likely time estimate

The time of an activity will take, if executed under normal conditions is the modal value.

The experienced person who may be an engineer, foreman or worker having sufficient technical competence is asked to guess the various time estimate.

19.24 FREQUENCY DISTRIBUTION CURVE FOR PERT

When the data regarding the time required to complete a particular job are collected. The probability distribution function may be represented as figure shown below: It may be of three types:

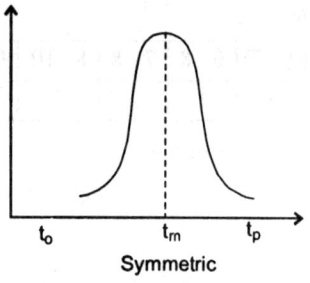

Symmetric

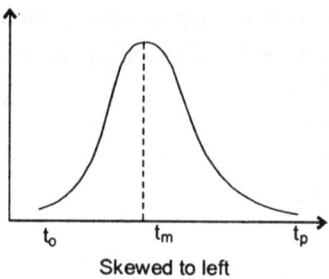

Skewed to left

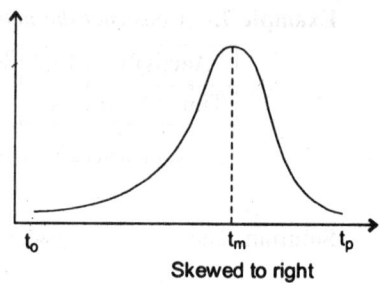

Skewed to right

$$\text{Variance} = \sigma^2 = \left(\frac{t_p - t_0}{6}\right)^2$$

$$\sigma^2 = \left(\frac{t_p - t_0}{6}\right)^2$$

$$\mu = \frac{t_0 + 4t_m + t_p}{6}$$

The expected time is then used as the activity duration and the critical path is obtained by analytical method.

The variance is used to find the probability of completing the whole project.

19.25 PROCEDURE

Find the variance of all activity durations of critical path. Add them and make the square root to find the standard deviation of the total project duration denoted by σ.

$$\sigma = \sqrt{\sigma_2^2 + \sigma_2^2 + \sigma_3^2 + \ldots\ldots\ldots}$$

A β-distribution curve represent the activity time frequency distribution i.e: normal distribution curve.

The area bounded by the curve and the z-axis is unity and standard deviation equal to one. First we calculate the standard normal variate.

$$Z = \frac{T - T_{cp}}{\sigma}, \text{ where } T_{cp} \text{ is critical path duration.}$$

Then the probability is read from the standard normal probability distribution table.

19.26 DIFFERENCES BETWEEN CRITICAL PATH METHOD AND PROGRAMME EVALUATION REVIEW TECHNIQUE
(Bihar SBTE, 2011, 2009)

CPM	PERT
1. CPM stands for critical path method.	1. PERT stands for project evaluation and review technique.
2. Time estimate to perform the activities of a project can be determined.	2. The time estimate can be longer.
3. There are four types of time require to perform an activity	3. It has four time estimates of the performance time of an activity.
(a) Earliest start time	(a) The optimistic time,
(b) Earliest completion time	(b) The pessimistic time
(c) Latest start time	(c) The most likely time
(d) Earlier completion time	(d) Expected time.

Example 7. *Construct the network with the help of table given below:*

Aactivity	1–2	1–3	2–4	3–4	3–5	4–9	5–6	5–7	6–8	7–8	8–10	9–10
Time (days)	4	1	1	1	6	5	4	8	1	2	5	7

Construct the network diagram and find critical path and its duration.

(Bihar SBTE, 2010)

Solution. The required network diagram is below:

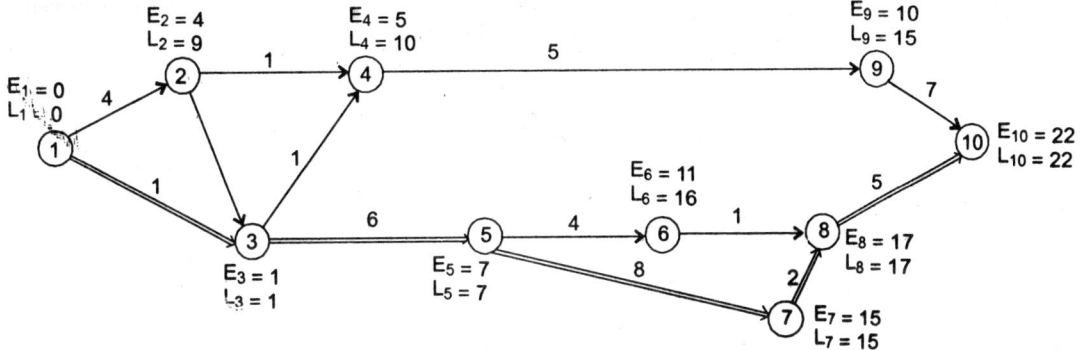

The earliest occurrence time (E) and the latest occurrence time (L) of each event are now computed by employing forward and backward pass calculations.

In forward pass computations.

$E_1 = 0$

$E_2 = E_1 + t_{12} = 0 + 4 = 4$

$E_3 = E_1 + t_{13} = 0 + 1 = 1$

$E_4 = \text{Max} [E_3 + t_{34}, E_2 + t_{24}] = \text{Max} [1 + 1, 4 + 1] = 5$

$E_5 = E_3 + t_{35} = 1 + 6 = 7$

$E_6 = E_5 + t_{56} = 7 + 4 = 11$

$E_7 = E_5 + t_{57} = 7 + 8 = 15$

$E_8 = \text{Max} [E_6 + t_{68}, E_7 + t_{78}] = \text{Max} [11 + 1, 15 + 2] = 17$

$E_9 = E_4 + t_{49} = 5 + 5 = 10$

$E_{10} = \text{Max} [E_8 + t_{8, 10}, E_9 + t_{9, 10}] = \text{Max} [17 + 5, 10 + 7] = 22$

In backward pass computations,

$L_{10} = E_{10} = 22$

$L_9 = L_{10} - t_{9, 10} = 22 - 7 = 15$

$L_8 = L_{10} - t_{8, 10} = 22 - 5 = 17$

$L_7 = L_8 - t_{78} = 17 - 2 = 15$

$L_6 = L_8 - t_{68} = 17 - 1 = 16$

$L_5 = \text{Min} [L_6 - t_{56}, L_7 - t_{57}] = \text{Min} [16 - 4, 15 - 8] = 7$

$L_4 = L_9 - t_{49} = 15 - 5 = 10$

$L_3 = \text{Min} [L_4 - t_{34}, L_5 - t_{35}] = \text{Min} [10 - 1, 7 - 6] = 1$

$L_2 = L_4 - t_{24} = 10 - 1 = 9$

$L_1 = \text{Min} [L_2 - t_{12}, L_3 - t_{13}] = \text{Min} [9 - 4, 1 - 1] = 0$

Here, we observe that the values of E and L are equal on the nodes (1), (3), (5), (7), (8) and (10). So, the slack values (L – E) are zero at the above nodes. Hence, the critical path is underlined by a line.

$$1 \xrightarrow{\ 1\ } 3 \xrightarrow{\ 6\ } 5 \xrightarrow{\ 8\ } 7 \xrightarrow{\ 2\ } 8 \xrightarrow{\ 5\ } 10$$

Project duration = 1 + 6 + 8 + 2 + 5 = 22 days.   **Ans.**

Example 8. *Find critical path for the following network:*   *(Bihar SBTE, 2010, 2009)*

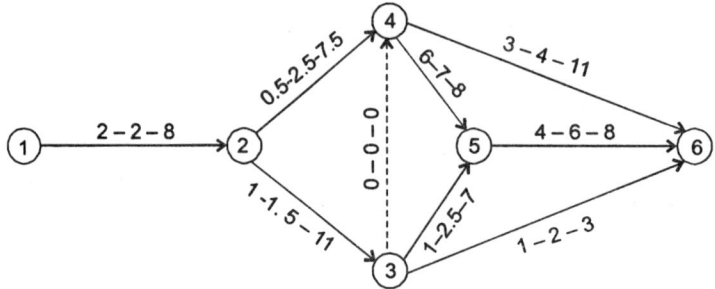

Solution.

Task (i, j)	t_0	t_m	t_p	t_θ
A (1, 2)	2	2	8	3
C (2, 3)	1	1.5	11	3
B (2, 4)	0.5	2.5	7.5	3
D (3, 4)	0	0	0	0
E (3, 5)	1	2.5	7	3
F (4, 5)	6	7	8	7
G (4, 6)	3	4	11	5
H (5, 6)	4	6	8	6
I (3, 6)	1	2	3	2

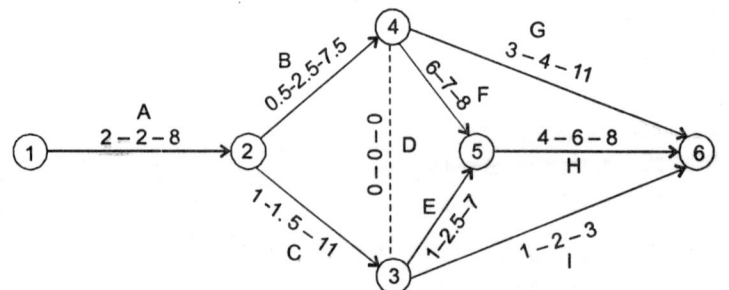

                                                      Ans.

Fourth column of the above table is filled by t_e as calculated below:

$$t_e = \frac{1}{6} \left[t_0 + 4\, t_m + t_p \right]$$

$$t_e(1, 2) = \frac{1}{6}(2 + 4 \times 2 + 8) = \frac{1}{6}(18) = 3$$

$$t_e(2, 3) = \frac{1}{6}(1 + 4 \times 1.5 + 11) = \frac{1}{6}(18) = 3$$

$$t_e(2, 4) = \frac{1}{6}(0.5 + 4 \times 2.5 + 7.5) = \frac{1}{6}(18) = 3$$

$$t_e(3, 4) = \frac{1}{6}(0 + 0 + 0) = 0$$

$$t_e(3, 5) = \frac{1}{6}(1 + 4 \times 2.5 + 7) = \frac{1}{6}(18) = 3$$

$$t_e(4, 5) = \frac{1}{6}(6 + 4 \times 7 + 8) = \frac{1}{6}(42) = 7$$

$$t_e(4, 6) = \frac{1}{6}(3 + 4 \times 4 + 11) = \frac{1}{6}(30) = 5$$

$$t_e(5, 6) = \frac{1}{6}(4 + 4 \times 6 + 8) = \frac{1}{6}(36) = 6$$

$$t_e(3, 6) = \frac{1}{6}(1 + 4 \times 2 + 3) = \frac{1}{6}(12) = 2$$

Let us assume that

$ES_1 = 0$

$ES_2 = ES_1 + t_e(1, 2) = 0 + 3 = 3$

$ES_3 = ES_2 + t_e(2, 3) = 3 + 3 = 6$

$ES_4 = \text{Max} [ES_2 + t_e(2, 4), ES_3 + t_e(3, 4)]$

$\qquad = \text{Max} [3 + 3, 6 + 0] = 6$

$ES_5 = \text{Max} [ES_3 + t_e(3, 5); ES_4 + t_e(4, 5)]$

$\qquad = \text{Max} [6 + 3, 6 + 7] = 13$

$ES_6 = \text{Max} [ES_3 + t_e(3, 6), ES_5 + t_e(5, 6), ES_4 + t_e(4, 6)]$

$\qquad = \text{Max} [6 + 2, 13 + 6, 6 + 5] = 19$

Now let us assume that $LF_6 = 19$

$LF_5 = LF_6 - t_e(5, 6) = 19 - 6 = 13$

$LF_4 = \text{Min} [LF_6 - t_e(4, 6), LF_5 - t_e(4, 5)]$

$\qquad = \text{Min} [19 - 5, 13 - 7] = 6$

$LF_3 = \text{Min} [LF_4 - t_e(3, 4), LF_5 - t_e(3, 5), LF_6 - t_e(3, 6)] = \text{Min} [6 - 0, 13 - 3, 19 - 2] = 6$

$LF_2 = \text{Min} [LF_3 - t_e(2, 3), LF_4 - t_e(2, 4)] = \text{Min} [6 - 3, 6 - 3] = 3$

$LF_1 = LF_2 - t_e(1, 2) = 3 - 3 = 0$

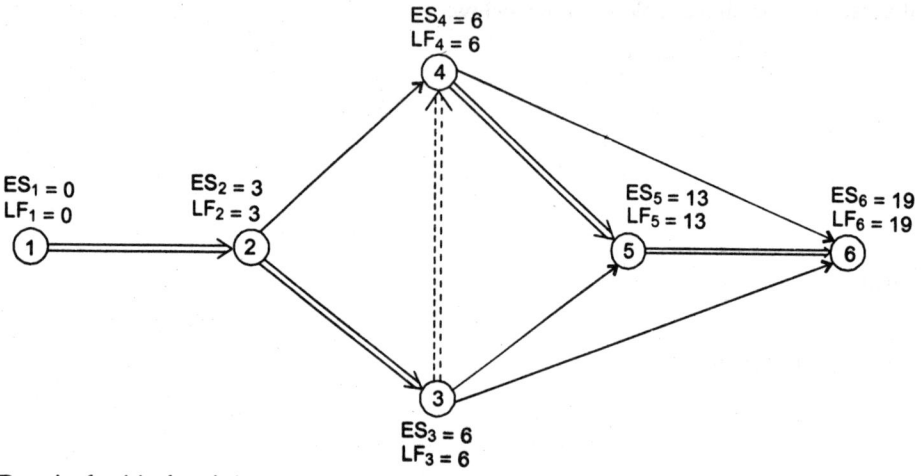

Required critical path is

Ans.

Example 9.   *The following table shows the jobs of a network alongwith their time estimates:*

Job	1 − 2	1 − 6	2 − 3	2 − 4	3 − 5	4 − 5	6 − 7	5 − 8	7 − 8
o (days)	1	2	2	2	7	5	5	3	8
m (days)	7	5	14	5	10	5	8	3	17
p (days)	13	14	26	8	19	17	29	9	32

Draw the network and find out the critical path.   *(Bihar SBTE, 2009)*

Solution.   Here t_p = Pessimistic time (It is the longest time to complete the activity).

t_m = Most likely time (It is the estimate of the normal activity time)

t_e = Expected time (It is known as average or, mean time for completion of the job).

t_0 = Optimistic time (It is the least possible time to complete the activity)

Job	t_0	t_m	t_p	$t_e = \dfrac{t_0 + 4t_m + t_p}{6}$	$\sigma = \dfrac{t_p - t_0}{6}$ (Standard deviation)
1 − 2	1	7	13	$te_{12} = \dfrac{1 + 4 \times 7 + 13}{6} = 7$	$\sigma_{12} = \dfrac{13 - 1}{6} = 2$
1 − 6	2	5	14	$te_{16} = 6$	$\sigma_{16} = 2$
2 − 3	2	14	26	$te_{23} = 14$	$\sigma_{23} = 4$
2 − 4	2	5	8	$te_{24} = 5$	$\sigma_{24} = 1$
3 − 5	7	10	19	$te_{35} = 11$	$\sigma_{35} = 2$
4 − 5	5	5	17	$te_{45} = 7$	$\sigma_{45} = 2$
6 − 7	5	8	29	$te_{67} = 4$	$\sigma_{67} = 4$
5 − 8	3	3	9	$te_{58} = 4$	$\sigma_{58} = 1$
7 − 8	8	17	32	$te_{78} = 18$	$\sigma_{78} = 4$

The network diagram for the above table is shown below:

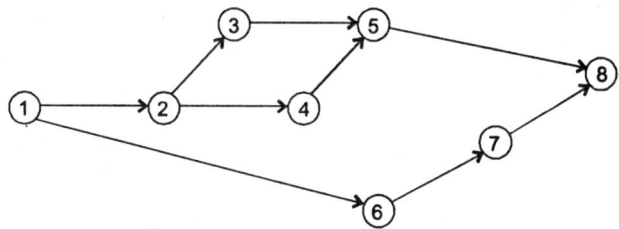

Let us assume that

$ES_1 = 0$

$ES_2 = ES_1 + t_e (1, 2) = 0 + 7 = 7$

$ES_3 = ES_2 + t_e (2, 3) = 7 + 14 = 21$

$ES_4 = ES_2 + t_e (2, 4) = 7 + 5 = 12$

$ES_5 = Max [ES_3 + t_e (3, 5), ES_4 + t_e (4, 5)] = Max [21 + 11, 12 + 7] = 32$

$ES_6 = ES_1 + t_e (1, 6) = 0 + 6 = 6$

$ES_7 = ES_6 + t_e (6, 7) = 6 + 11 = 17$

$ES_8 = [ES_5 + t_e (5, 8), ES_7 + t_e (7, 8)] = Max [32 + 4, 17 + 18] = 36$

Now let us assume that $LF_8 = 36$

$LF_7 = LF_8 - t_e (7, 8) = 36 - 18 = 18$

$LF_6 = LF_7 - t_e (6, 7) = 18 - 11 = 7$

$LF_5 = LF_8 - t_e (5, 8) = 36 - 4 = 32$

$LF_4 = LF_5 - t_e (4, 5) = 32 - 7 = 25$

$LF_3 = LF_5 - t_e (3, 5) = 32 - 11 = 21$

$LF_2 = Min [LF_4 - t_e (2, 4), LF_3 - t_e (2, 3)] = Min [25 - 5, 21 - 14] = 7$

$LF_1 = Min [LF_2 - t_e (1, 2), LF_6 - t_e (1, 6)] = Min [7 - 7, 7 - 6] = 0$

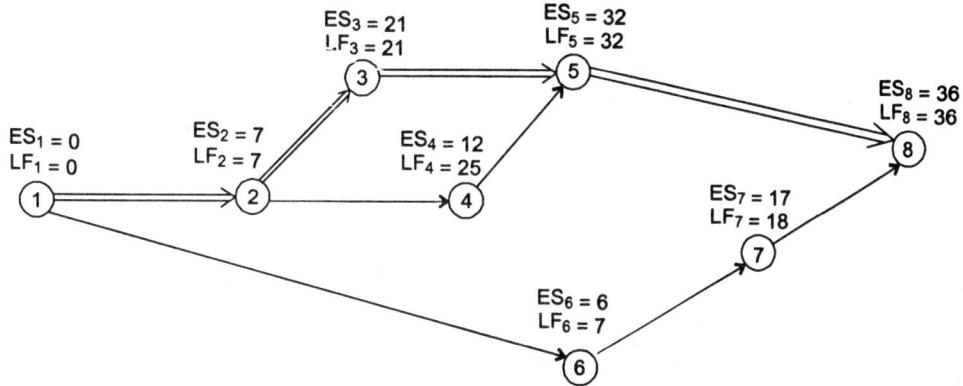

Here we observe that the values of ES and LF are equal on the nodes (1), (2), (3), (5) and (8). So the slack values (LF – ES) are zero at the above nodes. Hence, the critical path is underlined by a line:

Example 10. *A simple project is given below:*

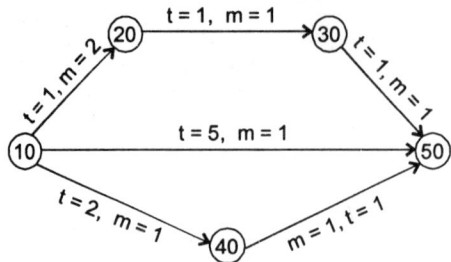

Here t is activity duration and m is labour requirement.

(i) Determine the maximum project schedule.

(ii) Find the project schedule if only two men are available.

(iii) Also write what type of resource scheduling it is?   *(Bihar SBTE, 2009)*

Solution.   *(i)* 5

(ii) 10

(iii) Maintenance project

Example 11. *Consider the network shown in the figure given below. The three time estimates, the expected activity durations and the variances are shown along the arrows. The earliest expected times and the latest allowable accurrence times are computed and put along the nodes. What is the probability of completing the project in (i) 12 days (ii) 14 days (iii) 10 days?*

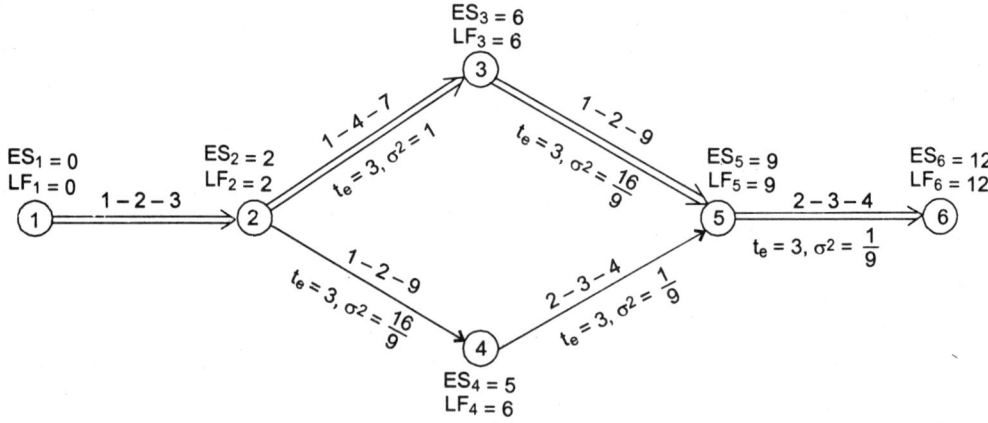

Solution.   We identify that the path $1 - 2 - 3 - 5 - 6$ is the critical path and expected project length is 12 days.

(i) Here T_{cp} = 12 days, T = 12 days

Standard deviation for the project length, $\sigma = \sqrt{\Sigma \sigma_{ij}^2}$ for all ij on the critical path.

$\therefore \quad \sigma = \sqrt{\dfrac{1}{9} + 1 + \dfrac{16}{9} + \dfrac{1}{9}} = 1.73$

$\therefore \quad$ Normal deviate, $z = \dfrac{T - T_{cp}}{\sigma} = \dfrac{12 - 12}{1.73} = 0$

$\therefore \quad$ Probability of completing the project (from table $C - 2$) = 50%

(ii) Here, T = 14 days

$$z = \frac{14-12}{1.73} = 1.16$$

∴ Corresponding probability = 87.7%

(iii) Here, T = 10 days

$$\therefore \quad z = \frac{10-12}{1.73} = -1.16$$

∴ Corresponding probability = 1 – 0.877 = 0.123 = 12.3%   **Ans.**

Example 12. *A project schedule has the following characteristics:*

Activity	Time	Activity	Time
1–2	2	4–5	5
1–4	2	4–8	9
1–7	1	5–8	4
2–3	4	6–9	3
3–6	1	7–8	3
		8–9	5

Construct CPM network and find critical path and time duration of the project.

(Bihar SBTE, 2008)

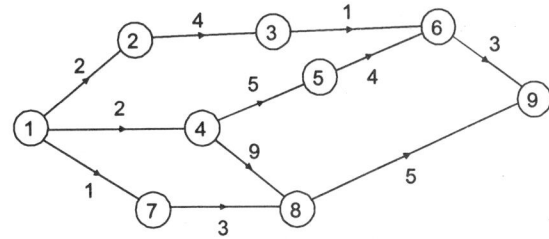

Solution.   The critical path is determined in two phases. In the first phase, one has to start calculations from the "start" node and move towards the "end" node by making use of the earliest start times. In the second phase, calculations are begined from the "end" node and move towards the "start" node by using the latest finish time.

Let us assume that the earliest start time of the first node be zero.

That is, $ES_1 = 0$

Now, $ES_2 = ES_1 + t_{12} = 0 + 2 = 2$

$ES_3 = ES_2 + t_{23} = 2 + 4 = 6$

$ES_4 = 0 + t_{23} = 2$

$ES_5 = ES_4 + t_{45} = 2 + 5 = 7$

$ES_6 = \max (2 + 4 + 1; 2 + 5 + 4) = 11$

$ES_7 = 1$

$ES_8 = \max (2 + 9; 1 + 3) = 11, 4$

$ES_9 = \max (11 + 3, 11 + 5) = 16$

Now, we assume that the target time to finish the last task of the project be 16 days. This
is, $LF_9 = 16$

Then, $LF_8 = 16 - 5 = 11$

$LF_7 = 16 - 5 (5 + 3) = 8$

$LF_6 = 16 - 3 = 13$

$LF_5 = 16 - (3 + 4) = 9$

$LF_4 = 16 - (3 + 4 + 5) = 4$

$LF_3 = 16 - (3 + 1) = 12$

$LF_2 = 16 - (3 + 1 + 4) = 8$

$LF_1 = 16 - (5 + 9 + 2) = 0$

Activity	Normal time	ES_1	EF_1	LF_1	Slatck $LF_1 - EF_1$
1–2	2	0	2	8	6
1–4	2	0	2	2	0
1–7	1	0	1	8	7
2–3	4	2	6	12	6
3–6	1	6	7	13	6
4–1	5	2	7	9	2
4–8	9	2	11	11	0
5–6	4	7	11	13	2
6–9	3	11	14	16	2
7–8	3	1	4	11	7
8–9	5	11	16	16	0

EXERCISE 19.1

1. Construct the network for a project in which activities have the following precedence relationships:
   A, C, D can start immediately.

   $E > B, C;$ $F, G > D;$ $H, I > E, F;$ $J > I, G;$ $K > H;$ $B > A.$

Ans.

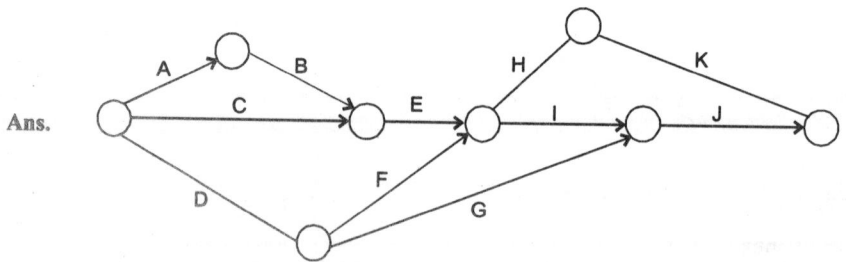

2. Using Fulkerson's rule number the events

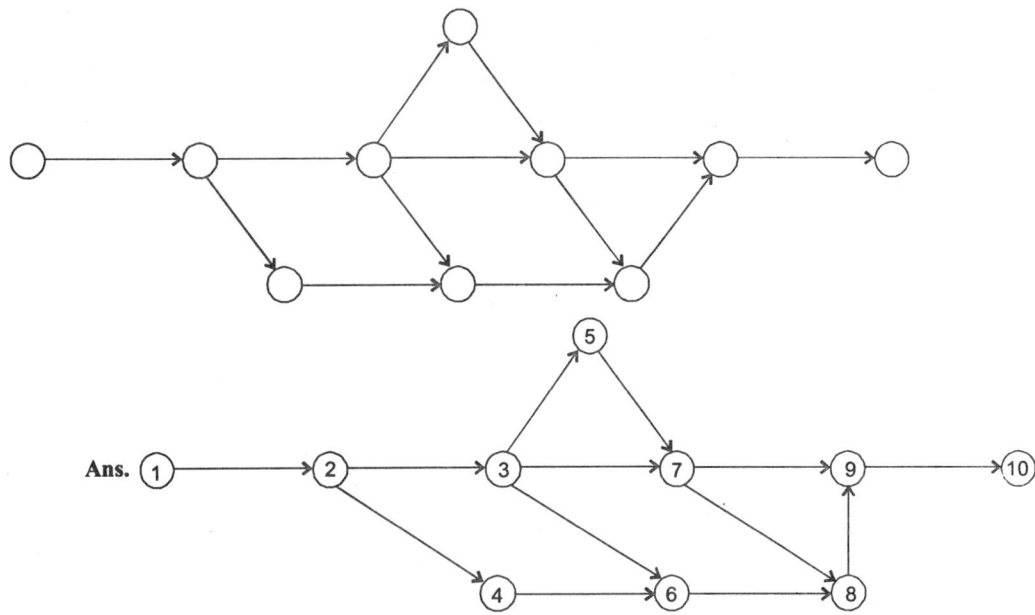

Ans.

3. The dependency relationships between the activities of a project are shown below. Draw the arrow diagram.

Activity Immediate	A	B	C	D	E	F
Predecessor	—	—	—	A, B	B	B, C

Ans.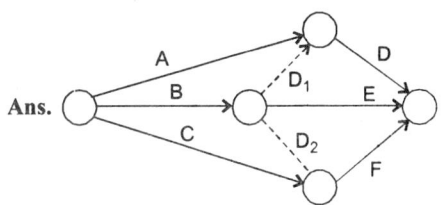

4. Construct the network for the following set of activities:

Activity Immediate	A	B	C	D	E	F	G	H	I	J
Predecessor	—	—	B	A, B	A, B	B	E, D, F	D, E	E, F	H, G, I, C

Ans.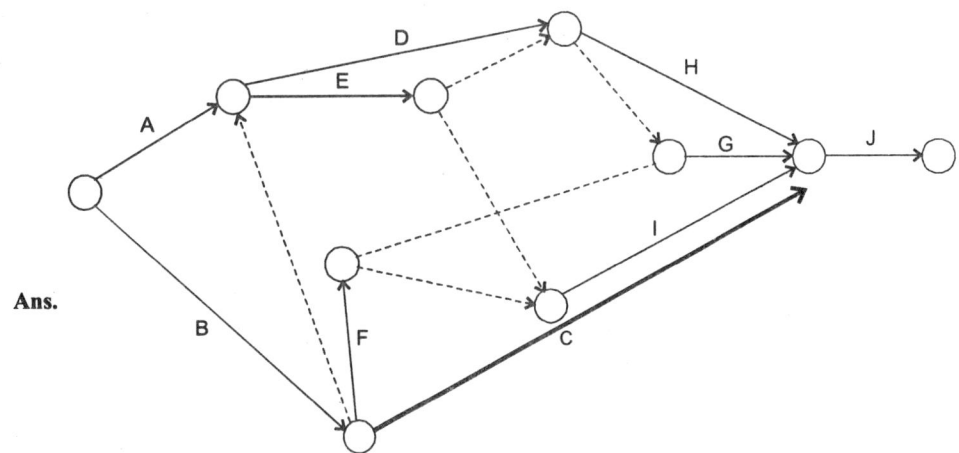

5. The following table shows the jobs of a network along their time estimates. The time estimates are in months.

Activity	A	B	C	D	E	F	G	H	I
Predecessor	–	A	A	–	D	B,CE	F	D	GH
Time (CPM) in months	5	7	6	4	9	8	10	11	8

Construct the network. Find critical path and total time of project. *(Bihar SBTE, 2012)*

6. Consider the network shown below. For each activity, the three time estimates t_o, t_m, and t_p are given along the arrows in the t_o-t_m-t_p order. Determine variance and expected time for each activity.

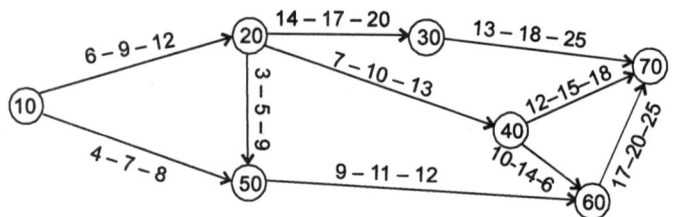

Ans. t_e: 9, 6.7, 17.0, 10.0, 5.33, 18.33, 13.67, 15.00, 10.83, 20.33
$\sigma = 1.00, 0.44, 1.00, 1.00, 1.00, 4.00, 1.00, 1.00, 0.25, 1.78$

7. Find the minimum time of completion of the project, when the time (in days) of completion of each task is written besides each activity of the network given below.

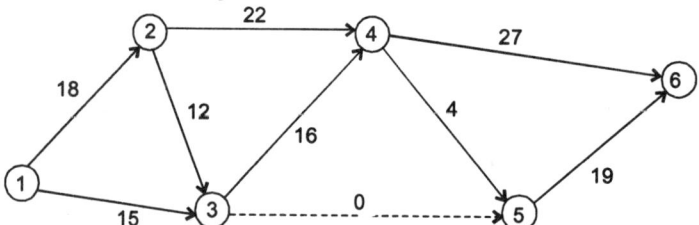

Determine the critical path.

Ans. CPM is

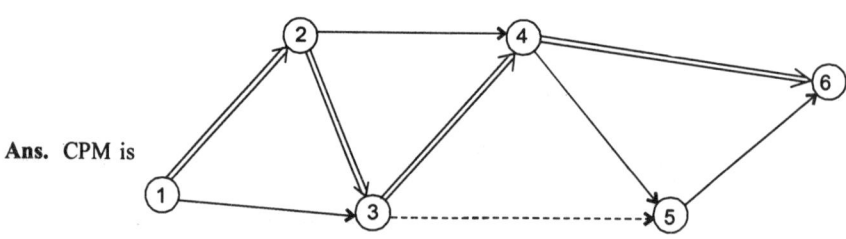

Minimum time of completion = 18 + 12 + 16 + 27 = 73 days.

8. Tasks A, B, C, ... H, I constitute a project. The precedence relationships are

A < D; A < E; B < F; D < F; C < G; C < H; F < I; G < I.

Draw a network to represent the project and find the minimum time of completion of the project when time, in days of each task is as follows:

Task	A	B	C	D	E	F	G	H	I
Time	8	10	8	10	16	17	18	14	9

Also identify the critical path.

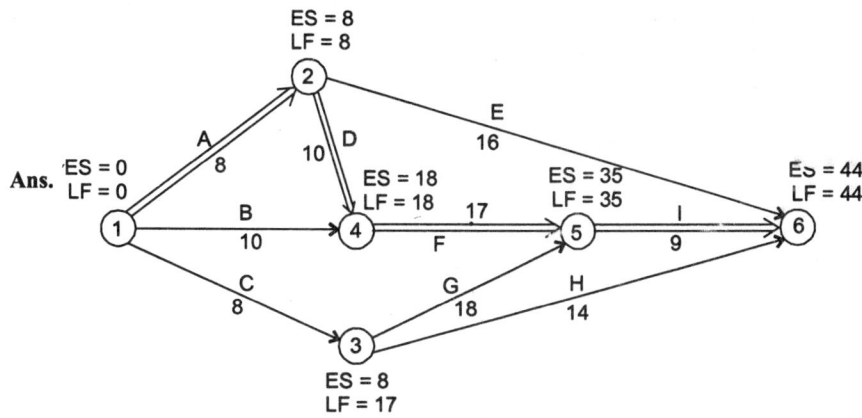

The longest in time involving = 44 days

9. The following table lists the jobs of a network along with their time estimates:

Job i − j	Duration t_o	t_m	t_p	Job i − j	Duration t_o	t_m	t_p
1 − 2	3	6	15	3 − 5	5	11	17
1 − 6	2	5	14	4 − 5	3	6	15
2 − 3	6	12	30	5 − 8	1	4	7
2 − 4	2	5	8	6 − 7	3	9	27
				7 − 8	4	19	28

(a) Draw the project network

(b) Calculate the length and variance of the critical path

(c) What is the probability that the jobs on the critical path will be completed in 41 days.

(d) What is the probability that the jobs on the next most critical path will be completed by the due date of 41 days?

Ans. (a)

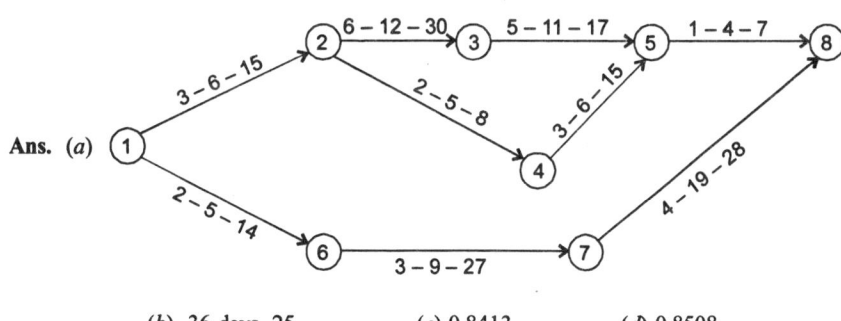

     (b) 36 days, 25,   (c) 0.8413   (d) 0.8508

Choose the correct alternative:

10. The time estimations in a CPM network are :

 (a) probabilistic   (b) Deterministic   (c) Both of these   (d) None of these

                      (Bihar SBTE, 2011, 2004) **Ans.** (b)

11. In CPM the critical path is the path of:

 (a) Minimum time   (b) Average time   (c) Maximum time   (d) Optimal time

                          (Bihar SBTE, 2009) **Ans.** (c)

12. A simple project is given below:

    Here t is activity duration and m is labour requirement.

    (i) Determine the mximum project schedule.

    (ii) Find the project schedule if only two men are available.

    (iii) Also write what type of resource scheduling it is?   *(Bihar SBTE, 2009)* **Ans.** *(i)*

         (i) 5   (ii) 10   (iii) Maintenance project

Fill in the blanks:

13. No. of time estimations in PERT network is ....

             (Bihar SBTE, 2005)   **Ans.** 3 (three)

Write true for correct and False for wrong statement:

14. In PERT network, the optimistic time estimates is more than the pessimistic time estimate.

             (Bihar SBTE, 2003) **Ans.** False

15. Dummy activity consumes time and other resources.   *(Bihar SBTE, 2003)* **Ans.** False

Objective Type Questions

16. Pert consists of .... time estimate

    (a) 01   (b) 02   (c) 03   (d) None of these

             (Bihar SBTE, 2014, 2011) **Ans.** ??

17. Critical path method consists of .... time estimate

    (a) One   (b) Two   (c) Three   (d) Four   *(Bihar SBTE, 2012)* **Ans.** *(a)*

18. In load levelling critical path is

    (a) fixed   (b) increased   (c) decreased   (d) any of these   *(Bihar SBTE, 2010)* **Ans.** *(a)*

19. If T_0 is optimistic time, T_m is most likely time. T_p is pessimistic time then expected time T_1 is equal to

    (a) $\dfrac{T_0 + T_m + T_p}{3}$ (b) $\dfrac{T_0 + 2T_m + T_p}{4}$ (c) $\dfrac{T_0 + 4T_m + T_p}{6}$ (d) $\dfrac{T_0 + 6T_m + T_p}{8}$

             (Bihar SBTE, 2008) **Ans.** *(c)*

20. If optimistic time = 2 hours, most likely time = 3 hours, passimistic time = 4 hours then expected time is equal to

    (a) 2 hours   (b) 3 hours   (c) 4 hours   (d) None of these

             (Bihar SBTE, 2014) **Ans.** *(??)*

20 Simulation

20.1 INTRODUCTION

Some time decision making problems arise in administration or business. Simulation is one of the important technique of solving a decision making problem with the help of modal.

Simulation, which can appropriately be called management laboratory, determines the effect of a number of alternate ways without disturbing the real system. It helps in selecting the best way with the prior assurances that its implementation will be beneficial.

20.2 SIMULATION   *(Bihar SBTE, 2009)*

Simulation is the imitation of reality which may be in the physical form or in the form of mathematical equations. For example; testing of an air craft model in a wind tunnel from which we determine the performance of the actual aircraft under real operating conditions.

Simulation can be defined as a representation of reality through the use of a model or other divices which will react in the same manner as reality under given set of conditions.

20.3 ADVANTAGES OF SIMULATION TECHNIQUE   *(Bihar SBTE, 2008)*

1. Some important managerial decision problems are solved by mathematical programming and experimentation with the actual system.
2. By simulation management we can foresee the difficulty and bottlenecks which may come due to new technique, equipment or process.
3. Operating personnels and non technical managers can understand easily the proposed plans.
4. It is flexible and can be modify according to changing conditions.
5. Computers simulation can express the performance of a system into few minutes.
6. It is easier than mathematical models.
7. Simulation has been used for training the managers.

20.4 LIMITATIONS

1. It does not produce optimum results.
2. By simulation we can not quantify all the variables that effect the behaviour of the system.
3. A large number of variables may exceed the capacity of the available computer.
4. Simulation is not a cheep method of analysis.
5. It is only useful to some simple problems.

20.5 APPLICATIONS OF SIMULATION

1. To evaluate the area under a curve.
2. To estimate the value of π

3. Industrial problem including the shop floor management, design of computer system etc.

4. In business, customer behaviour, price determination etc.

5. In social problem, population growth, effect of environment on health.

6. In biomedical, fluid balance, distribution of electrolyte in the human body.

20.6 MONTE-CARLO METHOD   (*Bihar SBTE, 2004*)

In world war II monte-Carlo method was developed by two mathematician. Johnvon newmann and Stainslaw ulam. How far neutrons, would travelling through different material. It was a successful technique. With this remarkable successs this technique became popular in business and industry now a days. The steps involved caring out Monte – Carlo simulation are:

1. Select the major of effectiveness (objective) function of the problem. It is either to be maximize or minimize.

2. Identify the variables that effect the major of objective function.

3. Determine the cumulative probability distribution of each variable. Plot these distributions with values of variable along x-axis and cumulative probability along y-axis.

4. Find a set of random numbers.

5. Consider the cumulative probability to a decimal value. Plot cumulative distribution along y-axis. Project this point horizontally so that it can meet distribution curve. Find the projection of this point on x-axis.

6. Record these values and substitute in the formula chosen for major of effectiveness.

7. Repeat steps 5 and 6 so that the decision maker may satisfy with the sample.

20.7 ADVANTAGES AND DIS-ADVANTAGES OF MONTE-CARLO METHOD   (*Bihar SBTE, 2003*)

Advantage

1. Monte carlo method helps in finding a solution of those problems which otherwise is not possible.

2. By this method the experiments are done on the simulated models rather on real system. Because on real system it will be inconvenient and time consuming.

3. Although the technique does not provide an optimal solution however it gives an insight for analysing the problem.

Disadvantage

1. Monte Carlo method does not provide an optimal solution.

2. By using real small model the results obtained are not good.

3. The computers are quite cumbersome even for a small problem.

4. It is costly affair as it consumes a lot of computer time.

Example 1. *Find the value of π experimentally by simulation.*   (*Bihar SBTE, 2009*)

Solution.   Draw coordinate axes OX and OY. With centre O draw an arc AB of unit radius and complete the square OACB equation of the circle is $x^2 + y^2 = 1$

Plot a point whose coordinates are (0.1,0.8). In this way plot 100 or 1000 of points. Suppose N is the total number of points considered out of which n are the points lies in / on the arc.Then

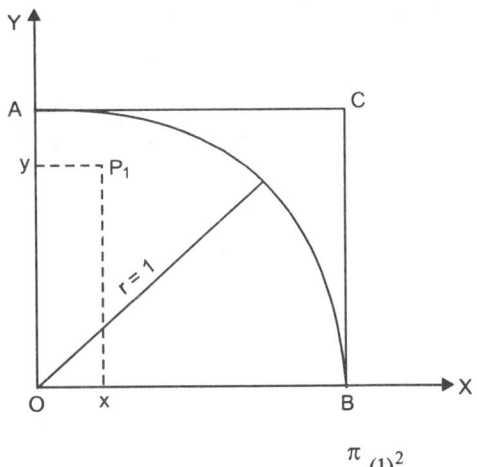

$$\frac{n}{N} = \frac{\text{area enclosed by the arc}}{\text{area of the square}} = \frac{\frac{\pi}{4}(1)^2}{1} = \frac{\pi}{4}$$

$$\Rightarrow \qquad \pi = \frac{4n}{N} \qquad\qquad\qquad \textbf{Ans.}$$

(Event Type Simulation)

Example 2. *Customers arrive at a service facility to get the required service. The interval and service times are 1.8 min and 4 min respectively. Simulate the system for 14 min. Determine the average waiting time of a customer and ideal time of the service facility.*

Solution. The arrival time of customer within 14 min periods will be

Customer	1	2	3	4	5	6	7	8
Arrival time	0	1.8	3.6	5.4	7.2	9	10.8	12.6

Customer	Service		Waiting time of Customer	Ideal time of service facility
	begins	ends		
1	0	4	0	0
2	4	8	4 – 1.8 = 2.2	0
3	8	12	8 – 3.6 = 4.4	0
4	12	16	12 – 5.4 = 6.6	0
5	14		14 – 7.2 = 6.8	
6			6.8 – 1.8 = 5.0	
7			5 – 1.8 = 3.2	
8			3.2 – 1.8 = 1.4	

The average waiting time of a customer.

$$= \frac{0 + 2.2 + 4.4 + 6.6 + 6.8 + 5 + 3.2 + 1.4}{8} = \frac{29.6}{8} = 3.7$$

Ideal time of facility = nill   **Ans.**

EXERCISE 20.1

1. What is simulation? Describe its advantages in solving the problems. Give its limitation with suitable examples?

2. Describe the simulation process. What are the reasons for using simulation.

3. What is simulation? Why simulation is useful in technical education?   (*Bihar SBTE, 2003, 2005*)

4. Explain why do you apply Monte carlo simulation technique for PERT network illustrate with example.

5. Explain with suitable examples the Monte Carlo method for solving theoretical problems.

6. Write short note on the simulation technique.   (*Bihar SBTE, 2003*)

Choose the correct alternative:

7. Monte Carlo refers to the name of a

   (a) Place   (b) Man   (c) Both of them   (*d*) None   (*Bihar SBTE, 2003*) **Ans.** (*a*)

8. Simulation is .... reality.

   (a) Duplication   (b) immitation   (c) improvement   (*d*) simplification

                                                                       (*Bihar SBTE, 2003*) **Ans.** (*a*)

2011 (Odd)

Bihar Board of Technical Education
THIRD SEMESTER
Professional Mathematics

Time : 3 Hrs

Full Marks : 80

Pass Marks : 26

Answer All 20 questions from Group A (each of 1 mark).

Group B consists of 10 questions out of which five questions are to be answered (each of 4 marks).

Group C consists of three parts I, II, III. Answer any five questions selecting not more than two from any part. (Each of 8 marks).

All parts of question must be answered at one place in sequence, otherwise they may not be evaluated.

The figures in the right hand margin indicate full marks.

GROUP – A

1. Select the most appropriate answer from the alternatives given under:

   (i) If x be the true value of a quantity and x_1 be the approximate value then the percentage error is

   (a) $\dfrac{x - x_1}{x} \times 100$ (b) $\dfrac{x - x_1}{x_1} \times 100$ (c) $\dfrac{x_1 - x}{x}$ (d) $\dfrac{x_1 + x}{x}$ **Ans.** (a)

   (ii) A first negative real root of the equation $x^3 + 3x^2 - 3 = 0$ lies between ....

   (a) $(0, -1)$ (b) $(-1, -2)$ (c) $(-2, -3)$ (d) $(-3, -4)$

   (See Example 8 on page 16) **Ans.** (b)

   (iii) If $a < b$ and $f(a)$ and $f(b)$ be of opposite signs for the equation $f(x) = 0$, then bu Regula falsi method, the 1st approximation of the root of the equation is ....

   (a) $\dfrac{af(b) - bf(a)}{n - a}$ (b) $\dfrac{af(b) - bf(a)}{f(a) - f(b)}$ (c) $\dfrac{af(b) - bf(a)}{f(b) - f(a)}$ (d) None of these

   (See Q. 10 on page 32) **Ans.** (c)

   (iv) Simultaneous linear algebraic equations may be solved by ....

   (a) Gauss-Jordan method   (b) Newton-Rapson method

   (c) Regula-Falsi method   (d) Hungarian method

   (See Q. 15 on page 46) **Ans.** (a)

   (v) Which of the following is true?

   (a) $E = e^{hD}$ (b) $E = e^{2hD}$ (c) $E^2 = e^{hD}$ (d) None of these

   (See Example 18 on page 63) **Ans.** (a)

(vi)   Newton Backward interpolation formula is useful for interpolation near the .... of Tabular values.

     (a) Beginning   (b) End   (c) Middle   (d) None of these

           Ans. (b)

(vii)   Which result is true?

     (a) Mean = Variance   (b) Mean = $\sqrt{\text{Variance}}$

     (c) Mean = Standard deviation   (d) Mean = $\sqrt{\text{Standard deviation}}$ **Ans. (a)**

(viii) $\Delta^2 y_0$ = ...

     (a) $y_1 - y_0$ (b) $y_2 + 2y_1 + y_0$ (c) $y_2 - 2y_1 + y_0$ (d) None of these

           (*See Example 15 on page 63*) **Ans. (c)**

(ix)   What is the probability of drawing one spade from a pack of 52 cards?

     (a) $\dfrac{1}{52}$ (b) $\dfrac{1}{4}$ (c) $\dfrac{4}{52}$ (d) $\dfrac{13}{51}$ **Ans. (c)**

(x)   A product of the form:

     x (x – 1) (x – 2) (x – 3) .... (x – 9) is denoted by ....

     (a) $[x]^9$ (b) $[x]^{10}$ (c) $[x]^{11}$ (d) None of these

           (*See Example 19 on page 63*) **Ans. (b)**

(xi)   A linear programming problem can be solved by ....

     (a) North-West corner method   (b) Simpson 1/3 Rule

     (c) Simplex method   (d) Newton-Rampson Method

           (*See Example 12 on page 255*) **Ans. (c)**

(xii)   Transportation problem in any company may be solved by...............

     (a) Newton Rapson Method   (b) Trepezoidal Method

     (c) Simpson 1/3 Method   (d) Least cost method

           (*See Example 9 on page 267*) **Ans. (d)**

(xiii)   In a critical path method, there is .... the estimate.

     (a) 03   (b) 02   (c) 04   (d) 01   **Ans. (a)**

(xiv)   In a given table the value of Mean. ....

Number	5	10	15	20	25
Frequency	2	2	3	2	1

     (a) 12   (b) 13   (c) 14   (d) 15   **Ans. (c)**

(xv)   Standard deviation of Normal Distribution is equal to ....

     (a) $\sqrt{np}$ (b) np   (c) npq   (d) $\sqrt{npq}$ **Ans. (d)**

(xvi)   What is the chance that a leap your, selecting at random, will contain 53 Monday?

(a) $\dfrac{1}{7}$ (b) $\dfrac{2}{53}$ (c) $\dfrac{1}{53}$ (d) $\dfrac{2}{7}$ **Ans. (d)**

(xvii)   A bag contains 3 white and 2 black balls. Find the probability of drawing white ball.

(a) $\dfrac{3}{5}$ (b) $\dfrac{2}{5}$ (c) $\dfrac{1}{5}$ (d) $\dfrac{4}{5}$

(See Q. (vi) on page 164) **Ans. (a)**

(xviii) If A and B are two independent events then ....

(a) $P(A \cap B) = P(A) \cdot P(B/A)$ (c) $P(A \cap B) = P(A) \cdot P(A/B)$

(c) $P(A \cap B) = P(A) \cdot P(B)$ (d) $P(A \cap B) = P(A) + P(B)$ **Ans. (c)**

(xix)   For difference equation $(E^2 + 5E + 6) Y_n = 0$, $Y_n = $ ....

(a) $C_1 2^n + C_2 3^n$ (b) $C_1 (-2)^n + C_2 (-3)^n$

(c) $C_1 (-2)^n + C_2 (3)^n$ (d) $C_1 (2)^n + C_2 (-3)^n$ **Ans. (b)**

(xx)   Which method is used to evaluate $\displaystyle\int_{x_0}^{x_0 + nh} f(x) \cdot dx$

(a) Gauss-Jordan Method   (b) Euler's method

(c) Simpson 1/3 Rule   (d) Hungarian Method

(See Example 22 on page 104) **Ans. (c)**

GROUP – B

2. **Answer any five questions (each of 4 marks):**

(i)   Solve the following equation by Gauss Jordan Method

$2x - 8y + z = -5$; $x - 2y + 9z = 8$; $3x + y - z = 3$ *(See Q. 13 on page 46)*

(ii)   Find a positive root of the equation $2x^3 - 3x^2 - 9 = 0$ by Newton-Rapson Method upto third approxmation of the root.   *(See Example 11 on page 21)*

(iii) Evaluate $\displaystyle\int_{-3}^{3} x^4 dx$ by using Trapezoidal Rule.   *(See Q. 10 on page 96)*

(iv)   For Poisson Distribution show that: $P(r) = \dfrac{m^r}{\lfloor r} e^{-m}$ *(See Art. 13.1 on page 192)*

(v)   Find the probability of drawing a card of Queen and a card of Heart from a pack of 52 cards when (i) 1st card is replaced (ii) 1st card is not replaced.   *(See Similar Example 9 on page 147)*

(vi)   Find the probability of getting 04 heads in 10 tosses of a coin.   **Ans. (210)** $\left(\dfrac{1}{2}\right)^{10}$

(vii)   find the total transportation cost for the following Table by Vogel's Approximaion Method (VAM):

Destination (cost in hundred)   *(See Example 16 on page 282)*

	D_1	D_2	D_3	D_4	Capacity
S_1	8	9	7	5	100
S_2	6	4	8	7	150
S_3	9	10	7	6	150
S_4	5	8	4	6	200
Demand	175	100	125	200	

(viii) White do you man by Network Analysis? Write the difference of CPM & PERT.

(See Art. 19.26 on page 318)

(ix)   Solve the following LPP by graphical method Maximum $z = 9x_1 + 6x_2$

Subject to:

$5x_1 + 10x_2 \le 50; \quad 8x_1 + 2x_2 \ge 16; \quad 3x_1 - 2x_2 \ge 6; \quad x_1, x_2 \ge 0$

(See Example 9 on page 221)

GROUP C
PART – I

3.   Answer any five questions selecting not more than two questions from each part (Each of 8 marks)

If

$x:$	4	8	12	16	20
$y:$	9	12	17	25	20

Find f (5), f (18) and f '' (18).   **Ans.** $f(5) = 10.256, f(18) = 24.789, f''(18) = -1.622$

4.   (a)   Establish Simpson $\frac{1}{3}$ Rule for numerical integration.   *(See Art. 17.4 on page 97)*

(b)   Find a real root of the $2x^3 + 5x - 9 = 0$ by Regula-Falsi Method upto third approximation of Root.

(See Example 25 on page 31)

5.   Solve any two of the following equations:

(i) $y_{n+2} - 5y_{n+1} - 6y_n = 12^n$ *(See Similar Example 8 on page 109)*
(ii) $(E^2 - 4E + 4) y_n = 2^n$ *(See Example 11 on page 110)*
(iii) $y_{n+1} - 2 \cos \alpha\, y_n + y_{n=1} = 0$ *(See Example 7 on page 108)*

PART – II

6.   Find Mean, Median, Mode and standard Deviation for the following table:

Class Interval	0 – 5	5 – 10	10 – 15	15 – 20	20 – 25
Frequency	5	7	6	10	2

(See Q. 30 on page 138)

7.   What is Binomial Probability Distributiobn? Find the Mean and Variance of the Binomial Probability Distribution.   *(See Page 179, 187, 188)*

8.   The probability that a Television manufactured by a company will defective is $\frac{1}{10}$. If 12 such Televisions are manufactured, find the probability that:

(a) Exactly two will be defective   (b) At least two will be defective
(c) None will be defective   *(See Example 16 on page 185)*

PART – III

9. Five jobs are to be done and five Technicians are available. Any Technician do any job and the Time taken (in hous) by each Technician to do the Jobs being given below in Table. Find the assignment of Jobs to the technicians that will minimize the Total time using Hungarian method.

Technicians

Jobs	T_1	T_2	T_3	T_4	T_5
J_1	32	38	40	28	40
J_2	40	24	28	31	36
J_3	41	27	33	30	37
J_4	22	38	41	36	36
J_5	29	33	40	35	39

(See Example 9 on page 302)

10. Solve the following LPP by using Simplex method

Max. $z = 3x_1 + 2x_2 + 5x_3$

Subject to

$$x_1 + x_2 + x_3 \le 9;$$
$$2x_1 + 3x_2 + 5x_3 \le 30;$$
$$2x_1 - x_2 - x_3 \le 0$$

and $x_1, x_2, x_3 \ge 0$ *(See Similar Example 8 on page 245)*

11. Construct a project Network and find critical path and Total project time for the following table of the activity of a network along with their CPM time estimate.

Activity	A	B	C	D	E	F
Predecessor	–	–	–	A	B	B

Activity	G	H	I	J	K
Predecessor	C	D	E	H,J	F,G

(See Q. 11 on page 339)

2012 (Odd Course)
Bihar Board of Technical Education
THIRD SEMESTER
Professional Mathematics

Time : 3 Hrs

Full Marks : 80
Pass Marks : 26

GROUP – A

2. **Select the most suitable answer from the given alternatives:**

(i) Round-off the number 6.0009 correct upto 4 significant figures is............ . **Ans.** 6.00

(ii) Define percentage error of a number. **(Ans.** *See (6) on page 4)*

(iii) $\Delta^5 (x^6 + 7x^5 + 9x^4 + 7)$ is a ....

    (a) Variable   (b) Constant   (c) Both of these   (d) None of these

    (See Example 17 on page 63) **Ans.** (a)

(iv) The function $f(x) = 2x^3 - x - 6 = 0$ has a real root lying between .... **Ans.** (a)

    (a) (1, 2)   (b) (2, 3)   (c) (3, 4)   (d) None of these   **Ans.** (a)

(v) Approximate root of the equation $f(x) = 0$ obtained by .... method.

    (a) Hungarian method   (b) Simpson 1/3 method

    (c) Interation method   (d) None of these   **Ans.** (c)

(vi) Complementary function of Difference equation $(E^2 - 1) Y_n = 0$ is ....

    (a) (–1, 1)   (b) (2, 1)   (c) (1, 1)   (d) (1, 0)   **Ans.** (a)

(vii) Which is correct relation?

    (a) $\log (1 + \Delta) = h^2 D^2$ (b) $\log (1 \div \Delta) = hD$

    (c) $\log (1 + \Delta) = hD$ (d) $\log (1 - \Delta) = h^2 D^2$ **Ans.** (c)

(viii) Transportation problem is solved by .... method

    (a) simpson 1/3 method   (b) Least cost method

    (c) Bisection method   (d) None of these

    (See Example 7 on page 267) **Ans.** (b)

(ix) Critical path method consists of .... time estimate

    (a) One   (b) Two   (c) Three   (d) Four

    (See Q. 17 on page 330) **Ans.** (a)

(x) The optimal solution of LPP:

    (a) Maximum   (b) Minimum   (c) Both (a) and (b)   (d) None   **Ans.** (c)

(xi) Linear programming problem can be solved by .... method.

    (a) Simpson 1/3 method   (b) N.W. corner method

    (c) Gauss-Jordan method   (d) None of these

    (See Example 16 on page 233) **Ans.** (b)

(xii) Hugarian method is used to solve ....

    (a) Transporation problem   (b) Assignment problem

    (c) Linear programming problem   (d) None of these   **Ans.** (b)

(xiii) If T_0 be optimistic time T_M be most likely time, Tp be pessimistic time then expected time T_E is equal to

(a) $\dfrac{T_0 + T_M + T_P}{4}$ (b) $\dfrac{T_0 + 2T_M + T_P}{6}$ (c) $\dfrac{T_0 + 4T_M + T_P}{4}$ (d) $\dfrac{T_0 + 4T_M + T_P}{6}$

Ans. (d)

(xiv) The variance of Binomial distribution $^nC_p P^r q^{n-r}$ $(r = 0, 1, 2.........n)$

(a) $\sqrt{npq}$ (b) $(npq)^2$ (c) npq   (d) np   **Ans. (a)**

(xv) A die is thrown. The probability that the digit coming up is greater than 5 is ....

(a) $\dfrac{1}{6}$ (b) $\dfrac{5}{6}$ (c) $\dfrac{1}{3}$ (d) $\dfrac{1}{2}$ **Ans. (a)**

(xvi) A card is drawn from a peak of 52 cards. What is the probability of getting queen of king?

(a) $\dfrac{1}{13}$ (b) $\dfrac{2}{13}$ (c) $\dfrac{3}{13}$ (d) $\dfrac{4}{13}$ **Ans. (b)**

(xvii) In Poisson Distribution standard deviation is equal to..............

(a) Mean   (b) Square of mean   (c) Square root of mean   (d) Variance

(See Example 25 on page 206) **Ans. (c)**

(xviii) Mode of a sample:

(a) Observed most frequently   (b) Occurs most frequently

(c) Both (a) and (b)   (d) None of these   **Ans. (d)**

(xix) The probability of event lies between ....

(a) $(-\infty$ and $\infty)$ (b) -1 and $+1)$ (c) $(0$ and $-1)$ (d) None of these   **Ans. (d)**

(xx) Which is correct relation?

(a) PA + P $(A^1) = \infty$ (B) $P(A) + P(A^1) = 0$ (c) $P(A) + P(A^1) = 1$ (d) $P(A) + P(A^1) = -1$

Ans. (c)

GROUP – B

2. Find the 1st positive root of $f(x) = 2x^3 - 3x - 6 = 90$ by Newton-Rapsion method up to third approximation of root.   *(See Example 5 on page 15)*

3. Establish Newton-Forward interpolation formmula.   *(See Art. 5.4 on page 64)*

4. If

x :	6	8	10	12	14
y :	4	20	16	30	40

find $f'(13)$.   *(See Example 10 on page 89)*

5. Solve the following LPP graphically:

Maximize $Z = 10x_1 + 25x_2$

Subject to condition $x_1 + 3x_2 \le 12$; $4x_1 + x_2 \ge 16$ and $x_1, x_2 \ge 0$ *(See Example 8 on page 220)*

6. Determine a solution to the following transportation problem using least cost method.

↓ Source	D_1	D_2	D_3	D_4	Capacity
S_1	8	9	6	10	400
S_2	12	15	17	8	300
S_3	9	5	20	7	500
Demand	250	250	450	250	1200

(See Example 7 on page 265)

7. What do you mean by newtwork analysis? How this is important in calculation of total project duration?   *(See Page 311)*

8. Bag A contains 10 red and 5 white balls. Bag B contains 8 red and 7 white balls. If any one ball (red or white) is transferred from A to bag B, find the probability of drawing one white ball from bag B.   *(See Example 23 on page 152)*

9. Find mean and standard deviation of poisson distribution.   (*See Art. 13.2, 13.3 on page 193*)

10. A card is drawn from an ordinary standard a gambler bets that it is a spade king. What is the probability of his not winning the bet?   (*See Example 10 on page 147*)

GROUP – C

11. Solve the following difference equations (Any two)

(a) $\left(E - 2\cos\alpha \dfrac{1}{E} \right) Y_n = 0$ (*See Example 7 on page 108*)

(b) $(E-5)(E-7)(E+6)(E-1) = 2^n$;   (*See Example 12 on page 111*)

(c) $Y_{n+2} - 4Y_{n+1} + 4y_n = 2^n$ (*See Example 11 on page 110*)

12. (a) Solve the following system of equations of Gauss Elimination Method.

$10x_1 + x_2 + x_3 = 12$;
$x_1 + 10x_2 + 2x_3 = 10$
$3x_1 + 3x_2 + 5x_2 = 8$ (*See Similar Example 1 on page 38*)

(b) Evaluate $\Delta^2 (2x^3 + 4x^2 + 7x - 6)$ (*See Example 14 on page 60*)

13. Eastablish Newton-Cote's formula and then deduce trapezoidal rule for numerical integration.

(*See page 93*)

14. Solve the following linear programming problem (LPP) using simplex method.

Maximum $Z = x_1 + x_2 + 3x_3$
Subject to condition $3x_1 + 2x_2 + x_3 \leq 3$
$2x_1 + x_2 + 2x_3 \leq 2$
and $x_1, x_2, x_3 \geq 0$ (*See Similar Example 9 on page 247*)

15. The following table shows the jobs of a network along their time estimates. The time estimates are in months.

Activity	A	B	C	D	E	F	G	H	I
Predecessor	–	A	A	–	D	B,C,E	F	D	G,H
Time (CPM) in Month	5	7	6	4	9	8	10	11	8

Construct the network. Find critical path and total time of project.   (*See Example 5 on page 328*)

16. Five jobs are to be processed and life machines are available. Any machine can process any job and the time taken (in hours) by each machine to do the job beings as follows:

Machines   (*See Example 9 on page 302*)

17. The following table shows the marks obtained by students in an examination. Calculate the mean median, mode and Standard Deviation of the distribution.

Marks	1–10	10–20	20–30	30–40	40–50	50–60
No. of Students	05	10	15	25	20	15

Ans. Mean = 35

S.D = 4.72

18. (a) What is the chance of getting 7 or 11 with 2 dice?   (*Bihar SBTE, 2012*) **Ans.** $p = \dfrac{2}{9}$

(b) If $P(A) = \dfrac{3}{8}, P(B) = \dfrac{1}{2}$ and $P(A \cap B) = \dfrac{1}{4}$ then find: **Ans.** (A/B) = $\dfrac{1}{2}$

(i) P (A/B) and (ii) P (A^1/B^1)

2014 (Odd Course)
Bihar Board of Technical Education
THIRD SEMESTER
Professional Mathematics

Time : 3 Hrs

Full Marks : 80
Pass Marks : 26

GROUP – A

2. **Choose the most suitable answer from the following options:**

(i) If 0.333 is the approximate value of 1/3 then relative error is

(a) 0.003   (b) 0.02   (c) 0.001   (d) None of these

(See Example 21 on page 7)

(ii) 1st positive root of the equaiton $x^4 + 3x^3 + 5x^2 + 7x - 6 = 0$ lies between ....

(a) (0, 1)   (b) (1, 2)   (c) (2, 3)   (d) (3, 4)   *(See Q. 7, on page 16)*

(iii) $\Delta^n (x^n + 5x^{n-1} + 4x^{n-2} + ... + 5) =$ ....

(a) constant   (b) n   (c) (n – 1)   (d) none of these   **Ans. (a)**

(iv) In Newton's forward Interpolation formula we use the operator:

(a) E   (b) ∇ (c) D   (d) None of these

(See Example 20 on page 73)

(v) Which relation is correct?

(a) $\nabla f(x) = f(x) + f(x - h)$ (b) $\nabla f(x) = f(x) - f(x - h)$

(c) $\nabla f(x) = f(x) - f(x + h)$ (d) None of these   **Ans. (b)**

(vi) Newton's Backward Interpolation formula is useful to interpolation near the .... of tabular form

(a) Beginning   (b) End   (c) Middle   (d) None of these

(See Example 12 on page 78) **Ans. (b)**

(vii) A card is drawn from a pack of 52 cards. What is the probability of getting a king

(a) 1/52   (b) 1/3   (c) 4/13   (d) None of these   **Ans. (d)**

(viii) Approximate root of the equation $f(x) = 0$ is obtained by ....

(a) Simplex Method   (b) Gauss-Jordan Method

(c) Graphical Method   (d) None of these *(See Q. 12 on page 16)* **Ans. (c)**

(ix) Order of difference equation $y_{n+2} + 3y_{n-1} + 2y_n = 0$ is.......

(a) 0   (b) 1   (c) 2   (d) None of these

(See Example 11 on page 109) **Ans. (c)**

(x) Pert consists of .... time estimate

(a) 01   (b) 02   (c) 03   (d) None of these

(See Q. 16 on page 330) **Ans. (b)**

(xi) Linear programming problem can be solved by

(a) Newton's Rapson Method   (b) Simplex method

(c) N.W. Corner Method   (d) None of these   **Ans. (b)**

(xii) Hungarian method is used to solve ....

(a) Linear Programming Problem   (b) Transportation Problem

(c) Assignment Problem   (d) None of these *(See Example 17 on page 309)* **Ans. (c)**

(xiii) Standard deviation of Binomial Distribution is equal to ....

(a) np   (b) np $(1-p)$ (c) $\sqrt{np(1-p)}$ (d) None of these   **Ans. (c)**

(xiv) Auxiliary equation of the difference equation $y_{n+2} + 5y_{n+1} + 6y_n = 2^n$ is.................

(a) $(E^2 + 6E + 5) = 0$ (b) $(E^2 + 5E + 6) = 0$

(c) $(E^2 - 5E + 6) = 0$ (d) None of these   **Ans. (b)**

(xv) A card is drawn from a pck of 52 cards What is the probability of getting a black card?

(a) 1/4   (b) 1/2   (c) 1/3   (d) 3/52   **Ans. (b)**

(xvi) If optimistic time = 2 hours, most likely time = 3 hours, passimistic time = 4 hours then expected time is equal to

(a) 2 hours   (b) 3 hours   (c) 4 hours   (d) None of these

(See Q. 20 on page 330)

(xvii) Feasible region of a linear programming problem lies in the ....

(a) First quadrant   (b) Second quadrant (c) Third quadrant   (d) Fourth quadrant   **Ans. (a)**

(xviii) Square of standard deviation is known as ....

(a) Mean deviation (b) Variance   (c) Quartile   (d) None of these   **Ans. (b)**

(xix) A graphical representation of nodes and connecting arrows represent activities is called ......

(a) Project   (b) Network   (c) Research   (d) None of these

(xx) There will be .... different solution in a n × n assignment problem

(a) n   (b) $n!$ (c) $(n-n)!$ (d) $(n+1)!$

(See Example 11 on page 309) **Ans. (b)**

GROUP – B

Answer all five questions.

2. Find the 1st positive root of the equation $5x^3 + 2x^2 + x - 9 = 0$ by Bisection Method up to third approximation of the root.   *(See Q. 6 on page 16)*

3. Solve the following equations by Gauss-Jordan Method

$3x + 4y + 5z = 12; \ x + 2y + 3z = 10; \ 2x + 3y + 2z = 8$ *(See 11. 6 on page 46)*

OR

Evaluate: $\Delta (5x^3 - 2x^2 + 3x + 9)$ *(See Example 2 on page 48)*

4. If

x	6	8	10	12	14	16
$y = f(x)$	15	10	20	25	22	35

(See Q. 2 on page 72)

then find y for $x = 15$

OR

Solve the following LPP by Graphical method.

Maximize $z = 10x_1 + 20x_2$

with condition

$2x_1 + 5x_2 \geq 10; \quad 7x_1 + 3x_2 \leq 49; \ 3x_1 + 7x_2 \geq 12 \ \& \ x_1, x_2 \geq 0$ *(See Example 10 on page 223)*

5. Determine a solution to the following transportation problem using VAM method.

Destination

Source	D_1	D_2	D_3	D_4	Capacity
S_1	20	25	30	40	500
S_2	22	40	35	50	500
S_3	25	20	38	20	500
Demand	300	450	350	400	

(See Example 14 on page 278)

OR

What is operational Research? Write the scope of operational research.   *(See Art. 14.1 on page 207)*

6. Bag 'A" contains 5 red balls and 7 black balls. Bag 'B" contains 6 red balls and 8 black balls. It any one ball (Red or Black) is transferred from bag A to Bag B. Find the probability of drawing one black ball frokm Bag B.   *(See Similar Example 14 on page 149)*

OR

A die tossed t times A success is getting 1 or 6 on a toss. Find the mean and variance of the number of success.   *(See Example 18 on page 189)*

GROUP – C

7. Find f′ (5) and f″ (5) for the following table:

x :	2	7	12	17	22
$y = f(x)$	8	15	10	25	30

(See Example 2 on page 81)

OR

evaluate: $\int_0^6 \frac{1}{1+x} dx$ by

(i) Trapezoidal rule and   *(See Similar Example 2 on page 95)*

(ii) Simpson 1/3 Rule   *(See Similar Example 7 on page 100)*

8. Solve the following difference equations:

(a) $(E^2 - 4E + 3y) y_n = 3^n$ *(See Example 10 on page 110)*

(b) $(y_{n+3} - 4y_{n+1} + y_n) = 5$ *(See Example 14 on page 111)*

OR

(a) Establish Newton Forward Interpolation formula.   *(See Art. 5.4 on page 64)*

(b) Prove that $E = 1 + \Delta = (1 - \nabla)^{-1}$ *(See Art. 4.5 on page 52 and Art. 4.6 on page 53)*

9. Find mean and standard deviation for Bionomial Distribution.

OR

Find mean, median, variance for the following table:

Class	0–10	10–20	20–30	30–40	40–50	50–60
Frequency	5	15	25	20	10	5

(See Example 21 on page 134)

10. Solve the following LPP using Simplex Method:

Maximize $Z = 10x_1 + 15x_2 + 20x_3$

with condition $2x_1 + 4x_2 + 6x_3 \leq 24$

$\qquad\qquad 3x_1 + 9x_2 + 6x_3 \leq 30$

$\qquad\qquad$ and x_1, x_2 or $x_3 \geq 0$ *(See Example 9 on page 247)*

OR

Solve by North-West corner method and tests its optimility for the following transportation problem:

INDEX